URBAN GROUNDWATER – MEETING THE CHALLENGE

SELECTED PAPERS ON HYDROGEOLOGY

8

INTERNATIONAL ASSOCIATION OF HYDROGEOLOGISTS

Urban groundwater – meeting the challenge

Selected papers from the 32nd International Geological Congress (IGC), Florence, Italy, August 2004

Matthias Eiswirth Memorial Volume

Edited by

Ken W.F. Howard
Vice-President of the International Association of Hydrogeologists (IAH), Chair of the IAH Commission on Groundwater in Urban Areas and Professor of Hydrogeology at the University of Toronto, Canada

LONDON / LEIDEN / NEW YORK / PHILADELPHIA / SINGAPORE

United Nations
Educational, Scientific and
Cultural Organization

International Hydrological Programme

UNESCO's International Hydrological Programme (IHP) and the International Association of Hydrogeologists (IAH) coordinated the preparation of this book as an IAH contribution to the implementation of the Groundwater and Urban Water Management components of the 6th phase of the IHP Programme (2002–2007) – Water Interaction Systems at Risks and Social Challenges

Taylor & Francis is an imprint of the Taylor & Francis Group, an informa business

Typeset by Charon Tec Ltd (A Macmillan Company), Chennai, India
Printed and bound in Great Britain by Antony Rowe

Published by: Taylor & Francis/Balkema
P.O. Box 447, 2300 AK Leiden, The Netherlands
e-mail: Pub.NL@tandf.co.uk
www.balkema.nl, www.taylorandfrancis.co.uk, www.crcpress.com

British Library Cataloguing in Publication Data
A catalogue record for this book is available from the British Library

Library of Congress Cataloging in Publication Data
A catalog record for this book has been requested

ISBN13 978-0-415-40745-8 (Hbk)

Contents

ASSESSMENT OF CONTAMINANT IMPACTS

URBAN RECHARGE MANAGEMENT

URBAN AQUIFER MANAGEMENT

Introduction: Urban groundwater – meeting the challenge

Ken W.F. Howard
Vice-President of the International Association of Hydrogeologists (IAH), Chair of the IAH Commission on Groundwater in Urban Areas, and Professor of Hydrogeology at the University of Toronto
1265 Military Trail, Toronto, Ontario M1C 1A4, Canada

1 URBAN GROUNDWATER AS A PRIORITY ISSUE

The world's human population is increasing at an unprecedented rate with much of this growth taking place in urban areas. At the turn of the century, 2.86 billion people lived in urban areas, representing close to half the world's inhabitants. This number is currently increasing at the rate of almost 200,000 per day and by 2030 is expected to reach almost 5 billion or 60% of the projected global population of 8.6 billion. Urban areas are the economic power houses of the world but maintaining healthy and sustainable living conditions will be a major undertaking. Groundwater – "out of the public sight" and thus "out of the political mind" (Foster, 1996), has emerged as a priority issue.

From a resource perspective, groundwater represents the world's largest and most important source of fresh potable water. Where available, groundwater is generally favoured over surface water since it is normally well protected from surface contaminants, is less susceptible to drought, and can be introduced incrementally to meet growing private, municipal and industrial demand with minimal upfront capital expenditure. Problems begin when groundwater's utility becomes compromised by urban growth, which intensifies demand (Shahin, 1990) and degrades water quality through the subsurface release of contaminants. The overall effect is to significantly increase water-supply costs which, without an appropriate response, can negatively affect human health, undermine economic stability and lead to socio-economic and environmental decline. In some parts of the world, problems associated with ill-managed water supplies are seen as a major threat to social and political stability. Globally, new technologies and carefully planned management and protection strategies are required to make more efficient use of an essentially finite resource and provide adequate protection for water quality and the environment. A major challenge will be to meet the growing demand for safe water supplies in the face of competing political, societal and economic issues and limited financial resources for technological development and essential infrastructure (Howard and Gelo, 2002).

In terms of water resources, the most severe problems are faced by low- and middle-income countries (Morris et al., 1997; Foster et al., 1999; Howard, 2002), notably in drier regions where groundwater represents the only viable resource for much of the year. However, this does not mean that cities in high-income countries are immune from the myriad of problems that can arise if groundwater is not managed with appropriate regard. In many cities, for example, leaking water distribution networks and sewers generate rising groundwater levels which, in turn lead to unstable land slopes as well as flooded basements, tunnels and electrical utilities. Moreover, groundwater polluted by contaminated urban runoff, and leaking sewers and underground storage tanks ultimately discharges to urban wetlands, springs and streams and can represent a serious threat to human health. In many respects, such problems are most severe in cities that are entirely reliant on surface water as a supply and have failed to monitor the underlying aquifer system.

2 URBAN GROUNDWATER SCIENCE

Major interest in the relationship between urban development and water began during the 1950s and 1960s when accelerating urban growth following World War II, notably in Europe and North America, began to generate a wide range of hydrological problems. Most were related to urban run-off and flooding, and within a few short years the discipline of "urban hydrology" became firmly established. As a science, urban "groundwater" took somewhat longer to emerge but during the past 20 years has attracted increasing attention from a wide range of disciplines. To date, remediation and problem resolution have taken priority over more proactive measures such as urban planning, resource management and groundwater protection. Nevertheless, significant progress has been made on a number of key problems and a wealth of knowledge has been gained.

Urban groundwater first came to the world stage at Urban Water '88, a UNESCO symposium that dealt with hydrological processes and water management in urban areas. In 1992, it was recognised once again at the "United Nations Conference on Environment and Development" in Rio, where Agenda 21 responded to a growing concern for rapid urban population growth by specifying the need to protect the quality and supply of freshwater resources through an integrated approach to the development, management and use of water in a sustainable way. These were timely events, as urban groundwater issues were becoming prominent in scientific literature throughout the 1980s and early 1990s. For example,

- Baxter (1982) documented the impacts of sewage effluent disposal on groundwater in the United Kingdom.
- Cavallaro et al. (1986) and Rivett et al. (1989, 1990), revealed the presence of organic contaminants in aquifers underlying industrialized urban areas in Europe.
- Flipse et al. (1984) and Morton et al. (1988) demonstrated the detrimental impacts of lawn fertilizer on urban groundwater quality.
- Eisen and Anderson (1980), Locat and Gelinas (1989) and Pilon and Howard (1987) raised awareness for the impacts of road de-icing chemicals on groundwater.
- Lerner (1986, 1990a, b, c) and Foster (1990) drew attention to the enhancement of urban recharge by leaking septic systems, sewage mains and water distribution pipes.

Research in urban groundwater accelerated throughout the 1990s. The IAH Commission for Groundwater in Urban Areas was formed in Oslo, Norway in 1993, and in 1997, IAH dedicated its XXVII Congress in Nottingham, UK, to the topic of "Groundwater in Urban Areas". The proceedings of this meeting were published in two volumes (Chilton et al., 1997; Chilton, 1999).

In the past five years urban groundwater issues have become a regular theme at groundwater conferences throughout the world. In May 2001, a NATO Advanced Research Workshop on "Current Problems of Hydrogeology in Urban Areas, Urban Agglomerates and Industrial Centres" was held in Baku, Azerbaijan (Howard and Israfilov, 2002). Three years later, a NATO Advanced Study Institute was held in the same city to further promote the urban groundwater research agenda. The proceedings of this meeting will shortly be published under the editorial direction of Dr. John Tellam of the University of Birmingham in England. Urban groundwater also featured strongly at the 3rd World Water Forum held in Kyoto, Osaka and Shiga, Japan in 2003 (Howard, 2004).

The most recent international urban groundwater meeting was organized with the support of the IAH Commission on Urban Groundwater at the 32nd International Geological Congress (IGC). This meeting took place in Florence, Italy in August, 2004 and attracted excellent speakers and a large audience. Research papers presented at this meeting form the basis of this book.

3 THE BOOK AND ITS DEDICATION

The book on "challenges in urban groundwater" is premised on a growing recognition that most urban groundwater problems are not uniquely associated with any particular region or hydrogeological environment, and that much can be learned by understanding and sharing in the successes and challenges of others. Urban groundwater issues are complex. As demonstrated here, valuable scientific and technological progress has been made in areas such as the urban water balance, contaminant source characterisation and transport, aquifer vulnerability mapping, artificial recharge and the control of groundwater in deep excavations, roads and tunnels. The challenge is to integrate these advances into effective, holistic, plans for pro-active, sustainable urban groundwater management.

The book showcases the best urban groundwater papers presented at the 32nd IGC and is supplemented by contributions solicited from other world experts active in urban groundwater research. The book is broadly divided into five sections dealing with key aspects of urban groundwater. They include:

- Recharge Mechanisms and the Urban Water Balance
- Sewer Exfiltration
- Assessment of Contaminant Impacts
- Urban Recharge Management
- Urban Aquifer Management

The book is dedicated to the late Dr. Matthias Eiswirth. Matthias was President of the German Chapter of the International Association of Hydrogeologists (IAH), and Vice-Chair of the IAH Commission on Groundwater in Urban Areas. Matthias was co-organizer of the urban groundwater session at IGC before the tragic events of December 30th, 2003 brought his outstanding career to a premature end. He was an exceptional research scientist and the book's contents reveal directly and indirectly, the immense contribution he made to urban groundwater science. He continues to be sorely missed by his friends in the urban groundwater research community.

REFERENCES

Baxter, K.M. 1982. The effects of the disposal of sewage effluents on groundwater quality in the United Kingdom. Ground Water 17: 429–437.

Cavallaro, A., Corradi, C., De Felice, G. and Grassi, P. 1986. Underground water pollution in Milan and the Province by industrial chlorinated organic compounds, in J.F. de, L.G. Solbe (eds.), Effects of Land Use on Fresh Waters, Ellis Horwood, Chichester, pp. 68–84.
Chilton, J. et al. (eds.). 1997. Groundwater in the Urban Environment: Volume 1: Problems, Processes and Management; Proc. of the XXV11 IAH Congress on Groundwater in the Urban Environment, Nottingham, UK, 21–27 September 1997, Rotterdam, Balkema, 682 pp.
Chilton, J. (ed.). 1999. Groundwater in the Urban Environment: Selected City Profiles, Rotterdam, Balkema, 342 pp.
Eisen, C. and Anderson, M.P. 1980. The effects of urbanization on groundwater quality, Milwaukee, Wisconsin, USA, in R.E. Jackson (ed.), Aquifer Contamination and Protection SRH no. 30, UNESCO Press, Paris, pp. 378–390.
Flipse, W.J. Jr., Katz, B.G., Lindner, J.B. and Markel, R. 1984. Sources of nitrate in ground water in a sewered housing development, Central Long Island, New York. Ground Water 22: 418–426.
Foster, S.S.D. 1990. Impacts of urbanisation on groundwater, International Association of Hydrological Sciences (IAHS) Publ. No. 198, pp. 187–207.
Foster, S.S.D. 1996. Ground for Concern. October, 1996. Water. Our Planet 8(3).
Foster, S., Morris, B., Lawrence, A. and Chilton, J. 1999. Groundwater impacts and issues in developing cities – An introductory review, in J. Chilton (ed.), Groundwater in the Urban Environment: Selected City Profiles, Rotterdam, Balkema, pp. 3–16.
Howard, K.W.F. 2002. Urban groundwater issues – an introduction, in K.W.F. Howard, R. Israfilov (eds.), Current problems of hydrogeology in urban areas, urban agglomerates and industrial centres. NATO Science Series IV Earth and Environmental Sciences vol. 8, pp. 1–15.
Howard, K.W.F. and Gelo, K. 2002. Intensive Groundwater Use in Urban Areas: the Case of Megacities, in R. Llamas, E. Custodio (eds.), Intensive Use of Groundwater: Challenges and Opportunities Balkema, pp. 35–58.
Howard, K.W.F. and Israfilov, R.G. 2002. Current problems of hydrogeology in urban areas, urban agglomerates and industrial centres. NATO Science Series IV, Earth and Environmental Sciences Vol. 8, Kluwer. 500 pp.
Howard, K.W.F. 2004. Groundwater for socio-economic development – the role of science. UNESCO IHP-VI series on Groundwater. No. 9, published as CD. ISBN 92-9220-029-1.
Lerner, D.N. 1986. Leaking pipes recharge ground water. Ground Water 24(5): 654–662.
Lerner, D.N. 1990a. Recharge due to urbanization, in D.N. Lerner, A.S. Issar, I. Simmers (eds.), Groundwater Recharge: A Guide Book for Estimation Natural Recharge, International Association of Hydrogeologists, International Contributions of Hydrogeology Vol. 8, Hannover, Heise, pp. 201–214.
Lerner, D.N. 1990b. Groundwater recharge in urban areas. Atmospheric Environment 24B(1): 29–33.
Lerner, D.N. 1990c. Groundwater recharge in urban areas, International Association of Hydrological Sciences (IAHS) Publ. No. 198, pp. 59–65.
Locat, J. and Gélinas, P. 1989. Infiltration of de-icing salts in aquifers: the Trois-Rivires-Ouest case, Quebec, Canada. Canadian Journal Earth Sciences 26: 2186–2193.
Morris, B.L., Lawrence, A.R. and Foster, S.D. 1997. Sustainable groundwater management for fast-growing cities: Mission achievable or mission impossible? in J. Chilton et al. (eds.), Groundwater in the Urban Environment: Problems, Processes and Management; Proc. of the XXV11 IAH Congress on Groundwater in the Urban Environment, Nottingham, UK, 21–27 September 1997, Rotterdam, Balkema, pp. 55–66.
Morton, T.G., Gold, A.J. and Sullivan, W.M. 1988. Influence of over watering and fertilization on nitrogen losses from home lawns. J. Environ. 17: 124–130.
Pilon, P. and Howard, K.W.F. 1987. Contamination of sub-surface waters by road de-icing chemicals. Water Pollution Research Journal of Canada 22(1): 157–171.
Rivett, M.O., Lerner, D.N., Lloyd, J.W. and Clark, L. 1989. Organic contamination of the Birmingham aquifer. Report PRS 2064-M. Water Research Centre, Marlow, UK.
Rivett, M.O., Lerner, D.N., Lloyd, J.W. and Clark, L. 1990. Organic contamination of the Birmingham aquifer, UK. Journal of Hydrology 113: 307–323.
Shahin, M. 1990. Impacts of urbanization of the Greater Cairo area on the groundwater in the underlying aquifer. International Association of Hydrological Sciences (IAHS) Publ. No. 198, pp. 243–249.

Dedication to Matthias Eiswirth

PD Dr. Matthias Eiswirth (1965–2003)

The IAH Symposium on Urban Groundwater at the 32nd International Geological Congress in Florence 2004 was dedicated to Matthias Eiswirth, the Co-Initiator and former Vice-Chairman of the IAH Commission on Urban Groundwater. Matthias died together with his 2-year old son in a tragic avalanche accident in the Swiss Alps on 30 December 2003, leaving his wife and parents. His death was a great shock for many in the hydrogeological community, and especially for his long-standing colleagues at the University of Karlsruhe – Department of Applied Geology.

Matthias was born in Karlsruhe in January 1965 and grew up nearby in Durmersheim. Since his childhood days he had been a very active sportsman, joining the German Alpine Society where he participated regularly in mountain hiking and skiing tours. One of his last major climbs was of Aconcagua (6960 masl) on the Argentina/Chile frontier in January 2003. His fascination for mountains was also a major motivation to pursue a career in geological sciences. Following university-entrance certification in 1984 he obtained the Professional Geologist Diploma in 1992. While he loved travelling and contacts with the international scientific community, he remained with Karlsruhe University. In 1995 he obtained his Doctorate with magna-cum-laude and there followed several years of renowned international research, including a 1-year scholarship in Australia.

Matthias was not an academic scientist who stayed in the "ivory tower". He always aimed to identify environmental problems, develop strategic solutions and put them into practice. His scientific work focused on hydrogeological topics – such as the protection of groundwater resources, groundwater risk evaluation from various contaminant sources, and subsurface contaminant transport. His contribution was essential for the success of a series of important research projects.

Earlier than most colleagues, he recognised the importance of sustainable water management in urban areas. From his Diploma thesis onward he investigated leaky sewer systems and his Doctorate thesis was on the emission and transport of contaminants from leaky sewers and landfill sites. Through this avenue he looked beyond the borders of the classical hydrogeology and acquired substantial knowledge in urban drainage systems and the possibilities offered by alternative technologies like grey water reuse. This was the basis for his 1-year research scholarship with CSIRO-Australia, where he joined a team working on sewer monitoring and tools for the quantification of urban water and solute balances. This work led to a thesis entitled "Balancing mass fluxes in the urban water cycle – pathways to sustainable use of urban water resources (Bilanzierung der Stoffflüsse im urbanen Wasserkreislauf – Wege zur Nachhaltigkeit urbaner Wasserressourcen)".

With this consolidated knowledge and many international contacts, he initiated and coordinated the AISWURS Project of the EC 5th Framework Programme with partners in the CSIRO (Australia), the British Geological Survey, University of Surrey (UK), IRGO (Slovenia) and GWK Consult (Germany) from November 2002. AISUWRS (Assessing and Improving the Sustainability of Urban Water Resources and Systems), which expanded mass balance approaches, added new models for the description of leaky sewer networks and combined these with numerical models of unsaturated and saturated zone transport. Being a passionate field worker, Matthias applied the approach to 4 case-study cities with large hydrochemical sampling programmes, in-situ measurements and innovative test sites – he was the heart and soul of this project. At the European level, Matthias was also active in the CityNet Research Cluster, the ASIA-Link Programme and the COST Action C19.

His scientific work and strong professional commitment were acknowledged by national and international colleagues. Due to his detailed expertise, Matthias was invited to join a number of panels and boards related to groundwater management and protection. From 1997 he acted as Vice-President of the IAH Commission on Urban Groundwater. In 2001 he was elected as President of the IAH German Chapter, and worked on the management board of the German Geological Society-Hydrogeology Section and the International Association of Tracer Hydrology. He was also a member of the German Science Foundation-Water Forum and evaluator for the EC 6th Framework Programme.

Matthias was an exceptionally well qualified colleague. With his profound scientific knowledge, personal commitment and dynamic style, he convinced and motivated students and senior colleagues around him alike. Despite his young age, he had already achieved much and experienced remarkable successes, but how much more was still to come! His absence creates large gaps in our community. We remember him with gratitude for all the things he initiated and pushed to success with energy and skill. His ideas will remain with us and provide guidance to young colleagues. But besides his scientific achievements he will also remain in our memory as someone who cared about the people around him, providing support wherever possible and appropriate, and taking on many responsibilities without hesitation. We have lost a dear friend.

Prof. Dr. Heinz Hötzl
IAH Chair Commission on Karst Hydrogeology

Dr. Stephen Foster
IAH President

SELECTED REFERENCES

Eiswirth, M. 1993. Leckstellendetektion und Nachweis der Schadstoffausbreitung aus Abwasserkanälen. – WLB Wasser, Luft und Boden, 9: 34–38.

Eiswirth, M., Hötzl, H. and Merkler, G.-P. 1995. The detection and simulation of contaminant transport from leaky sewerage systems and waste disposal sites by combined hydrogeological and geophysical methods. – in: Kovar, K. and Krásn, J. (eds.): Groundwater Quality: Remediation and Protection GQ'95. IAHS-Publ., 225: 337–346.

Eiswirth, M. and Hötzl, H. 1995. Tracer techniques and soil gas surveys for the detection of contaminant transport in the unsaturated zone. – in: Leibundgut, C. (ed.): Comparison of tracer technologies for hydrological systems. IAHS-Publ., 229: 31–39.

Eiswirth, M. 1995. Charakterisierung und Simulation des Schadstofftransports aus Abwasserkanälen und Mülldeponien. – Schr. Angew. Geol. Karlsruhe, 38: 258 S.; Karlsruhe.

Eiswirth, M., Ohlenbusch, R. and Schnell, K. 1997. Use of artificial and natural tracers for the estimation of urban groundwater contamination by chemical grout injections. – in: Kranjc, A. (ed.): Tracer Hydrology 97: pp. 313–320; Rotterdam (Balkema).

Eiswirth, M. and Hötzl, H. 1997. The impact of leaking sewers on urban groundwater. – in: Chilton, J. et al. (eds.): Groundwater in the urban environment. Vol. 1: Problems, Processes and Management, pp. 399–404.

Eiswirth, M. and Hötzl, H. 1998. Attenuation and biodegradation processes in leachate plumes within different aquifers. – in: Van Brahana et al. (eds.): Gambling with groundwater – Physical, chemical and biological aspects of aquifer-stream relations: 249–255.

Eiswirth, M., Hötzl, H., Reichert, B. and Weber, K. 1998. Grundwasser- und Bodenluftuntersuchungen im Testfeld Süd. – Grundwasser, 4: 151–158.

Eiswirth, M., Hötzl, H., Jentsch, G. and Krauthausen, B. 1998. Contamination of a karst aquifer by a sanitary landfill, SW-Germany. – in: Drew, D. and Hötzl, H. (eds.): Karst hydrogeology and human activities – impacts, consequences and implications. – Int. Contr. Hydrogeology, 20: 151–158.

Eiswirth, M., Ohlenbusch, R. and Schnell, K. 1999. Impact of chemical grout injection on urban groundwater. – in: Ellis, B. (ed.): Impacts of Urban Growth on Surface Water and Groundwater Quality, Proc. of IUGG 99 Symposium 5, Birmingham, July 1999, IAHS Publ., 259: 187–194; Wallingford, Oxfordshire.

Eiswirth, M., Hötzl, H., Burn, L.S., Gray, S. and Mitchell, V.G. 2001. Contaminant loads within the urban water system – Scenario analyses and new strategies. – in: Seiler, K.-P. and Wohnlich, S. (eds.): New approaches to characterising groundwater flow, Balkema, Vol. 1: 493–498.

Eiswirth, M. 2001. Hydrogeological factors for sustainable urban water systems. – in: Howard, K. and Israfilov, R. (eds.): Current problems of Hydrogeology in Urban Areas, Urban Agglomerates and Industrial Centres, Kluwer, pp. 159–183.

Weber, K., Eiswirth, M. and Hötzl, H. 2002. Soil-gas investigations for the delineation of VOC contaminations in the subsurface. – in: Breh, W., Gottlieb, J., Hötzl, H., Kern, F., Liesch, T. and Niessner, R. (eds.): Field Screening Europe 2001, 14–16 May 2001, Karlsruhe, pp. 119–124.

Eiswirth, M. 2002. Bilanzierung der Stoffflüsse im urbanen Wasserkreislauf – Wege zur Nachhaltigkeit urbaner Wasserressourcen. – An der Fakultät für Bio- und Geowissenschaften der Universität Karlsruhe angenommene Habilitationsschrift.

Eiswirth, M., Wolf, L. and Hötzl, H. 2003. Assessing the sustainability of urban water resources, Diffuse input of chemicals into soil and groundwater – Assessment and Management, 26–28 Feb 2003, Mitteilung Institut Grundwasserwirtschaft TU Dresden, Bd. 3, pp. 205–215.

Eiswirth, M., Hötzl, H., Cronin, A., Morris, B., Veselič, M., Bufler, R., Burn, S. and Dillon, P. 2003. Assessing and Improving Sustainability of Urban Water Resources and Systems. – RMZ Materials and Geoenvironment, 50.

Eiswirth, M., Wolf, L. and Hötzl, H. 2004. Balancing the contaminant input into urban water resources. Environmental Geology, 46, 2, 246–256.

Wolf, L., Held, I., Eiswirth, M. and Hötzl, H. 2004. Environmental impact of leaky sewers on groundwater quality. Acta hydrochimica et hydrobiologica, 32, S.361–373.

Recharge mechanisms and the urban water balance

CHAPTER 1

Urban-enhanced groundwater recharge: review and case study of Austin, Texas, USA

B. Garcia-Fresca
The John A. and Katherine G. Jackson School of Geosciences – The University of Texas at Austin
P.O. Box B, University Station – Austin, TX 78713-8902

ABSTRACT: Cities and urban populations are growing at a high pace and as consequence, so are the local scale anthropic impacts on the hydrologic cycle. The shallow urban underground is an intricate network of tunnels, conduits, utilities, and other buried structures comparable to a natural karstic system, except that the "urban karst" is generated much more rapidly. Urbanisation also introduces new sources of water that increase groundwater recharge. These sources include irrigation of parks and lawns, leakage from water mains and sewers, and infiltration structures. Geologic, land use, and utilities information for the city of Austin, Texas, was compiled and processed by means of a GIS in order to make a water balance for the city. The areal extent of Austin, has increased five-fold since the 1960s. Direct recharge from rainfall has decreased, due to impervious pavements, from 53 mm/a under pre-urban conditions to 31 mm/a in the year 2000. However, 85 mm/a of treated tap water never reaches wastewater treatment plants and potentially contributes to recharge. A conservative estimate yields 63 mm/a of recharge from urban sources and a total recharge rate that nearly doubles that of pre-urban conditions.

1 INTRODUCTION

The magnitude of anthropic impacts upon their environment makes humans the major geologic agent on the surface of the planet (e.g. Heiken et al., 2003). These effects are most severe in areas of high population concentration and, today, half of the world's population lives in urban areas.

Urban development alters all aspects of the water cycle: the climate; the quantity, quality, and regime of surface water and groundwater; and the land surface and subsurface. Urbanisation affects the local climate by altering surface temperatures, albedo, precipitation, evaporation and transpiration rates, as well as the atmospheric energy balance (e.g. Changnon, 1976; Bornstein and Lin, 2000). Urban development and population growth increase water demand. This imposes increased demands on surface and groundwater resources, often requiring interbasinal transfers, which affect the natural water budget of the area. Water quality is a key issue in urban settings as shallow aquifers and surface waters in cities are subject to pollution by a multitude of point and non-point sources, some of which are still poorly understood. Urbanisation affects stream regimes by modifying both base flow and flood discharge, bank erosion, sedimentation, land-sliding, declines in water quality, and increased flooding (Leopold, 1968, 1973). Changes in surface water systems are commonly visible and apparent even to casual observers, while effects on groundwater systems that may be equally significant are not always obvious. Human effects on groundwater in urban areas include overexploitation, subsidence, seawater intrusion, groundwater contamination, changes in recharge and discharge, alteration of the permeability structure, and destruction of important environmental resources, including wetlands and urban streams (e.g. Chilton et al., 1997; Garcia-Fresca and Sharp, *in press*; Howard, 2002).

The covering and replacement of natural rocks, soils, and vegetation by pavements, foundations, buildings, metallic structures, dams, tunnels, and other structures has a profound impact on the hydrology of an area. The urban underground is an intricate and rapidly changing network of tunnels, buried utilities, garages, and other buried structures that disturb the natural structure of the ground and alter its porosity and hydraulic conductivity. Based on the studies of porosity of karstic aquifers by Worthington (2003), and the volume of underground tunnels and installations catalogued for Quebec City by Boivin (1990), Garcia-Fresca and Sharp (*in press*) conclude that the urban underground has secondary porosities and perhaps permeability distributions comparable to those of a karstic system (Table 1).

Boivin (1990) did not provide estimates for the porosity created by smaller utility lines, trenches, pipes, and conduits. Krothe (2002) and Krothe et al. (2002) documented orders of magnitude increases in field permeability measurements along utility trenches and showed by finite-difference numerical modeling that high permeability utility trenches alter groundwater flow direction and velocity. Thus, the urban underground is comparable to a shallow karstic system (Sharp et al., 2001; Krothe et al., 2002; Sharp et al., 2003; Garcia-Fresca, 2004; Garcia-Fresca and Sharp, *in press*). The following characteristics make the urban underground similar to karst:

- Utility trenches are analogous to naturally fractured systems;
- Larger underground openings, excavations, and tunnels are analogous to natural conduits, caves, and channels;
- Permeabilities are highly anisotropic and heterogeneous;
- Storm drains are analogous to dolines, swallets, and sink holes;
- Rain water can be stored in the shallow underground just as in the epikarst;
- Recharge can be from both diffuse (precipitation and irrigation return flows) and discrete sources (i.e. leaky pipes).

This "urban karstification" is in continuous evolution as new structures are built over the older ones, buried structures are abandoned, and as existing geological structures, lithofacies,

Table 1. Porosity values for four karstic aquifers (after Worthington, 2003) and estimated porosity from human construction in Quebec City (after Boivin, 1990).

	Porosity (%)		
	Matrix	Fractures	Conduits/channels
Smithville, Ontario, Canada	6.6	0.02	0.003
Mammoth Cave, Kentucky, USA	2.4	0.03	0.06
Chalk, England	30	0.01	0.02
Nohoch Nah Chich, Mexico	17	0.1	0.5
Quebec City, Canada	n/a	Unknown	0.06

and other features are leveled and buried by construction. However, the development of the urban karst takes place at a much faster rate than the natural karst.

In the following sections urban effects on groundwater recharge and the mechanisms of recharge are discussed. Then the urban water balance and an estimate of recharge for the city of Austin are presented.

2 GROUNDWATER RECHARGE IN THE URBAN ENVIRONMENT

The hydrologic community has recognized that natural groundwater recharge can be inhibited in urban areas as impervious cover enhances runoff and limits infiltration (i.e. Leopold, 1968; Coldewey and Meßer, 1997). However urban development introduces new sources of recharge: leakage from water and wastewater distribution and collection systems, leaks from storm sewers and irrigation return flow from lawns, parks, and golf courses (Lerner, 1986). Quantifying groundwater recharge in urban areas is especially challenging because the urban environment is complex, not only due to the heterogeneity of the shallow underground, but also because a large variety of land uses coexist and overlap. The uncertainties intrinsic in quantifying different sources of groundwater recharge and discharge make it desirable to simplify with water balances based on the amount of groundwater abstractions, imports, water use, and wastewater outflows.

Numerous examples of significant water table-rise and increase on recharge to the groundwater have been reported in the last decade (e.g. Foster et al., 1994; Chilton et al., 1997; Chilton, 1999; Howard and Israfilov, 2002). A compilation of groundwater recharge data for various cities is portrayed in Figure 1 as a function of aridity as expressed by the mean annual rainfall of each location. The figure is expanded from Foster et al. (1994), who proposes ranges of natural recharge for non-urban environments, probable minimum recharge rates for comprehensively sewered and drained cities, and the probable maximum recharge rates for unsewered and undrained cities. These have been revised after adding nineteen data points to the original four found in Foster et al. (1994). In all cases, except for Birmingham, UK, the total recharge to the groundwater is increased by urban development. In the case of Birmingham, Lerner (1997) estimates a 4% loss in recharge, and is expressed as a downward pointing arrow in Figure 1. Urban-enhanced recharge is most significant in arid climates and in cities in developing countries. In a broader sense, urbanisation introduces new sources and pathways of recharge (Lerner, 1986) and affects water quality.

3 MECHANISMS OF RECHARGE IN URBAN AREAS

Simmers (1998) and Garcia-Fresca (2004) describe four types of recharge: (1) direct recharge: vertical percolation of rainwater through the unsaturated zone; (2) indirect recharge: water losses from surface water bodies and from water and sewage distribution systems; (3) localized recharge: percolation through preferential pathways (desiccation cracks, burrows, lithologic contacts, faults, fractures, and karstic features); and (4) artificial recharge: return flows from irrigation of parks and lawns, and designed infiltration systems. The four mechanisms of recharge generally combine to increase recharge with urbanisation, but the categories can overlap and are not mutually exclusive.

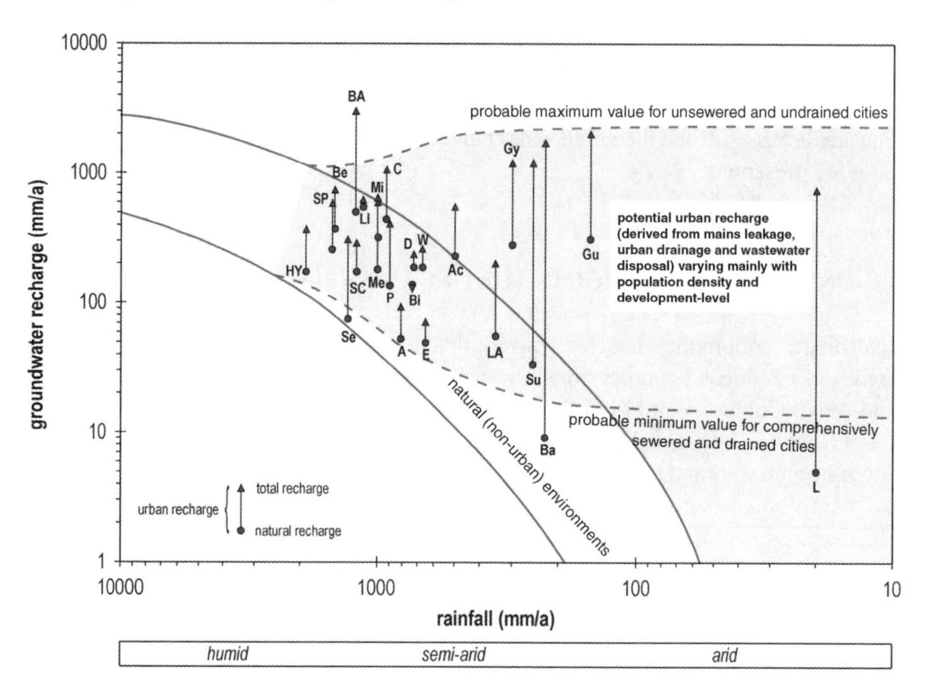

Figure 1. Urban-enhanced groundwater recharge in twenty-three cities around the world (modified from Foster et al., 1994). **HY**: Hat Yai, Thailand (Foster et al., 1994); **SP**: São Paulo, Brazil (Menegasse et al., 1999); **Be**: Bermuda, UK (Lerner, 1990a); **Se**: Seoul, Korea (Kim et al., 2001); **BA**: Buenos Aires, Argentina (Foster, 1990); **SC**: Santa Cruz, Bolivia (Foster et al., 1994); **LI**: Long Island (New York), USA (Ku et al., 1992); **Mi**: Milan, Italy (Giudici et al., 2001); **Me**: Mérida, México (Foster et al., 1994); **C**: Caracas, Venezuela (Seiler and Alvarado Rivas, 1999); **P**: Perth, Australia (Appelyard et al., 1999); **A**: Austin (Texas), USA (Garcia-Fresca, 2004); **Bi**: Birmingham, UK (Knipe et al., 1993); **D**: Dresden, Germany (Grischek et al., 1996); **W**: Wolverhampton, UK (Hooker et al., 1999); **E**: Évora, Portugal (Duque et al., 2002); **Ac**: Aguascalientes, México (Lara and Ortiz, 1999); **LA**: Los Angeles (California), USA (Geomatrix, 1997): **Ba**: Baku, Azerbaijan (Israfilov, 2002); **Su**: Sumgayit, Azerbaijan (Israfilov, 2002); **Gy**: Gyandja, Azerbaijan (Israfilov, 2002); **Gu**: Gulistan, Uzbekistan (Ikramov and Yakubov, 2002); **L**: Lima, Perú (Foster et al., 1994).

3.1 *Direct recharge*

Direct recharge in cities takes place by percolation in unpaved areas, and to a lesser extent through paved surfaces that are not always perfectly "impervious". The significance of direct recharge decreases as the aridity of the climate or the amount of impervious cover increases. Direct recharge can be estimated by assessing the amount of pervious cover in the city. Precipitation and potential evapotranspiration data are transformed into effective precipitation (e.g. Lerner et al., 1993) with a daily soil moisture balance. This method uses root constants and wilting points to account for different crops and soil types. A proportion of the impervious cover should be treated as permeable, as some infiltration does take place through asphalt, concrete, bricks etc. According to Lerner (2002) roughly 50% of the impervious cover should be treated as permeable.

3.2 *Indirect recharge*

Indirect recharge is the sum of the recharge coming from seepage out of surface water bodies, leakage from water mains, wastewater and storm sewers, and on-site sanitation systems.

Table 2. Compilation of water main or distribution system losses in various cities of the world. Some general rates are given in *italics*.

City	Water main loss [%]	Reference
Hull, UK	5	Chastain-Howley, *pers. comm.*
Los Angeles, USA	6–8	Geomatrix, 1997, *unpub.*
Hong Kong, China	8	Lerner, 1997
San Antonio, USA	8.5	Austin American Statesman, 1998
Évora, Portugal	8.5	Duque et al., 2002
Milan, Italy	10	Giudici et al., 2001
Austin, USA	12	Austin American Statesman, 1998
N. Auckland, NZ	12.3	Farley and Trow, 2003
Toronto, Canada	14	City of Toronto, 2001, *pers. comm.*
Calgary, Canada	15	Grasby et al., 1997
US average	*16*	*Thornton, 2002*
Dresden, Germany	18	Grischek et al., 1996
São Paulo, Brazil	16	Menegasse et al., 1999
UK general rates	*20–25*	*Lerner, 1997*
Göteborg, Sweden	26	Norin et al., 1999
Round Rock, USA	26	Austin American Statesman, 1998
Tomsk, Russia	15–30	Pokrovsky et al., 1999
Amman, Jordan	30	Salameh et al., 2002
Kharkiv, Ukraine	30	Jakovljev et al., 2002
Sana'a, Yemen	30	Alderwish and Dottridge, 1998
Brushy Creek, USA	33	Austin American Statesman, 1998
Calcutta, India	36	Basu and Main, 2001
San Marcos, USA	37	Austin American Statesman, 1998
St. Petersburg, Russia	~30	Vodocanal, 2000, *unpub.*
Developing countries	*30–60*	*Foster et al., 1998*
Lusaka, Zambia	45	Nkhuwa, 1999
Mérida, México	~50	Foster et al., 1994
Lima, Perú	45–60	Lerner, 1986
Cairo, Egypt	>60	Amer and Sherif, 1997
Some Italian systems	*>80*	*Farley and Trow, 2003*

Recharge from losing streams in urban areas changes as stream flows are altered by urbanisation. Decline in aquifer heads caused by pumpage can alter the hydraulic gradients between the surface and the aquifer and between adjacent formations to enhance recharge.

A simple way to assess the water available for recharge is to make a balance of the water served versus the wastewater treated. Yang et al. (1999) quantified the recharge in the city of Nottingham, UK, with a groundwater flow model calibrated with solute balances for chloride, sulfate, and nitrogen. They concluded current recharge to the aquifer is less than prior to urbanisation; however, water main leakage is the main current source of recharge in Nottingham.

3.2.1 *Leakage from water mains*

Water mains are pressurized to avoid infiltration of contaminants and to ensure distribution to the far reaches of the water system. Pressure is the main cause of leakages in water distribution systems. A review of the literature shows that typical values of water loss from the distribution system are around 20 to 30% (Table 2). The most efficient cities report loses around 10%, and values of 30 to 60% are common in the less developed countries. In arid climates, the amount of water distributed in a city is often significantly greater than rainfall (Foster et al., 1994). Thus, mains leakage is a consistent source of indirect groundwater recharge.

Lerner et al. (1990) proposed several indirect methods to estimate leakage from water distribution networks as direct measurements are often not practical. One method is to assume a certain percentage of the water supplied; Thornton (2002) suggests that about 60% of unaccounted water can be attributed to leakage. Other methods include mass balances of inputs and outputs to the network. External losses on consumers' premises (the "consumer's side of the water meter") are not accounted by water supply authorities. These losses may be reflected as legitimate use per property, but this can be the leakiest part of the system. Leakage rates vary spatially depending on the pressure of the water, the age and the material of the pipes, and the maintenance of the system.

3.2.2 *Leakage from wastewater sewers*

Reports of groundwater contamination by sewage or wastewater are numerous (e.g. Eiswirth and Hötzl, 1997; Blarasin et al., 1999; Rieckermann et al., 2003) and indicate that leakage from sewers is common and widespread. When sewer lines are located below the water table, they may infiltrate groundwater, and when located above the water table they may leak. Because flows in these pipes are not under pressure, it is reasonable to assume they leak less than water mains. Many cities lack sewerage networks and rely on septic tanks or similar systems to dispose of grey water. In these cases, most of the supplied water is recharged to the subsurface (Foster et al., 1994).

Albeit scarce within the literature, increasing efforts have recently been made to quantify wastewater leakage from sewers. The few published estimations seem to agree on leakage rates of 5% of the sewage flow through the network; these include Barcelona (Vázquez-Suñé, 2003), Nottingham (Yang et al., 1999), Munich (Lerner, 1997), Dresden (Grischek et al., 1996), and several other German cities (Foster et al., 1994). However, Giudici et al. (2001) report 20% losses from the sewage network in Milan.

Recently, more sophisticated methods to quantify leakage from sewage networks have been developed. For instance, Eiswirth et al. (2004) propose a software model to simulate the urban water, wastewater and stormwater systems. Another method adds artificial tracers to the network and analyzes the composition downflow in order to make a mass-balance of the introduced solutes (Rieckermann et al., 2003).

3.2.3 *Leakage from storm sewers*

Recharge from stormwater occurs under transient high-flow conditions, making it very difficult to measure and model. Lerner (2002) proposes to use an empirical approach, or to assume some proportion of the surface of the city is not impermeable, to account for this water.

3.2.4 *Septic tank infiltration*

On-site wastewater treatment systems can be assumed to recharge all the water they receive, except for some small losses to evapotranspiration, and perhaps stream baseflow. Thus, about 90% of the water supplied in unsewered cities can recharge the groundwater (Foster et al., 1994).

3.3 *Localized recharge*

Localized recharge takes place through faults, fractures, etc. and thus, it depends mainly on the geologic materials and structures, as well as the soil types in each particular area. As defined above, localized recharge is not directly related to urbanisation although it can be affected by urbanisation. Numerous approaches exist for modeling flow through fractures and conduits (e.g. Sharp, 1993; Halihan et al., 1999; Zahm, 1998).

3.4 *Artificial recharge*

Artificial recharge consists of water intentionally applied to the subsurface and includes devices designed to enhance infiltration, as well as irrigation water in excess of plant needs.

3.4.1 *Designed infiltration structures*

A variety of man-made structures are constructed to reduce flooding, relieve the sewerage networks, and promote groundwater recharge. Such structures include recreational lakes and ponds, soakways, runoff detention ponds, retention basins, artificial infiltration ponds, spreading basins, recharge ditches, and injection wells.

It can be assumed that infiltration devices recharge all the water they receive, except for some losses to evapotranspiration and stream interflows, as is the case of septic tanks. The importance of such recharge sources depends on their abundance in a city, their location with respect to the aquifers and the particular design characteristics of each device. Maintenance plays an important role, and when clogging takes place they may become ineffective and minimize recharge.

3.4.2 *Irrigation return flow*

The water directly applied to parks and lawns, in excess of the plant requirements, will percolate and recharge the groundwater, except for some loss to evaporation and interflow. What makes this source of recharge different from effective precipitation is the intentionality of its application, as well as the uncertainties related to its quantification.

This source of recharge can be especially significant in arid and semi-arid climates. La Dell (1986) and Lerner (1990b) illustrate this with the example of Doha (Qatar), where the water table rise is directly related to the excessive irrigation of parks and lawns.

Recharge from excess irrigation can be quantified by mass balancing water supply, water use, the physical properties of the soils, and evapotranspiration (e.g. Berg et al., 1996). In arid and semi-arid areas, variations in these parameters should be obvious when comparing the drier and wetter months.

4 URBAN-ENHANCED RECHARGE IN AUSTIN, TEXAS

Austin is located in central Texas (Figure 2) and has a subtropical humid climate with a mean annual temperature of 20°C and a mean annual precipitation of 813 mm/a. It is situated within the Colorado River basin, the main source of water supply. Austin sits over a major fault zone which juxtaposes a variety of geologic materials (Rose, 1972; Garner and Young, 1976) including the Edwards aquifer, one of the most prolific karstic aquifers in the world, and minor hydrogeologic units within Quaternary fluvial deposits. The population has increased steadily since 1985 and at exponential rates since the 1960s, reaching 656,562 people in the year 2000.

This section analyzes the effects of urban development on groundwater recharge in the city of Austin. Estimations of direct recharge from precipitation were carried out by means of spatial analysis using a GIS. Contributions to groundwater recharge from urban sources were estimated by means of a water balance. The results are summarized in Table 3 and fully described in Garcia-Fresca (2004).

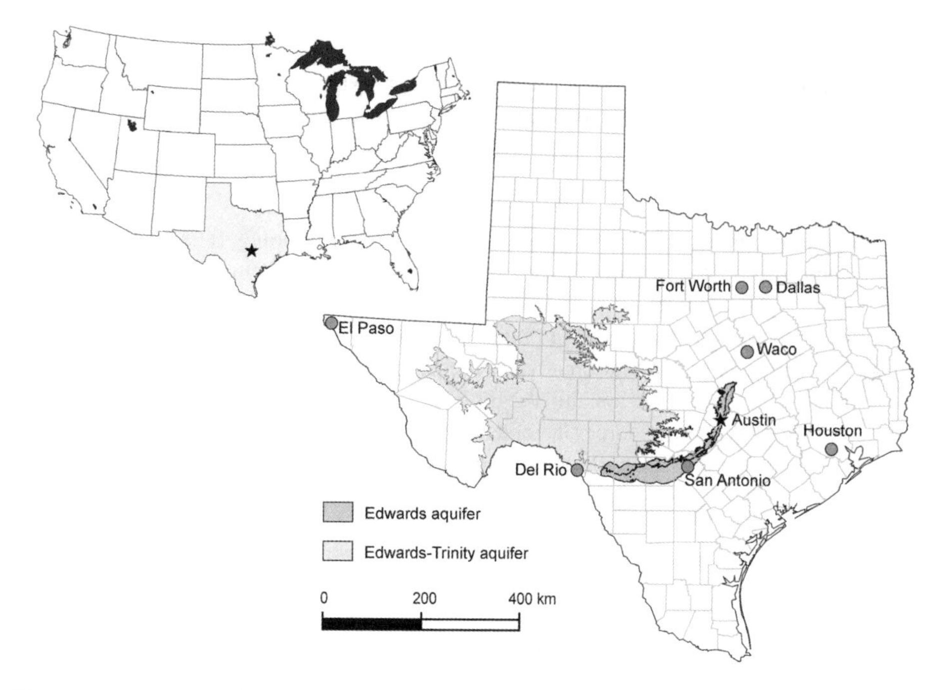

Figure 2. Location of the City of Austin and some of the major aquifers in central Texas.

4.1 *Direct recharge*

A comparison of direct recharge from precipitation before and after urbanisation, was conducted in order to assess the effects of development on this type of recharge. The spatial distribution of land-uses prior and after urban development was conducted by means of a GIS (ArcGIS).

4.1.1 *Direct recharge under pre-urban conditions*

Direct recharge from effective precipitation prior to urban development was estimated based on the hydraulic properties of the different lithologies cropping out within the city limits. Each hydrogeological unit was isolated and an infiltration coefficient assigned, as a percentage of precipitation. Values of the infiltration coefficients were compiled from the literature for the particular outcrops, or for similar units in Texas. The infiltration coefficient for clays and shales was assumed to be 0%. Figure 3 illustrates the outcrop analysis process and two of the resulting separated outcrops: the Barton Springs segment of the Edwards aquifer, and the Quaternary deposits, with infiltration coefficients of 8 and 9%, respectively. Pre-urban recharge is estimated at 53 mm/a.

4.1.2 *Direct recharge under urban conditions (year 2000)*

In this case direct recharge from precipitation was assessed in a similar fashion, as a function of the type of outcrop and the amount of impervious cover for the different urban land uses in the year 2000 (Figure 4). Direct recharge in the year 2000 is estimated to be 31 mm/a and, thus, it has decreased with increasing urban development.

4.2 *Urban sources of recharge*

A water balance of urban water supply, uses, and sewage volumes assesses the recharge available from strictly urban sources. In the year 2000, Austin put into the distribution system

Table 3. Water and wastewater statistics and water balance for Austin for the year 2000. Sources of data: (1) US Census Bureau, *online*; (2) City of Austin, *online*; (3) NOAA, *online*; (4) Garcia-Fresca, 2004; (5) City of Austin Water and Wastewater Utility, *personal communication*; (6) Austin American Statesman, 1998; (7) TexasET, *online*.

			mm/a	Data sources
Population		656,562		*(1)*
Area		704 km^2		*(2)*
Population density		933 p/km^2		*(4)*
Mean annual precipitation			813	*(3)*
Direct recharge (preurban)			53	*(4)*
Direct recharge (urban)			31	*(4)*
Served water (w)	Population served	738,229		*(2)*
	Area served	710 km^2		*(2)*
	Average	541,000 m^3/d	278	*(2)*
	Peak	856,000 m^3/d	440	*(2)*
	Max capacity	984,000 m^3/d	506	*(2)*
Treated wastewater (ww)	Population served	685,783		*(2)*
	Area served	601 km^2		*(2)*
	Average	318,000 m^3/d	193	*(2)*
	Max capacity	492,000 m^3/d	299	*(2)*
Excess urban water	Avg w – max ww		−21	*(4)*
	Avg w – avg ww		85 avg	*(4)*
	Max w – max ww		207 max	*(4)*
Gross unbilled water	12%	64,920 m^3/d	33	*(5,6)*
Mains leakage rate	7.7%	41,657 m^3/d	21	*(4)*
Sewer leakage rate	5%	16,737 m^3/d	10	*(4)*
Irrigation		Not area weighted	54 avg	*(4)*
			175 max	*(4)*
		Area weighed to adjust for	90 avg	*(4)*
		Pervious/impervious cover	291 max	*(4)*
Plant water requirement (PWR)		Not area weighted	364	*(7)*
		Area weighed	219	*(4)*
		From irrigation only	22	*(4)*
Irrigation return flow			32	*(4)*
Total recharge		ET not accounted	116	*(4)*
		After subtracting PWR	94 avg	*(4)*

an average of 541,000 m^3/d, and an average of 318,000 m^3/d was treated at the wastewater treatment plants. Of interest is the fact that the maximum wastewater treatment capacity (492,000 m^3/d) cannot accommodate even the average volume of water supplied. "Excess urban water" is estimated by balancing the drinking water supplied and the wastewater treated, 85 mm/a on average, and represents the amount of water of exclusively urban origin potentially available for recharge, both as indirect and artificial recharge (Figure 5). A fraction of the excess urban water is lost to leakage from the utility networks, and the rest is assumed to be used to irrigate parks and lawns.

The City of Austin estimates water losses as the difference between served water and billed consumption (Pedersen, 2003, *personal communication*). As summarised in Figure 6, 12% of the water usage is "unaccounted for" or "gross unbilled" treated water. Unaccounted-for water can be broken down into "unbilled uses" and "losses". Unbilled uses are estimated at 6.8% of

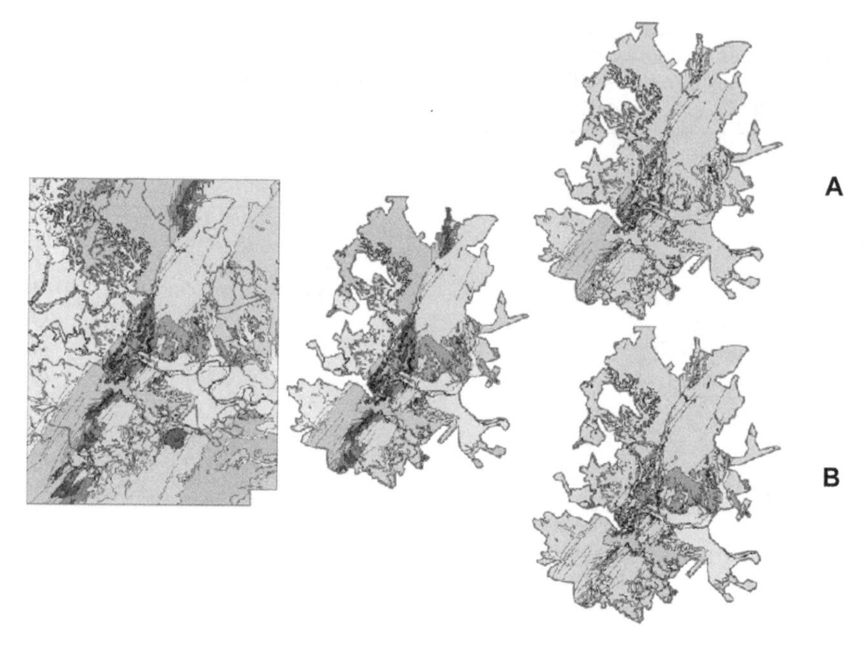

Figure 3. Outcrop analysis. The geology of the area is cropped to match the shape of the full service jurisdiction of the City of Austin (i.e. the portion of the city fully served by mains and sewers). As an example, two of the resulting separated outcrops are presented: (A) the Barton Springs segment of the Edwards aquifer (infiltration coefficient, 8%) and (B) Quaternary deposits (infiltration coefficient, 9%).

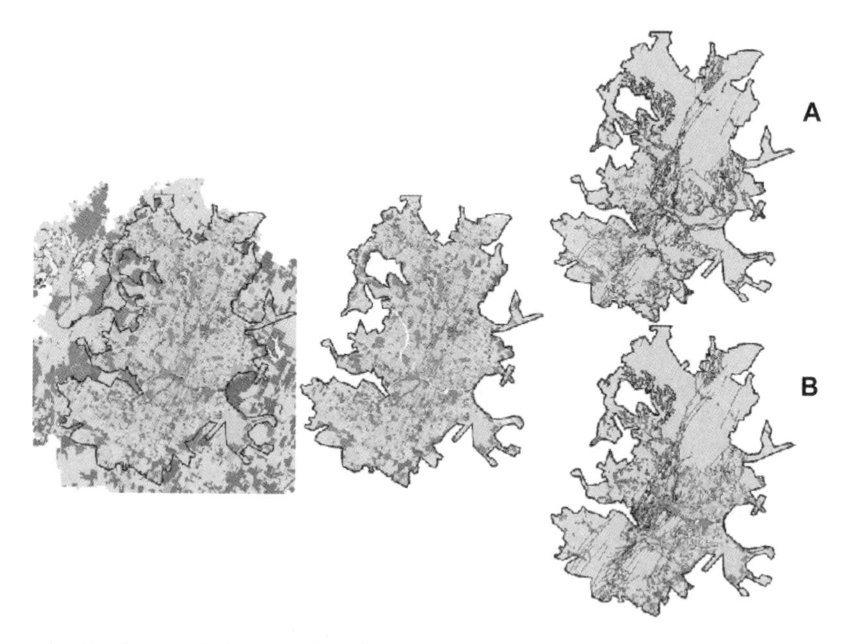

Figure 4. Land use analysis, consisting of an assessment of the different types of land cover and their percentage of impervious cover, within each outcrop. (A) the Barton Springs segment of the Edwards aquifer (impervious cover, 58%) and (B) Quaternary deposits (impervious cover, 2%). Data for year 2000.

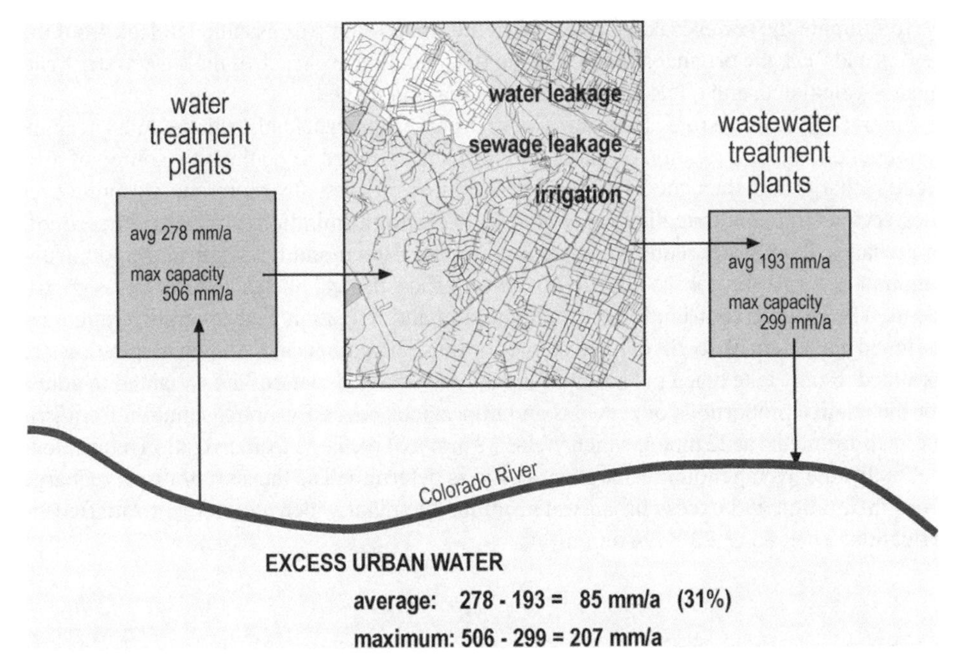

Figure 5. Excess urban water is defined as the difference between the amount of drinking water treated and the amount of wastewater treated. For the city of Austin it was a total of 85 mm for the year 2000.

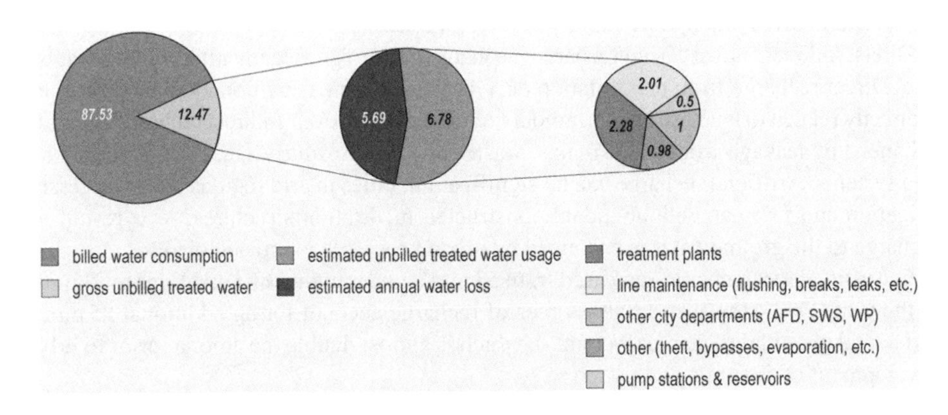

Figure 6. Water demand, usage and losses in the city of Austin, year 2000.

the total treated water; these include fire fighting water, thefts, municipal swimming pools, leakage and water mains breakages (the last two represent less than 2.01% of the total treated water). The other 5.7% of the water is simply "lost". Accordingly, the water potentially available for recharge is approximately 7.7% of the water treated, which for 2000 amounts 21 mm/a. This amount is consistent with Thornton's (2002) estimate of leakage rates of approximately 60% of gross unbilled water.

To estimate the sewer leakage rate, 5% of the wastewater was assumed to leak from the sewers and thus, the original amount of water that should have reached the wastewater treatment is calculated, and a leakage rate of 10 mm/a established.

Subtracting leakage rates from the excess urban water, artificial recharge must account for about 54 mm/a. Irrigation of parks and lawns is assumed to be the only source of artificial recharge and, thus, the 54 mm/a over the entire city actually represents 90 mm/a over the pervious fraction. Irrigation water not lost to evapotranspiration turns into either runoff or recharge. Plant water requirements are computed from monthly reference evapotranspiration rates for Austin for the year 2003, crop coefficients, and allowable plant stress coefficients. The relative contributions of precipitation and irrigation to evapotranspiration are assumed to be equal to their relative proportions. Irrigation and evapotranspiration are assumed to only take place in "pervious areas" and thus calculations are weighted to adjust for the relative proportions of pervious and impervious cover. Evapotranspiration from irrigation is estimated at 22 mm/a, which yields 32 mm/a of recharge from irrigation return flow.

Finally, the average total recharge in Austin is determined as the sum of direct recharge from infiltration and excess urban water, minus the plant water requirement satisfied by irrigation: $31 + 85 - 22 = 94$ mm/a.

5 CONCLUSIONS

The complex network of utility trenches, tunnels, and other buried structures below cities alter the natural permeability field and affect groundwater flow and transport. Thus, the shallow urban underground can be compared to a karstic system. The "urban karst" is generated and evolves much more rapidly than natural karsts.

Direct, indirect, and artificial recharge mechanisms are significantly affected by urbanisation. Direct recharge from precipitation decreases with increasing urban development and is directly related to land use and the amount of impervious cover. Indirect recharge is greatly enhanced by leakage from water mains, wastewater and storm sewers, and on-site sanitation systems. Artificial recharge can be significant in cities in arid regions due to excessive irrigation and in areas with abundant constructed infiltration structures. As a result, net recharge to the groundwater is enhanced by urban processes in cities globally.

In Austin, direct recharge decreased from 53 mm/a under pre-urban conditions to 31 mm/a in the year 2000. However urban sources of recharge account for an additional 85 mm/a, and a total potential recharge (94 mm/a), which is almost double the amount prior to urban development of 53 mm/a.

ACKNOWLEDGEMENTS

Thanks to professor J.M. Sharp Jr., who inspired this work, and to the kind anonymous reviewers who helped improve this paper. The Geology Foundation of the University of Texas at Austin, the Travel Grants Program of the 28th International Geological Congress Organizing Committee, and the Chevron Professorship at the Department of Geological Sciences of the University of Texas generously contributed to finance different aspects of the research presented here.

REFERENCES

Alderwish, A.M. and Dottridge, J. 1998. Recharge components in a semi-arid area: the Sana'a Basin, Yemen. In Robins NS (ed) *Groundwater pollution, aquifer recharge and vulnerability*. London, Geological Society Special Publication No 130, p 169–177.

Amer, A.M. and Sherif, M.M. 1997. Groundwater rise in Greater Cairo: Cause and effects on antiquities. In Chilton J (ed) *Groundwater in the Urban Environment: Problems, Processes and Management*. Rotterdam, Balkema, p 213–217.

Appelyard, S.J., Davidson, W.A. and Commander, D.P. 1999. The effects of urban development on the utilisation of groundwater resources in Perth, Western Australia. In Chilton J (ed) *Groundwater in the Urban Environment – Selected City Profiles*. Balkema, Rotterdam, p 97–104.

Austin American Statesman. 1998 16 July 1998 edition.

Basu, S.R. and Main, H.A.C. 2001. Calcutta's water supply: demand, governance and environmental change. *Applied Geography*, v 21, p 23–44.

Berg, A., Byrne, J. and Rogerson, R. 1996. An urban water balance study, Lethbridge Alberta: estimation of urban lawn overwatering and potential effects on local water tables. *Canadian Water Resources Journal*, v 21, No 4, p 17–27.

Blarasin, M., Cabrera, A., Villegas, M., Frigerio, C. and Bettera, S. 1999. Groundwater contamination from septic tank system in two neighborhoods in Rio Carto City, Cordoba, Argentina. In Chilton J (ed) *Groundwater in the Urban Environment – Selected City Profiles*. Balkema, Rotterdam, p 31–38.

Boivin, D.J. 1990. Underground space use and planning in the Quebec City area. *Tunnelling and Underground Space Technology*, v 5, No 1–2, p 69–83.

Bornstein, R. and Lin, Q. 2000. Urban heat islands and summertime convective thunderstorms in Atlanta: tree case studies. *Atmospheric Environment*, v 34, p 507–526.

Changnon, S.A. Jr. 1976. Inadvertent weather modification. *Water Resources Bulletin*, v 12, p 695–718.

Chilton, J. et al. (eds). 1997. *Groundwater in the Urban Environment: Problems, Processes and Management*. Balkema, Rotterdam, 682 p.

Chilton, J. 1999. *Groundwater in the Urban Environment – Selected City Profiles*. Balkema, Rotterdam, 356 p.

City of Austin (online) Water and Wastewater Utility. http://www.ci.austin.tx.us/water/ [accessed February 2003].

Coldewey, W.G. and Meßer, J. 1997. The effects of urbanization on groundwater recharge in the Ruhr region of Germany. In Chilton J et al. (eds) *Groundwater in the Urban Environment: Problems, Processes and Management*. Balkema, Rotterdam, p 115–119.

Duque, J., Chambel, A. and Madeira, M. 2002. The influence of urbanisation on groundwater recharge and discharge in the City of Evora, south Portugal. In Howard KWF and Israfilov RG (eds) *Current Problems of Hydrogeology in Urban Areas, Urban Agglomerates and Industrial Centres*. NATO Science Series. IV Earth and Environmental Sciences, v 8, p 127–137.

Eiswirth, M. and Hötzl, H. 1997. The impact of leaking sewers on urban groundwater. In Chilton J et al. (eds) *Groundwater in the Urban Environment: Problems, Processes and Management*. Balkema, Rotterdam, p 399–404.

Eiswirth, M., Wolf, L. and Hötzl, H. 2004. Balancing the contaminant input into urban water resources. *Environmental Geology*, v 46, p 246–256.

Farley, M. and Trow, S. 2003. Losses in water distribution networks: A practitioner's guide to assessment, monitoring and control. *International Water Association*, IWA Publishing, London, UK, 282 p.

Foster, S.S.D. 1990. Impacts of urbanisation on groundwater. In Massing H, Packman J and Zuidema FC (eds) *Hydrological Processes and Water Management in Urban Areas*. International Association of Hydrological Sciences, Publication No 198, p 187–207.

Foster, S.S.D., Morris, B.L. and Lawrence, A.R. 1994. Effects of urbanization on groundwater recharge. In Wilkinson WB (ed) *Groundwater Problems in Urban Areas*. Tomas Telford, London, p 43–63.

Foster, S.S.D., Lawrence, A.R. and Morris, B.L. 1998. *Groundwater in urban development – Assessing management needs and formulating policy strategies*. Washington DC, World Bank Technical Paper 390, 55 p.

Garcia-Fresca, B. 2004. *Urban effects on groundwater recharge in Austin, Texas*. MS thesis, The University of Texas, Austin, 173 p.

Garcia-Fresca, B. and Sharp, J.M. Jr. (in press). Hydrogeologic considerations of urban development – urban-induced recharge. In Ehlen J, Haneberg B and Larson R (eds) *Humans as Geologic Agents*. Geological Society of America Reviews in Engineering Geology.

Garner, L.E. and Young, K.P. 1976. *Environmental geology of the Austin area: an aid to urban planning*. The University of Texas at Austin, Bureau of Economic Geology Report of Investigations No 86, 39 p.

Geomatrix, Inc. 1997. *Conceptual hydrogeologic model, Charnock wellfield regional assessment*. Unpublished report, Los Angeles.

Giudici, M., Colpo, F., Ponzini, G. and Romano, E. 2001. Calibration of groundwater recharge and hydraulic conductivity for the aquifer system beneath the city of Milan, Italy. International Association of Hydrological Sciences, Publication No 269, p 43–50.

Grasby, S.E., Hutcheon, I. and Krouse, H.R. 1997. Application of the stable isotope composition of SO_4 to tracing anomalous TDS in Nose Creek, southern Alberta, Canada. Applied Geochemistry, v 12, p 567–575.

Grischek, T., Nestler, W., Piechniczek, K. and Fischer, T. 1996. Urban groundwater in Dresden, Germany. *Hydrogeology Journal*, v 4, p 48–63.

Halihan, T., Sharp, J.M. Jr. and Mace, R.E. 1999. Interpreting flow using permeability at multiple scales. In Palmer AR, Palmer MV and Sasowsky ID (eds) *Karst Modeling*. Karst Waters Institute Special Publication No 5, Charlottesville, p 82–96.

Heiken, G., Fakundiny, R. and Sutter, J. 2003. Earth Science in the City. *American Geophysical Union Special Publication Series*, v 56, 440 p.

Hooker, P.J., Mc Bridge, D., Brown, M.J., Lawrence, A.R. and Gooddy, D.C. 1999. An integrated hydrogeological case study of a post-industrial city in the West Midlands of England. In Chilton J (ed) *Groundwater in the Urban Environment – Selected City Profiles*. Balkema, Rotterdam, p 145–150.

Howard, K.W.F. 2002. Urban groundwater issues – an introduction. In Howard KWF and Israfilov RG (eds) *Current Problems of Hydrogeology in Urban Areas, Urban Agglomerates and Industrial Centres*. NATO Science Series. IV Earth and Environmental Sciences, v 8, p 1–15.

Howard, K.W.F. and Israfilov, R.G. 2002. *Current Problems of Hydrogeology in Urban Areas, Urban Agglomerates and Industrial Centres*. NATO Science Series. IV Earth and Environmental Sciences, v 8, 504 p.

Ikramov, R.K. and Yakubov, K.H.I. 2002. A survey of groundwater level rise and recommendations for high water table mitigation for the city of Gulistan, Republic of Uzbekistan. In Howard KWF and Israfilov RG (eds) *Current Problems of Hydrogeology in Urban Areas, Urban Agglomerates and Industrial Centres*. NATO Science Series. IV Earth and Environmental Sciences, v 8, p 425–436.

Israfilov, R.G. 2002. Groundwater anomalies in the urban areas of Azerbaijan. In Howard KWF and Israfilov RG (eds) *Current Problems of Hydrogeology in Urban Areas, Urban Agglomerates and Industrial Centres*. NATO Science Series. IV Earth and Environmental Sciences, v 8, p 17–37.

Jakovljev, V.V., Svirenko, L.P., Chebanov, O.J.U. and Spirn, O.I. 2002. Rising groundwater levels in North-Eastern Ukraine: Hazardous trends in urban areas. In Howard KWF and Israfilov RG (eds) *Current Problems of Hydrogeology in Urban Areas, Urban Agglomerates and Industrial Centres*. NATO Science Series. IV Earth and Environmental Sciences, v 8, p 221–241.

Kim, Y.Y., Lee, K.K. and Sung, I.H. 2001. Urbanization and the groundwater budget, metropolitan Seoul area, Korea. *Hydrogeology Journal*, v 9, p 401–412.

Knipe, C.V., Lloyd, J.W., Lerner, D.N. and Greswell, R. 1993. *Rising groundwater levels in Birmingham and the engineering implications*. CIRIA Special Publication No 92, 114 p.

Krothe, J.N. 2002. *Effects of urbanization on hydrogeological systems: the physical effects of utility trenches*. MS thesis, The University of Texas, Austin, 190 p.

Krothe, J.N., Garcia-Fresca, B. and Sharp, J.M. Jr. 2002. Effects of urbanization on groundwater systems. *Abstracts for the International Association of Hydrogeologists XXXII Congress, Mar del Plata, Agentina*, p 45.

Ku, H.F.H., Hagelin, N.W. and Buxton, H.T. 1992. Effects of urban storm runoff control on groundwater recharge in Nassau County, New York. *Ground Water*, No 30, p 507–514.

La Dell, T. 1986. Rising groundwater in Doha, Qatar. Geological Society of London News, v 15, abstract.

Lara, G.F. and Ortiz, F.G. 1999. Feasibility study for the attenuation of groundwater exploitation impacts in the urban area of Aguascalientes, Mexico. In Chilton J (ed) *Groundwater in the Urban Environment – Selected City Profiles*. Balkema, Rotterdam, p 181–187.

Leopold, L.B. 1968. *Hydrogeology for urban land planning – A guidebook on the hydrologic effects of urban land use*. US Geological Survey Circular 554, 18 p.
Leopold, L.B. 1973. River channel change with time; an example. *Geological Society of America Bulletin*, v 84, No 6, p 1845–1860.
Lerner, D.N. 1986. Leaking pipes recharge ground water. *Ground Water*, v 24, p 654–662.
Lerner, D.N. 1990a. Groundwater recharge in urban areas. In Massing H, Packman J and Zuidema FC (eds) *Hydrological Processes and Water Management in Urban Areas*. International Association of Hydrological Sciences, Publication No 198, Wallingford, UK, p 69–76.
Lerner, D.N. 1990b. Groundwater recharge in urban areas. *Atmospheric Environment*, v 24B, p 29–33.
Lerner, D.N., Issar, A.S. and Simmers, I. 1990. Groundwater recharge – A guide to understanding and estimating natural recharge. *IAH International Contributions to Hydrogeology*, v 8. Heise, Hannover, 345 p.
Lerner, D.N., Burston, M.W. and Bishop, P.K. 1993. Hydrogeology of the Coventry region (UK): an urbanized, multilayer, dual porosity aquifer system. *Journal of Hydrology*, v 149, p 111–135.
Lerner, D.N. 1997. Too much or too little: Recharge in urban areas. In Chilton J et al. (eds) *Groundwater in the Urban Environment: Problems, Processes and Management*. Balkema, Rotterdam, p 41–47.
Lerner, D.N. 2002. Identifying and quantifying urban recharge: a review. *Hydrogeology Journal*, v 10, p 143–152.
Menegasse, L.N., Duarte, U. and Pereira, P.R.B. 1999. Effects of urbanisation on aquifer recharge and superficial runoff in the Sumare and Pompeia basins, Sao Paulo Municipality, Brazil. In Chilton J (ed) *Groundwater in the Urban Environment – Selected City Profiles*. Balkema, Rotterdam, p 195–199.
NOAA – National Oceanic and Atmospheric Administration (online) http://www.noaa.gov/ [accessed July 2003].
Nkhuwa, D.C.W. 1999. The need for consideration of groundwater resources in urban planning in Lusaka, Zambia. In Chilton J (ed) *Groundwater in the Urban Environment – Selected City Profiles*. Balkema, Rotterdam, p 201–208.
Norin, M., Hulten, A.M. and Svensson, C. 1999. Groundwater studies conducted in Goteborg, Sweden. In Chilton J (ed) *Groundwater in the Urban Environment – Selected City Profiles*. Balkema, Rotterdam, p 209–216.
Pokrovsky, D.S., Rogov, G.M. and Kuzevanov, K.I. 1999. The impact of urbanization on the hydrogeological conditions of Tomsk. In Chilton J (ed) *Groundwater in the Urban Environment – Selected City Profiles*. Balkema, Rotterdam, p 217–223.
Rieckermann, J., Kracht, O. and Gujer, W. 2003. How "tight" is our sewer system? Swiss Federal Institute for Environmental Science and Technology, *EAWAG News* No 57, p 29–31.
Rose, P.R. 1972. *Edwards Group, surface and subsurface, central Texas*. The University of Texas Bureau of Economic Geology Report of Investigations No 74, 198 p.
Salameh, E., Alawi, M., Batarseh, M. and Jiries, A. 2002. Determination of trihalomethanes and the ionic composition of groundwater at Amman City, Jordan. *Hydrogeology Journal*, v 10, p 332–339.
Seiler, K.P. and Alvarado Rivas, J. 1999. Recharge and discharge of the Caracas Aquifer, Venezuela. In Chilton J (ed) *Groundwater in the Urban Environment – Selected City Profiles*. Balkema, Rotterdam, p 233–238.
Sharp, J.M. Jr. 1993. Fractured aquifers/reservoirs: approaches, problems, and opportunities. In Banks D and Banks S (eds) Hydrogeology of Hard Rocks. *Memoir of the XXIV Congress, International Association of Hydrogeologists*, Oslo, v 24, No 1, p 23–38.
Sharp, J.M. Jr., Hansen, C.N. and Krothe, J.N. 2001. Effects of urbanization on hydrogeological systems: the physical effects of utility trenches. In Seiler K-P and Wohnlich S (eds) *New Approaches Characterizing Groundwater Flow, supplement volume*. Proceedings of the International Association of Hydrogeologists XXXI Congress, Munich, Germany, 4 p.
Sharp, J.M. Jr., Krothe, J., Mather, J.D., Garcia-Fresca, B. and Stewart, C.A. 2003. Effects of urbanization on groundwater systems. In Heiken G, Fakundiny R and Sutter J (eds) *Earth Science in the City*. American Geophysical Union Special Publication Series, v 56, p 257–278.
Simmers, I. 1998. Groundwater recharge: an overview of estimation 'problems' and recent developments. In Robins NS (ed) *Groundwater Pollution, Aquifer Recharge and Vulnerability*. Geological Society Special Publication No 130, London, p 107–115.
TexasET – Texas Evapotranspiration network (online) http://texaset.tamu.edu [accessed March 2004].
Thornton, J. 2002. *Water Loss Control Manual*. McGraw-Hill, New York, 526 p.

US Census Bureau (online) http://www.census.gov/ [accessed November 2003].
Vázquez-Suñé, E. 2003. *Urban groundwater. Barcelona City case study*. Doctoral Thesis, Universitat Politécnica de Catalunya, Departament d'Enginyeria del Terreny, Cartográfica I Geofísica, unpublished, 134 p.
Worthington, S.R.H. 2003. A comprehensive strategy for understanding flow in carbonate aquifers. In Palmer AN, Palmer MV and Sasowsky ID (eds) *Speleogenesis and Evolution of Karst Aquifers*. Karst Modeling Special Publication No 5, The Karst Waters Institute, Charles Town, p 30–37.
Yang, Y., Lerner, D.N., Barrett, M.H. and Tellam, J.H. 1999. Quantification of groundwater recharge in the city of Nottingham, UK. *Environmental Geology*, v 38, p 183–198.
Zahm, C.K. 1998. *Use of outcrop fracture measurements to estimate regional groundwater flow: Barton Springs segment of Edwards Aquifer, central Texas*. MS thesis, The University of Texas, Austin, 154 p.

CHAPTER 2

Flow and solute transport monitoring at an urban lysimeter at the Union Brewery, Ljubljana, Slovenia

B. Trček[1] and A. Juren[2]
[1]*Geological Survey of Slovenia, Dimičeva ul. 14, SI–1000 Ljubljana, Slovenia*
[2]*Geosi, Geological Institute d.o.o., Kebetova 24, SI–1000 Ljubljana, Slovenia*

ABSTRACT: An urban lysimeter has been constructed within a highly urbanised area at Union Brewery in Ljubljana (the capital of Slovenia), with the intention of monitoring and controlling the environmental impacts of industry and traffic on groundwater within a Pleistocene alluvial gravel aquifer. The physico-chemical and isotopic properties of sampled groundwater have already produced general information on the hydrodynamic functioning of the study area and on solute transport – the main flow components, the flow hierarchy and the environmental response to the flow system have been indicated. Two important flow types are identified – lateral and vertical flow. The former has an important role in groundwater protection, whilst the latter is the main influence on contaminant transport towards the saturated zone.

1 INTRODUCTION

Groundwater from a Pleistocene alluvial gravel aquifer is becoming an increasingly important drinking water source for the Ljubljana area. It is also an invaluable water source for Union Brewery, which is located within a highly urbanised and industrialised area near the centre of Ljubljana and supplies high quality groundwater from four production wells (Figure 1). The managers of the brewery are aware that this water should be protected. Therefore, investigations of the environmental impacts of industry and traffic on groundwater of the Pleistocene alluvial gravel aquifer have been undertaken. Flow and solute transport monitoring was conducted in numerous piezometers within the brewery and in its vicinity (some of them are illustrated in Figure 1), as well as at the lysimeter, which is the topic of this paper. The main purpose of the piezometric monitoring is to investigate groundwater quality, whereas the main purpose of the lysimeter monitoring (Frank et al., 2001; Klotz et al., 1999; Raimer et al., 2003) is to study possible contamination in the vicinity of the brewery and with that to evaluate the role of the unsaturated zone in groundwater protection.

2 DESCRIPTION OF THE STUDY AREA

The urban lysimeter at Union Brewery has been constructed adjacent to the brewery (Figure 1). 42 boreholes were drilled into the right and left walls of the 8.5 m deep construction (Figure 2).

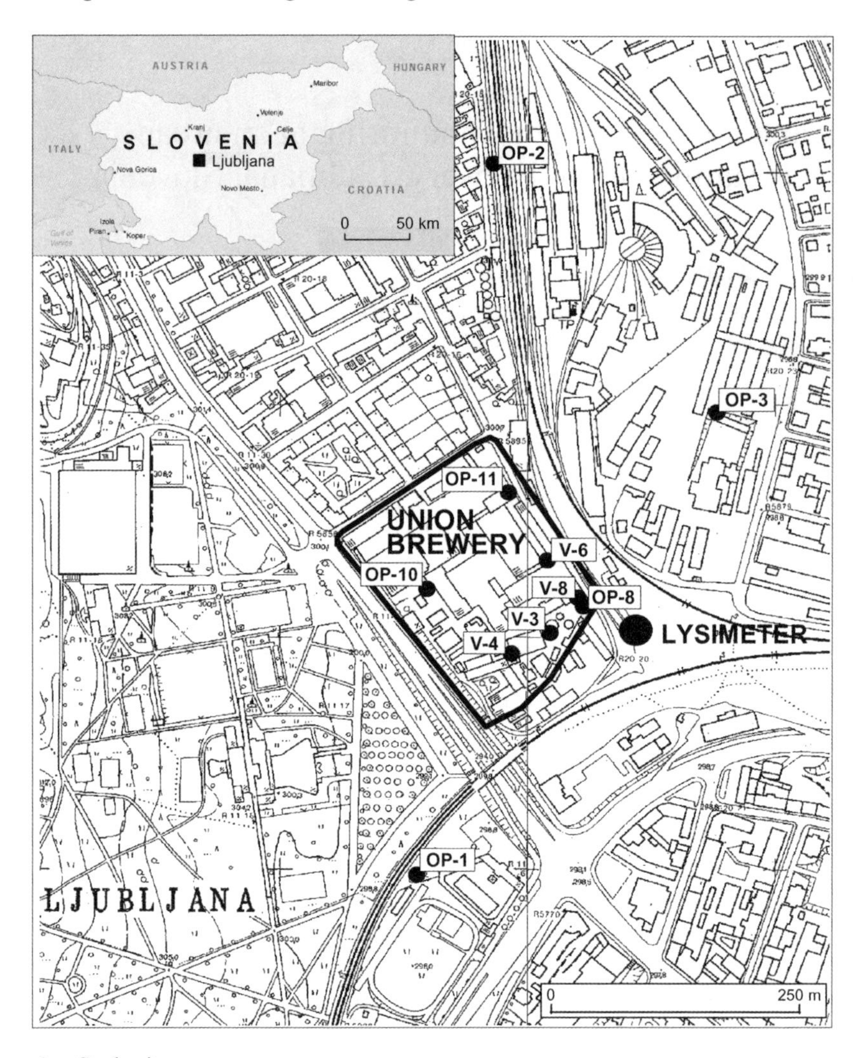

Figure 1. Study site.

To illustrate this, projections of boreholes in the right upper level wall are depicted in Figure 2.

The right wall of the lysimeter, which is located under industrial railway tracks, contains 36 boreholes that are up to 8 m long. They are distributed in six columns (1–6) and six levels (I–VI) at depths of 0.3, 0.6, 1.2, 1.8, 3.0 and 4.0 m (Figures 2, 3). The boreholes are defined by their distribution: RI-1, RI-2, …, RVI-5, RVI-6.

A further six boreholes were drilled beneath the asphalt surface into the left wall of the lysimeter (Figure 2). They are distributed in six columns (1–6) and three levels (I–III) at depths of 0.60, 1.20 and 1.80 m. Hence, they are referred to as LI-1, LI-2, …, LIII-5, LIII-6.

By the beginning of January 2003, the lysimeter was completely equipped with a UMS environmental monitoring system – the recording and sampling system. 15 tensiometers, 9 Time Domain Reflectometry (TDR) probes and 21 suction cups were installed into the boreholes, as listed in Table 1.

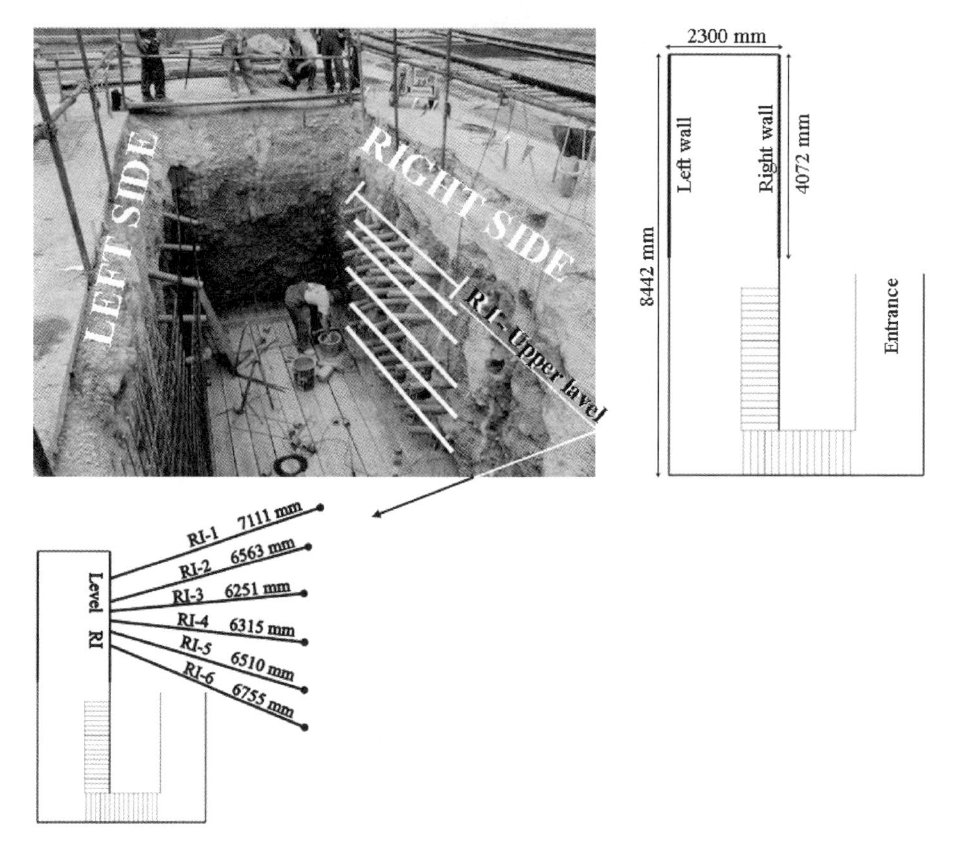

Figure 2. Lysimeter construction with the projection of the upper right level boreholes RI/1–6.

The boreholes penetrate four layers: sandy gravel, silt-sandy gravel, clayey silt-sandy silt with gravel and gravel with sand and silt. The upper three layers are artificial, whereas the fourth layer consists of river deposits. A detailed geological cross-section of the ends of the boreholes in the right wall of the lysimeter is presented in Figure 3.

4 METHODS AND TECHNIQUES

Monitoring of flow and solute transport processes in the lysimeter commenced in June 2003. During the first year of research, continuous measurement of hydrodynamic parameters (capillary pressure and water content), of water balance and of physico-chemical water parameters (pH and electro-conductivity) were carried out to obtain basic information on the study area. In addition, monthly water sampling for analysis of the ^{18}O and ^{2}H isotopic composition was undertaken to obtain additional information about mixing processes and groundwater residence times in the unsaturated zone (Berg et al., 1987; Burgman et al., 1987; Trček, 2003).

Groundwater was sampled with suction cups installed at the end of the boreholes. A total of 18 sampling points were established on the right side of the lysimeter: RI-1 to RI-3, RII-1 to RII-3, etc. (Table 1, Figure 3), whereas on the left side of the lysimeter only 3 sampling

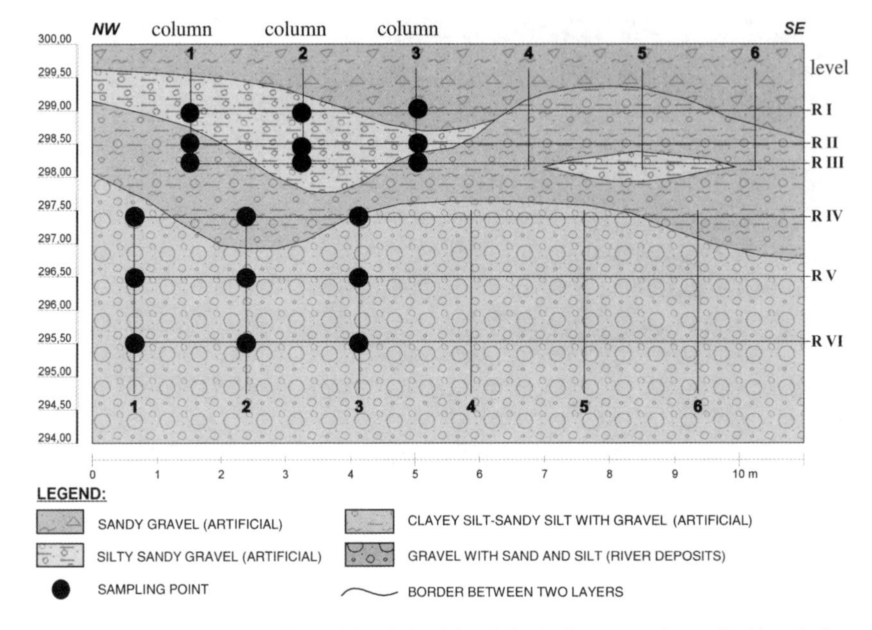

Figure 3. Geological cross-section of the right side of the lysimeter at the end of boreholes, with sampling points indicated (modified after Juren et al., 2003).

Table 1. Position of measuring probes installed in the lysimeter boreholes (■ – installed suction cups, ◘ – installed TDR probes, ○ – installed tensiometers).

Right side	Column						Left side	Column					
	1	2	3	4	5	6		6	5	4	3	2	1
RI	■	■	■	◘	○	○	LI			■○			◘
RII	■	■	■	◘	○	○	LII		■○			◘	
RIII	■	■	■	◘	○	○	LIII	■○			◘		
RIV	■	■	■	◘	○	○							
RV	■	■	■	◘	○	○							
RVI	■	■	■	◘	○	○							

points were established: LI-4, LII-5 and LIII-6 (Table 1). In addition, precipitation was sampled near the entrance to the lysimeter.

Groundwater was sampled and preserved based upon the method described by Clark and Fritz (1997). The characteristics of the isotopic composition of natural substances are described in numerous publications (e.g. Clark and Fritz, 1997; Kendall and McDonnell, 1998). The ^{18}O composition is expressed relative to the standard SMOW (Standard Mean Ocean Water), conventionally reported in terms of a relative value δ,

$$\delta_x (‰) = (R_x/R_{st} - 1)1000 \quad (1)$$

where R_x is the isotope ratio (e.g. $^{18}O/^{16}O$) in the substance x, R_{st} is the isotope ratio in the corresponding international standard substance, and δ is expressed in parts per thousand.

Water samples were analysed for $\delta^{18}O$ in the isotope laboratory at the GSF-Institute of Groundwater Ecology in Neuherberg (Germany), with a standard analytical error of ±0.05‰.

5 RESULTS

The water balance for the lysimeter sampling points during the first phase of the research is presented in Table 2. There is an absence of data for sampling points RIII-2 and RIII-3 for the first part of the monitoring period, because a proper measuring system was only established in April 2004. Nevertheless, it can be observed in Table 2 that these two sampling points discharged the highest volumes, and that on both the right and left side of the lysimeter, the bulk of the water is discharged to sampling points on level III. It is important to note that a low discharge occurs under the asphalt surface (LIII-6, during this period, drainage water was not observed at the other sampling sites).

Figure 3 illustrates that the sampling points on level III are located near the contact between two structurally different layers: silty-sandy gravel and underlying clayey silty-sandy silt with gravel grains. The hydraulic conductivity of the upper layer is higher than that of the lower layer. Therefore it is presumed that the greater volumes discharged from level III result from the development of a lateral flow component. Figures 4 and 5 demonstrate that the discharges of level III are strongly dependent on precipitation levels and intensity. Figure 5 also indicates the occurrence of vertical flow from level III, which results in increased volume of discharge from sampling points at the lower levels, particularly from RIV-2 (Oct. 2003, Apr. and Jun. 2004).

Statistical characteristics of the electro-conductivity of sampled water are presented in the form of boxplots ranges from 180 to 615 μS/cm (Figure 6). Significantly higher values were recorded on the left side of the lysimeter – up to 4000 μS/cm. These most probably result from winter contamination.

Lowest electro-conductivity values in the lysimeter are connected with levels I and II (Figures 6 and 7), whereas highest values are connected with level III (Figures 6 and 7), which reflects the important role of the lateral flow component near this level. In addition, Figure 7 also illustrates when and where the vertical flow component dominated. Vertical breakthrough of water from level III into level IV is particularly highlighted for April 04.

Boxplots of the ^{18}O isotopic composition of sampled water (precipitation and groundwater) are presented in Figure 8. Precipitation values range between −4.1 and −15.2‰ with a mean value of −8.9‰. Groundwater values vary between −4.5 and −14.7‰, whilst the means of single sampling points are between −8 and −10.7‰. The means as well as the spread of $\delta^{18}O$ for the various lysimeter sampling points differ significantly. These differences most probably reflect different residence times of the seepage water. Comparison with precipitation indicates that the ranges of groundwater for the upper two levels (I and II) are highest, reflecting the intensive groundwater dynamics and short residence times. On the other hand, the range of groundwater values for the lower levels (III, IV, V and VI) are relatively small, which reflects less intensive dynamics and longer residence times.

The $\delta^{18}O$ characteristics of the sampled water are also illustrated in Figures 9 and 10, which present the parameter time-trends for groundwater of the upper and lower lysimeter levels, respectively.

Table 2. Water balance of lysimeter sampling points.

	Volume (ml)													Precipitation (mm)
	RII 1	RV 1	RI 2	RII 2	RIII 2	RIV 2	RVI 2	RI 3	RIII 3	RIV 3	RV 3	RVI 3	LIII 6	
10.07.03	280	340	86	41		70	19	455		110	45	160	5	57.7
27.08.03	385	490	38	45		95	38	370		45	65	38		71.6
17.09.03	175	175	21	20		50		220			40		38	44.5
16.10.03	380	200	110	24		890	29	190		100	24	40	37	110.7
12.11.03	190	180	100	20		60	55	180			20	30		121.4
09.12.03	190		60	20		60		120					20	73.8
20.01.04	280		20			80		90						150.3
17.02.04	180	10	35	20		48		25		27	7	7	10	12.7
25.03.04	230	30	190			50		40				35	20	122.5
15.04.04	420		110		49936	620			75580	40			25	94.3
12.05.04	190	23	25	27	79590	40		20	92550		20	35	5	64.8
15.06.04	520	10	30	20	76880	510		110	136210	50	20	30	25	83.1
13.07.04	210	40	100	20	81140	50		150	125330		30	30	15	133.0
11.08.04	220	25	80	22	89320	70			132530		70	50	20	89.2
Total volume	3420	1485	825	237		2573	141	1820		372	241	375	185	1007.4

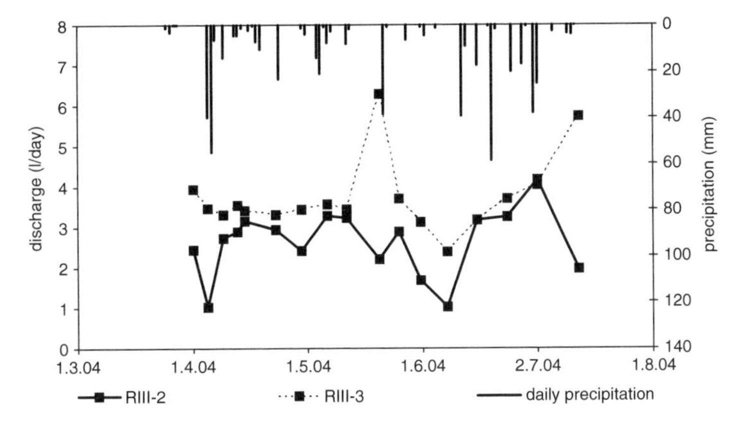

Figure 4. Daily discharge collected at lysimeter sampling points RIII-2 and RIII-3.

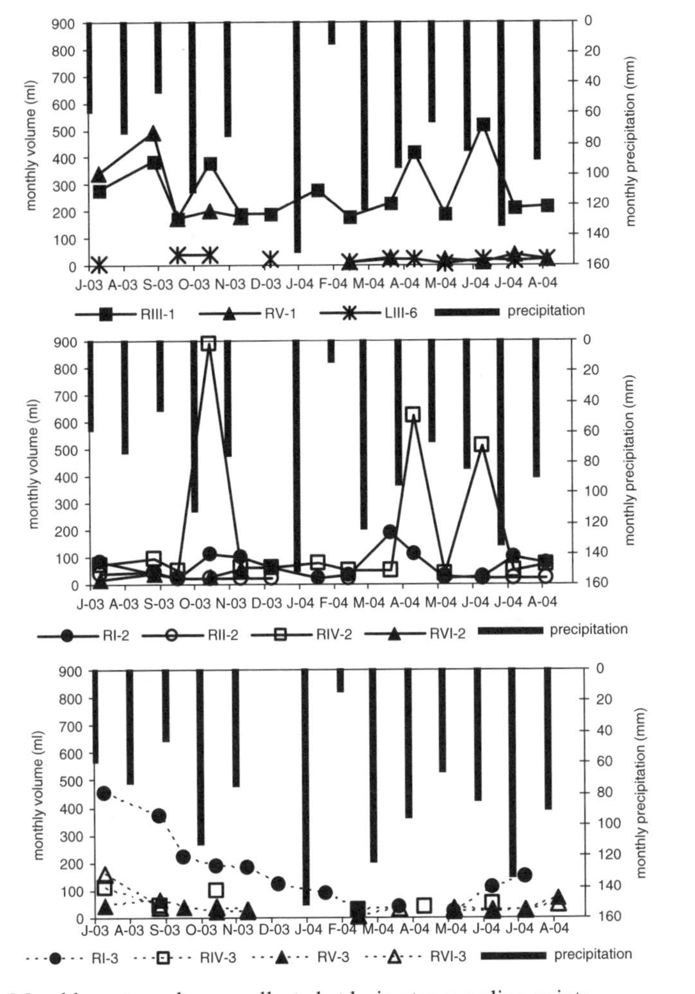

Figure 5. Monthly water volumes collected at lysimeter sampling points.

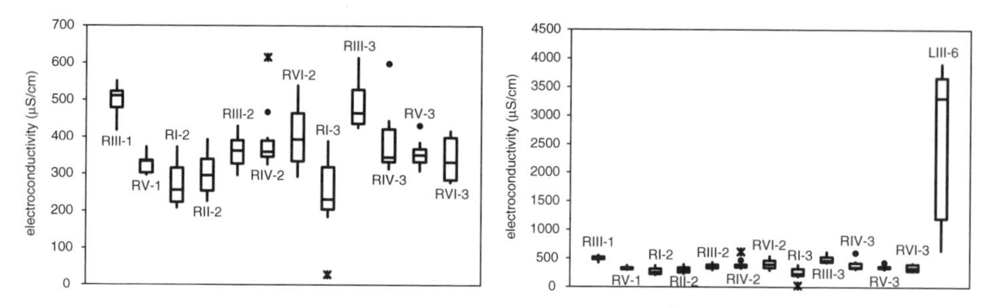

Figure 6. Boxplots of electroconductivity values for water sampled on the right and left side of the lysimeter beneath the industrial railway tracks and the asphalt surface, respectively.

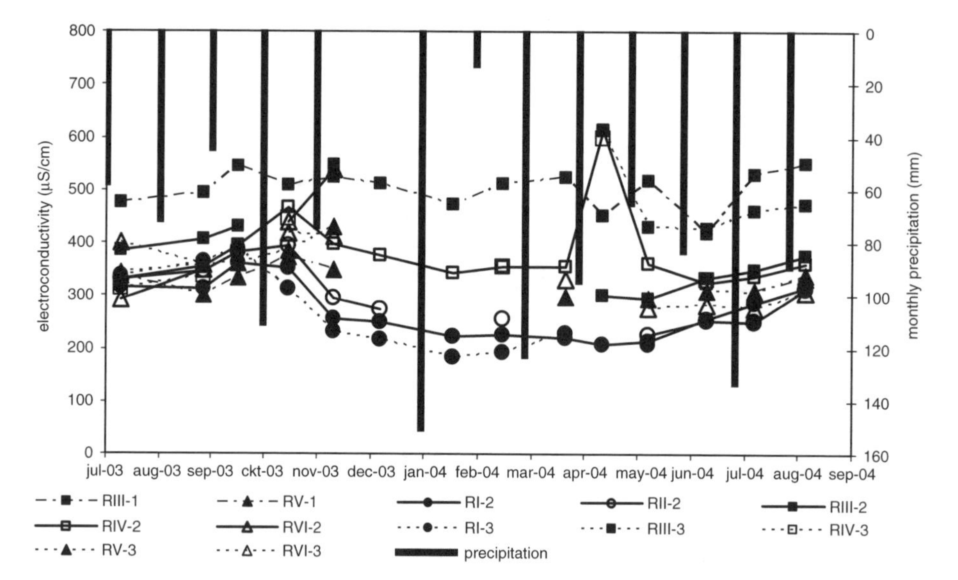

Figure 7. Time-trend plot of electroconductivity values for water sampled on the right side of the lysimeter beneath the industrial railway tracks.

Comparison of $\delta^{18}O$ trends in precipitation and groundwater in Figures 9 and 10 demonstrates that variations in this parameter are much more attenuated in the lower levels of the lysimeter, which probably reflects a longer average groundwater residence time. Peak values in both figures indicate vertical flow and solute transport in the aquifer during the main hydrological events, i.e. October 2003 and April 2004. For example, in April 2004, precipitation pushed low $\delta^{18}O$ water into the lower lysimeter levels (Figure 10). It is presumed that these values may have resulted from snowmelt. During the previous month, the influence of snowmelt is readily observed in the upper levels of the lysimeter (Figure 9).

6 DISCUSSION AND CONCLUSIONS

Results of the first phase of the research at the Union Brewery lysimeter has produced general information on the hydrodynamic functioning of the study area and on solute transport.

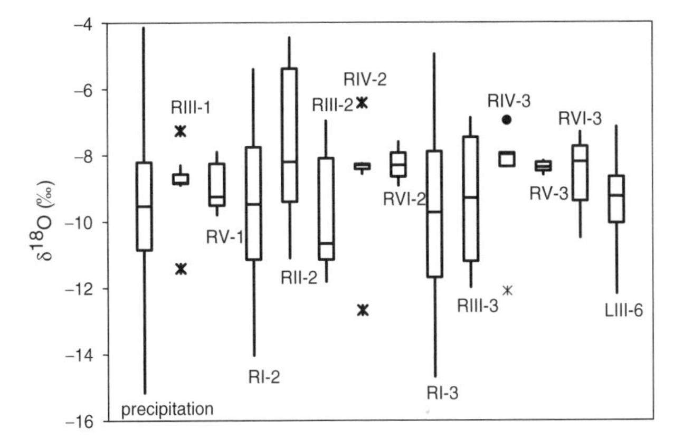

Figure 8. Boxplots of $\delta^{18}O$ values in sampled water.

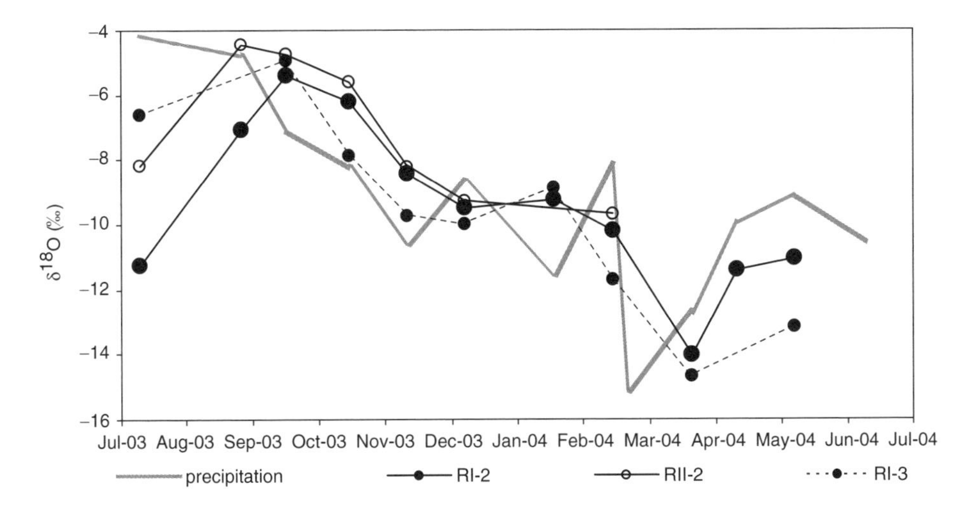

Figure 9. Time-trend plot of $\delta^{18}O$ values in water sampled from the lysimeter upper levels.

The synthesis of one-year of monitoring data has revealed the basic characteristics of flow and solute/contaminant transport, since the main flow components, the flow hierarchy and the environmental response to the flow system are all indicated. Two important flow types were identified – lateral and vertical flow. Lateral flow has an important role in the protection of groundwater of the Pleistocene alluvial gravel aquifer (Seiler et al., 2000). However, the role of vertical flow is quite the opposite, because it is the main factor controlling contaminant transport towards the saturated zone. Hence, investigation of the occurrence and frequency of rapid recharge events represents one of the main themes of the next research phase.

The presented results will help us to design and upgrade the methodology for the future research, which will also include tracer experiments and short-term monitoring of characteristic hydrological events, such as storm events and snowmelt.

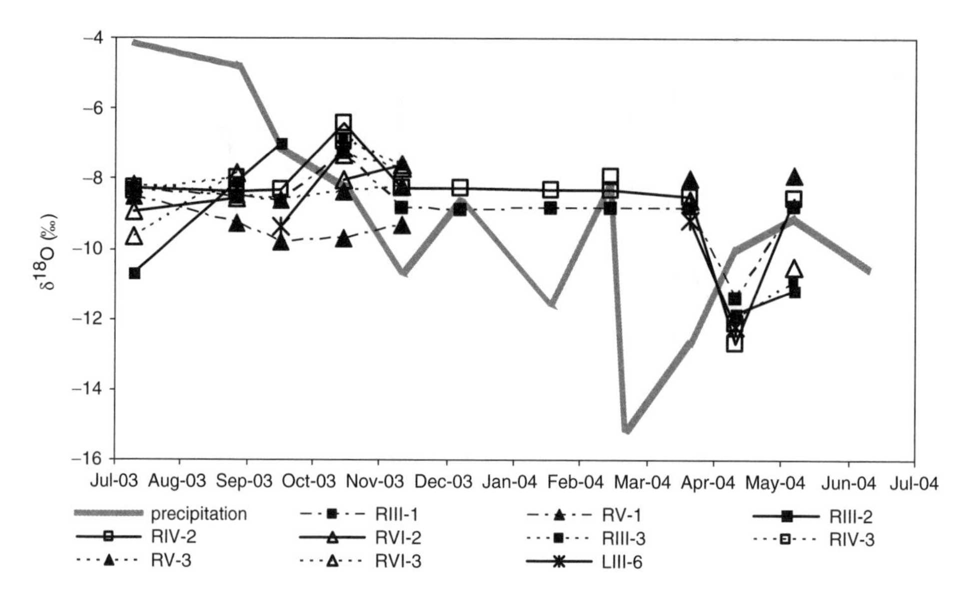

Figure 10. Time-trend plot of $\delta^{18}O$ values in water sampled from the lysimeter lower levels.

REFERENCES

Berg, W., Čenčur Curk, B., Frank, J., Feichtinger, F., Nützmann, G., Papesch, W., Rajner, V., Rank, D., Schneider, S., Seiler, K.P., Steiner, K.H., Stenitzer, E., Stichler, W., Trcek, B., Vargay, Z., Veselič, M. and Zojer, H. 2001. Tracers in the unsaturated zone. *Steir. Beitr. Hydrogeol.*, 52: 1–102.

Burgman, J.O., Calles, B. and Westman, F. 1987. Conclusions from a ten year study of oxygen-18 in precipitation and runoff in Sweden. In: *Publication of IAEA Symposium 299 on Isotope Techniques in Water Resources Development*. Vienna: 579–590. Vienna: IAEA.

Clark, I.D. and Fritz, P. 1997. *Environmental Isotopes in Hydrogeology*. New York: Lewis Publishers.

Fank, J., Ramspacher, P. and Zojer, H. 2001. Art der Sickerwassergewinnung und Ergebnisinterpretation. In: *Bericht der BAL über die Lysimetertagung, Gumpenstein, 16–17 April 1991*: 55–62. Gumpenstein: BAL.

Juren, A., Pregl, M. and Veselič, M. 2003. Project of an urban lysimeter at the Union brewery, Ljubljana, Slovenia. *RMZ-Materials and Geoenvironment* 50(3): 153–156.

Kendall, C. and McDonnell, J.J. 1998. *Isotope tracers in catchment hydrology*. Amsterdam: Elsevier.

Klotz, D., Seiler, K.P., Scheunert, I. and Schroll, R. 1999. Die Lysimeteranlagen des GSF-Forschungscentrums fuer Umwelt und Gesundheit. In: *Bericht ueber die 8. Lysimetertagung*: 157–160. Gumpenstein: BAL.

Raimer, D., Berthier, E. and Andrieu, H. 2003. Development and results of an urban lysimeter. *Geophysical Research Abstracts* 5: 37–92.

Seiler, K.P., Loewenstern, S. and Schneider, S. 2000. The role of by-pass and matrix flow in the unsaturated zone for groundwater protection. In: Sililo et al. (eds.), *Groundwater: Past Achievements and Future Challenges:* 43–76. Rotterdam: Balkema.

Trček, B. 2003. *Epikarst zone and the karst aquifer behaviour, a case study of the Hubelj catchment, Slovenia*. Ljubljana: Geološki zavod Slvenije.

CHAPTER 3

Urban infrastructure and its impact on groundwater contamination

S. Burn[1], M. Eiswirth[2], R. Correll[3], A. Cronin[4], D. DeSilva[1], C. Diaper[1], P. Dillon[5], U. Mohrlok[6], B. Morris[7], J. Rueedi[4], L. Wolf[2], G. Vizintin[8] and U. Vött[9]

[1]CSIRO Manufacturing & Infrastructure Technology, Highett, Australia
[2]Department of Applied Geology, Karlsruhe University, Karlsruhe, Germany
[3]CSIRO Mathematical & Information Sciences, Urrbrae, Australia
[4]University of Surrey, Robens Centre for Public & Environmental Health, Guildford, UK
[5]CSIRO Land & Water, Glen Osmond, Australia
[6]Institute for Hydromechanics, Karlsruhe University, Karlsruhe, Germany
[7]British Geological Survey, Wallingford, Oxon, UK
[8]IRGO – Institute for Mining, Geotechnology and Environment, Ljubljana, Slovenia
[9]GKW Consult GmbH, Mannheim, Germany

ABSTRACT: This paper describes a major European/Australian initiative on Assessing and Improving the Sustainability of Urban Water Resources and Systems (AISUWRS). The project aimed to allow cities to assess the impact of leaking urban water infrastructure on the long-term viability of their groundwater supplies for potable water use. By understanding contaminant fluxes and the movement of contaminants from the infrastructure into the underlying aquifer, water utility managers are able to predict whether urban contaminants and the utilities own urban water management systems could adversely affect aquifer sustainability. Decision support models, which feed socio-economic models have been developed in the project, which allows the assessment of different scenarios to determine the most sustainable method of managing water/sewer/stormwater infrastructure to minimise aquifer pollution effects. One of the major outcomes will be a set of guidelines for the sustainable development of urban water systems that take into account the likely future effects of groundwater contamination. For the verification and validation of the model, detailed field studies have been carried out in the four case study cities of Mt Gambier (Australia), Doncaster (UK), Rastatt (Germany) and Ljubljana (Slovenia). The models will be directly applicable to other European and Australian cities that are dependent on groundwater for their potable water supply, and will allow the long-term cost estimation of city-specific scenarios and practices to be assessed.

1 INTRODUCTION

Traditional water management strategies for major cities have assumed that water supplies and the corresponding water demands are generally in balance, and that cities have enough water to meet demand in the short to medium term. However, in Australia, recent droughts and increased demands have forced authorities to reassess the situation, and many cities

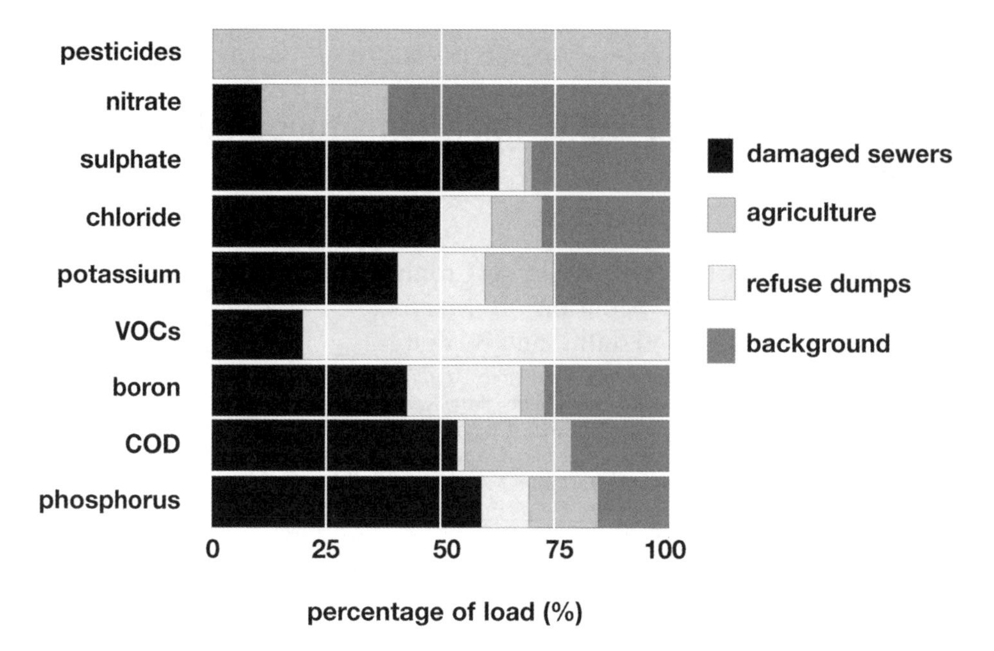

Figure 1. Example of the contaminant loads in urban groundwater from various sources (after Eiswirth et al., 2001): VOCs = volatile organic compounds, COD = chemical oxygen demand.

are now forecast to have significant water deficits. These developments raise questions concerning traditional planning concepts that assume unlimited supplies of potable water. In the future, cities will either have to significantly reduce their water demand or find new sources of potable water. This is especially true if urban population densities increase and, with environmental constraints and limitations on funding major infrastructure such as dams, it is becoming increasingly evident that the present management of urban water resources and systems will not be a suitable model for service provision into the 21st Century. Consequently, increased emphasis will be placed upon the better usage of groundwater reserves, including the use of aquifer storage and recovery (ASR).

In Europe and Australia, groundwater is a major potable water source for many cities. However, in spite of strong efforts by the European Union and other international organisations in the past 20 years, groundwater pollution (inorganic, organic and microbial) from industry, traffic, sewers and peri-urban agriculture continues to threaten urban groundwater resources (Eiswirth et al., 2000, 2001). As detailed in Figure 1, a significant proportion of this contamination can be attributed to leaking urban water infrastructure such as sewers and stormwater pipes. Groundwater provides over 40% of the water supply in Europe and is becoming an important source for Australian water authorities. Thus, efficient and cost-effective management of groundwater is essential for maintaining quality of life, and measures must be taken to ensure the long-term sustainability of potable groundwater water supplies.

Water quality in urban aquifers cannot currently be reliably predicted, and as a result the health and safety of people who depend on urban groundwater as potable water cannot be assured. An adequate knowledge base of the current status of urban water resources, together with an understanding of the processes involved (including the effects of leaking urban water infrastructure), is required to address this situation.

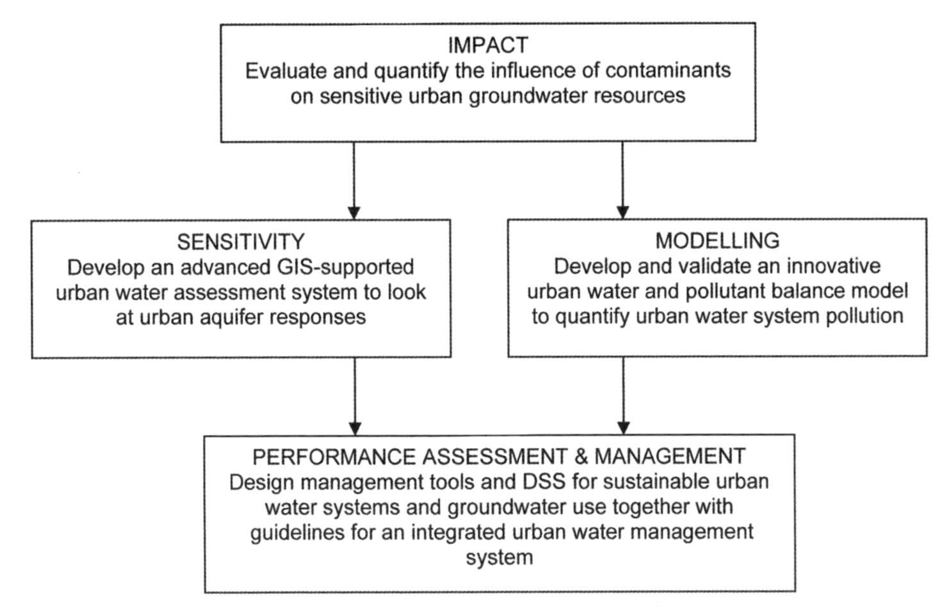

Figure 2. The ethos of the AISUWRS initiative.

As illustrated in Figure 2, the objective of the project titled Assessing and Improving the Sustainability of Urban Water Resources and Systems (AISUWRS) was to develop an innovative system to help quantify and manage urban water systems, by assessing the important role of pollution originating from urban water infrastructure on the underlying groundwater resources. A decision support system (DSS) which links a suite of models has been employed to assess the impact of a range of operational or management scenarios on groundwater contamination. For example, the cost of rehabilitating urban water infrastructure could be compared with groundwater treatment costs, after groundwater contamination.

Because of the worldwide long-term concern over the sustainability of groundwater sources for potable applications, the European Commission, Australia's Commonwealth Scientific and Industrial Research Organisation (CSIRO) and Australia's Department of Education Science and Training have funded the AISUWRS initiative which involves five European partners and one Australian partner:

- Karlsruhe University, Germany,
- University of Survey, UK,
- British Geological Survey, UK,
- Institute for Mining, Geotechnology and Environment, Slovenia,
- GKW Consult, Germany, and
- CSIRO, Australia.

2 SCIENTIFIC AND TECHNICAL OBJECTIVES

The European Water Framework Directive requires member states to achieve good groundwater status in aquifers and in areas with a dynamic exchange between ground and surface water. Contaminants emanating from urban areas such as nitrogen, sulphur and volatile organic compounds (VOCs) threaten both the sustainability of drinking water supplies and the ecological integrity of surface and groundwater in sensitive areas.

The overall scope of the AISUWRS initiative is to use computer tools to assess different aspects of the sustainability of urban groundwater resources and systems by studying the effects of urban water infrastructure. This has been assessed by utilising a DSS and model suite in case studies based on the four cities of Mt Gambier (Australia), Doncaster (UK), Rastatt (Germany) and Ljubljana (Slovenia). These are cities or large towns where groundwater forms the principal potable water supply. While the three European cities utilised the full suite of models, a different approach was needed in Mt Gambier because the karstic nature of the aquifer with its double porosity flow cannot be accurately reproduced with the models used in the other studies.

In the AISUWRS project, a range of different urban water supply/sewerage service scenarios were analysed in these case study cities, and a critical component of the project involved assessing how each scenario differs in its handling of contaminants from the pipe infrastructure and the potential effect on groundwater contamination. Potential contaminant flow paths to the aquifer, their sinks and their transport through the aquifer were identified and quantified for each of the case study cities. In the AISUWRS project, a management system including a DSS and a socio-economic model (SEM) have been developed to control the developed models and allow the impact of the different scenarios to be assessed. Additionally, the SEM was used to assess the stakeholders' behavioural patterns, the stakeholders' acceptance of the present service situation, as well as their acceptance of future scenarios with the associated costs. This management system enabled the project to deliver guidelines and recommendations for the safeguarding and protection of urban groundwater resources.

While some of these conditions may be specific to the selected case study cities, it is expected that the approach and methods used in AISUWRS will be directly transferable to urban regions worldwide.

3 WORK PROGRAM

The work program of AISUWRS was based on a suite of six models that allow the prediction of the effects of urban water infrastructure on groundwater contamination, as shown in Figure 3. The six models are:

- An Urban Volume and Quality (UVQ) model.
- A pipeline leakage model.
- An analytical unsaturated flow and transport model to predict leaching to the aquifer beneath each pipeline leak (SLEAKI) and from direct infiltration from open spaces (POSI).
- A numerical unsaturated flow and transport model to inform and calibrate the analytical model.
- A numerical groundwater flow and transport models used to predict contaminant transport within the aquifer.
- An analytical transport model for use in karstic aquifers, based on tracer and contaminant fate studies.

The DSS has been developed to control the functionality of these models and to allow the assessment of a range of scenarios, whilst the SEM allows the financial implications

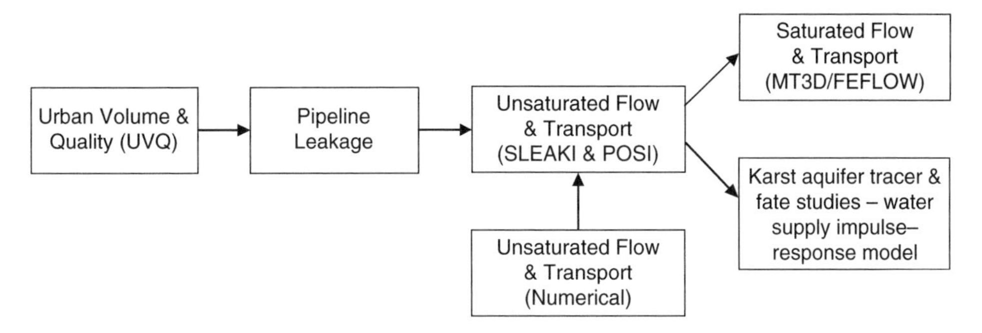

Figure 3. The six models informing the AISUWRS decision support system.

of these scenarios to be assessed, for example to compare the costs of infrastructure rehabilitation with those costs caused by groundwater contamination; it also analyses the stakeholders' acceptance and actual sensitivities and preferences. The graphical interaction of the model development and the interaction with other components of the program are illustrated in Figure 4.

The potential contaminant loads from different urban contaminant sources (sewer leakages, stormwater overflow, etc.) and their transport to the aquifer are illustrated in Figure 5. Each model in the array acts collectively and interacts with a database in order to simulate contaminant quantities and their flow to the aquifer. The final stage allows an assessment of scenarios such as the benefits associated with various management actions for example, infrastructure rehabilitation. All models were calibrated and validated in the four case study cities. A brief discussion of each component is given below.

3.1 *Urban volume and quality model*

As part of Australia's Urban Water Program, a tool has been developed to estimate the water flows and contaminant loads within an urban water system. It is a conceptual, daily time step, urban water and contaminant balance model called UVQ (Urban Volume and Quality) (Mitchell et al., 2003, 2004). This model represents water and contaminant flows through existing urban water, sewerage and stormwater systems, from source to discharge point. The UVQ modelling tool has been further developed in the AISUWRS project to permit the analysis of contaminant flows, and this has been tested at each case study site, by exploring alternative water supply reticulation, sewage collection and stormwater drainage scenarios. Current results from the Doncaster case study city are documented in Rueedi et al. (2004).

3.2 *Pipeline leakage model*

The pipeline leakage assessment model allows for an understanding of the impact of leakage from sewer, stormwater and water supply pipelines on groundwater contamination.

Whilst leakage from the water supply system can be estimated through district meter readings, customer water meter readings and night flow analysis, the level of exfiltration from sewer and stormwater pipelines is more difficult to measure accurately. The pipeline leakage model allows the quantification of leakage or exfiltration from pipelines and can also allow for the infiltration calculation where the groundwater level is above the pipe network

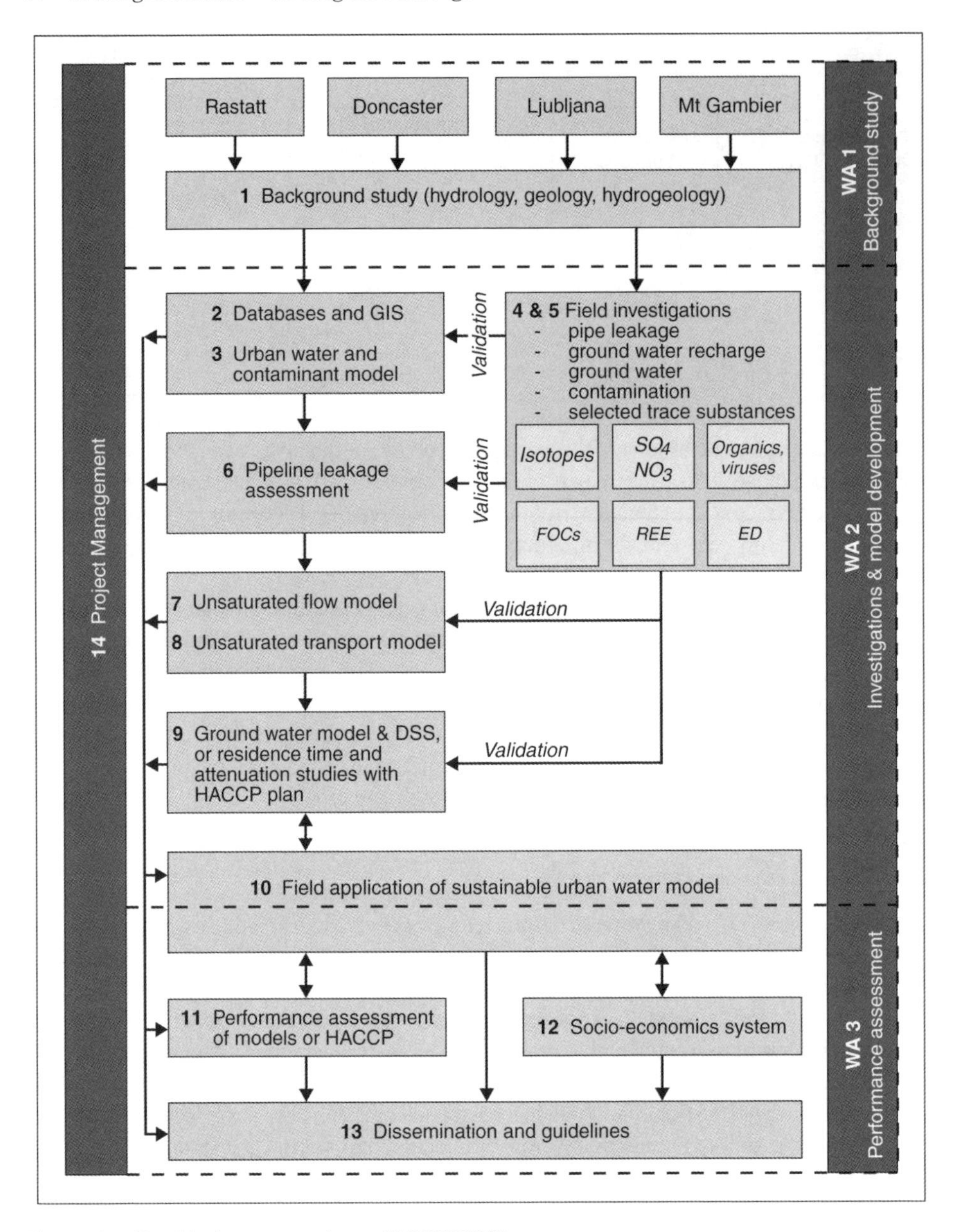

Figure 4. Graphical representations of AISUWRS's components.

(DeSilva et al., 2004). For water pipelines, the model assesses leakage from cast iron, ductile iron, polyvinylchloride, polyethylene and mild steel pipelines based on their size, age and pressure ratings, and it is based on leakage rates measured in current reticulation systems. For sewer and stormwater pipes, the leakage model relates closed circuit television (CCTV) damage assessed in pipes, (usually made of vitrified clay and concrete), and correlates these damage functions with variable flow volumes in pipelines, as illustrated in Figure 6,

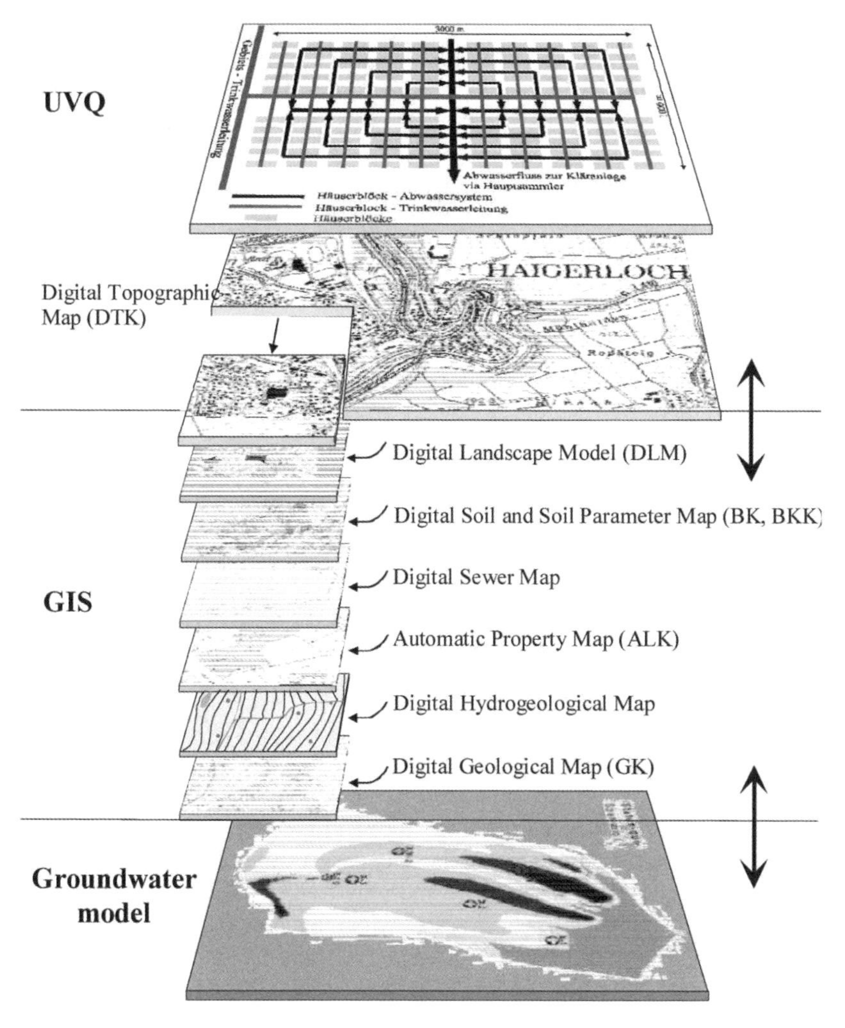

Figure 5. Model concept of the combined urban water model.

to calculate the point source leakage rate per asset for sewer pipelines of varying sizes, age, material and terrain/soil installation conditions.

The pipeline leakage model has been verified by CCTV analysis of selected sections of pipes, where leakage has been assessed. Leakage volumes were measured by isolating sections of pipeline, filling these pipelines with sewage and then measuring leakage rates. Pipe sections selected for the study cover the sizes, ages and materials located in soil types typically found at the case study sites. The distribution of leakage rates calculated for the different pipelines has been complemented, where possible, by estimates of exfiltration on a catchment-wide basis, by comparing gauged sewer flows with metered water consumption.

3.3 *Unsaturated flow model*

Infiltration rates to urban groundwater from different urban sources (precipitation, irrigation, rainwater infiltration, water system losses, sewer leakage, etc.) have been quantified

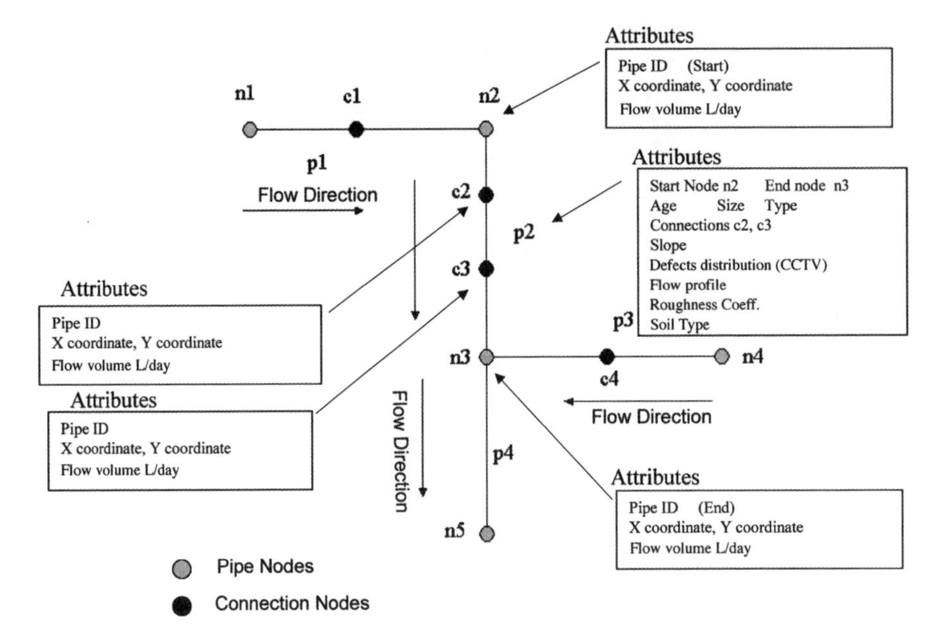

Figure 6. Schematic for calculation of flow volumes in pipelines.

in order to assess the urban groundwater recharge and seasonal and spatial variations in the unsaturated zone at the case study sites. Besides modelling the field data sets of the case study cities, sensitivity studies have been performed to investigate the effect of different soil characteristics and different hydraulic boundary conditions like infiltration rates, depth of groundwater table, soil moisture and pipe leakage rates. A recently developed simple mass balance approach has been tested with the numerical results collected in the case study cities. This approach has been used to generalise the effect of different recharge sources and different infiltration conditions. It is recognised that this modelling is difficult and cannot be carried out in the time frames used by the DSS, consequently this model has been used to verify a simplified unsaturated flow model included in the unsaturated transport model described below.

3.4 *Unsaturated transport model*

The unsaturated transport model estimates the likely fluxes of reactive solutes and selected microbial indicators of faecal contamination that are of importance to the case study cities, and that are expected to reach the aquifer from leaks in sewers, from stormwater infiltration, and from reuse of grey water and reclaimed water. The model developed, the Sewer Leakage Index (SLeakI) extended concepts from the Pesticide Impact Rating Index (PIRI) that has been successfully used to indicate pesticide leaching, and has been subsequently adapted by Miller et al. (2002) as an Aquifer Storage and Recovery Risk Index (ASRRI) to gauge attenuation of contaminants in aquifers following the injection of surface waters.

PIRI uses a one-dimensional (vertical) solute leaching program that models the unsaturated zone where a source of reactive or non-reactive contaminant is deposited at the soil surface, with the daily rainfall record being used to derive leaching rates. PIRI output appears as relative probabilities of contaminant concentrations leaching to the aquifer. In SLeakI which is based on PIRI, a new approach has been developed using 3-D flow from a point source, allowing for transfer through two homogenous soil layers to the aquifer. SLeakI derives mass fluxes from a series of point sources (at the surface or at the sewer leak) that are each modelled independently. SLeakI is a simple approach and is useful for broad-scale applications in the absence of detailed site-specific data that would be necessary for the effective use of more complex models. It is suited for use in small-scale analysis and has not been utilised before for city-wide analysis. In the case study cities, where more extensive data and models are available, the effectiveness of SLeakI (which uses soil parameters that can be mapped at city-wide scale) has been quantitatively evaluated. While SLeakI cannot hope to capture detail at every point source, it provides a reasonable estimate of the mass flux or total number of reactive solutes and microbial indicators (boron, *E. coli*, etc.) reaching groundwater as an average over a large number of point sources.

To predict the infiltration associated with urban green areas and other public open spaces a second simpler 1-D model has been developed called the Public Open Space Index (POSI). This software can also be applied to area sources such as septic tanks. POSI uses as input the UVQ output file for gardens and public open spaces that infiltrate into the soil. As with SLeakI, the DSS aggregates the POSI output to a neighbourhood scale.

3.5 *Groundwater flow and transport model*

For the three European case study cities of Rastatt, Doncaster and Ljubljana, the initial conceptual models have been adapted from the regional groundwater flow models, which cover an area of aquifer outcrop and sub-crop much larger than the urbanised extent of the cities themselves (or more precisely, of the detailed study areas). The existing regional/sub-regional groundwater flow models of the different urban areas have been used as a base, and expanded to populate one of the widely used proprietary codes such as MODFLOW or FEFLOW. For solute balance modelling, either Multiple Analytical Pathways (MAP) solute transport models or a proprietary code such as MT3D or FEFLOW have been used. Interaction with the urban water mass balance model (UVQ), especially for the complex recharge calculations typical of an urban area overlying an aquifer, is a particularly useful and novel feature of this input.

All project partners have used modelling approaches that facilitate interfacing the case study flow and contaminant transport models with the urban water models. This compatibility stage is the step future users of this research (water authorities/city managers/environmental regulators) will need to apply to their own groundwater case studies. Developing the groundwater model in the European case cities and linking it to the sophisticated estimate of urban recharge developed in the AISUWRS project, establishes inflows and outflows, and allows interpretation of groundwater flow patterns and the impact of contaminant fluxes on groundwater quality. In each of the European case studies, model calibration has occurred by (a) the use of a groundwater flow model to aid the quantification of urban recharge, and (b) a solute transport model to compare the predicted with measured concentrations of contaminant species. Current results for the groundwater model in the Rastatt case study are detailed in Wolf et al. (2004, 2005).

In the case of Mt Gambier, the karstic aquifer renders the above approach unsuitable for predicting contaminant breakthrough (if any) to the drinking water source (the Blue

Lake). In this instance, a Hazard Analysis and Critical Control Points (HACCP) approach has been adopted to determine the relative risk of contaminants from leaky sewers and stormwater drainage wells reaching the Blue Lake in concentrations that could adversely impact the city's water supply. Inputs to the HACCP plan include evaluations of potential contaminants and the measured quality of stormwater and sewage, travel times from drainage wells to the Blue Lake using tracers, and laboratory measurements of contaminant attenuation in the aquifer under the same environmental conditions (Vanderzalm et al., 2004).

3.6 *Decision Support System*

If the state of the urban water system is accurately known then the impact on groundwater quality can be assessed, e.g. by advanced condition monitoring, and decisions on how to operate asset management schemes and whether to undertake maintenance, repair, replacement or other management measures can be made. However, in the majority of cases, the current condition of the infrastructure is not accurately known, and consequently it is difficult to satisfactorily evaluate replacement or preventive maintenance priorities for sewers and water mains to combat the problem of pipe leakage. It is widely recognised that true asset management requires a balance between risk and planning, and thus it was essential to develop:

- A DSS that supported rational and objective comparisons between the engineering and environmental benefits of each scenario assessed.
- An SEM that reflects the stakeholders' acceptance of measures and associated costs, which can be used to predict stakeholders' behavioural changes in the future.

The DSS model developed in AISUWRS allows operational and rehabilitation procedures to be analysed, and facilitates the analysis of a range of scenarios developed in consultation with the stakeholders, for example:

- continuing with current (climate and demand) conditions,
- 25% reduction in rainfall,
- 25% increase in water usage (through an increase in population),
- groundwater contamination above World Health Organisation guidelines,
- the implementation of greywater recycling systems.

These scenarios are linked to a range of responses including water re-use strategies, proactive pipeline rehabilitation, relocation of current potable water extraction systems and upgrading current water restriction policies and treatment systems.

The DSS incorporates probability analysis of pipe failure utilising CCTV data collected from sewer systems, as well as techniques already established for water pipelines; and uses these data, as well as data from the UVQ, the SLeakI and POSI models, to predict the flow of contaminants to the aquifer. By controlling these models and the inputs needed for different scenarios, it allows the models to provide output to the groundwater flow model and the SEM for financial and socio-economic scenario assessment. The SEM incorporates estimates on whole-life costing and analyses the stakeholders' acceptance regarding financial changes for the different service levels provided for both Rastatt and Mt Gambier. Outputs are detailed to stakeholders via participatory GIS methodologies.

The DSS system has been applied to the case study cities, principally as examples of common settings (industrialised cities located on different aquifers), with the aim of widening its use into a generic system applicable to other cities worldwide.

4 CONCLUSIONS

Aquifer protection is becoming increasingly important as nations recognise the need to manage their water resources in a sustainable way. Nowhere is this more important than in groundwater-dependent cities that rely on underlying or nearby aquifers for domestic, commercial and industrial water supply. As municipal authorities in these cities develop their plans for sustainability, they will increasingly need to understand the interaction of groundwater with leaking urban water infrastructure and to proactively manage the possible consequences of aquifer contamination.

The AISUWRS initiative meets the needs of these municipalities in that it has developed an innovative urban water assessment and management system to allow quantification of the impact of urban water pollution and infrastructure degradation on groundwater resources. Using key indicators for a range of contaminants, the DSS developed allows the effectiveness of different scenarios such as the impact of climate change or urban water rehabilitation and their impact on groundwater contamination to be assessed. Combined with a GIS-supported assessment scheme, AISUWRS will allow users (water supply/drainage companies and water authorities) access to an urban water management and DSS for safeguarding and managing urban groundwater resources. The inclusion of the Mt Gambier case study provides pilot-study values for cities in karst or fractured rock environments, or where there is little information about flow and solute transport in aquifers. The HACCP approach developed for Mt Gambier is expected to allow key knowledge gaps to be identified more quickly and, by addressing these gaps, enable the consequences for water infrastructure planning to be more directly related to Australian and World Health Organization drinking water guidelines, both of which are currently in draft form.

A need for a multidisciplinary research effort to develop the AISUWRS approach has resulted in the AISUWRS initiative, which has established cooperation between specialists in urban water and groundwater systems in Europe and Australia.

ACKNOWLEDGMENTS

This paper is dedicated to the memory of Matthias Eiswirth, coordinator of the AISUWRS project, whose vision of urban groundwater utilisation in Europe led to the establishment of the AISUWRS project. AISUWRS is a research project supported by the European Commission under the Fifth Framework Programme, and contributes to the implementation of the key action "sustainable management and quality of water" within the Energy, Environment and Sustainable Development thematic program. The AISUWRS project is a member of the cluster CityNet – the network of European research projects on integrated urban water management. The financial support of the Australian Government through an IAP-International S&T Competitive Grant from the Department of Education, Science and Technology and of the UK Natural Environment Research Council is gratefully acknowledged. Case study city partners have also contributed significantly to this project and their support is greatly appreciated.

REFERENCES

DeSilva, D., Burn, S., Tjandraatmadja, G., Moglia, M., Davis, P., Wolf, L., Held, I., Vollertsen, J., Williams, W. and Hafskjold, L. 2004. Sustainable Management of Leakage from Wastewater Pipelines, IWA World Water Congress, Sept. 2004, Marrakech, Proceedings on CD-ROM.

Eiswirth, M., Hötzl, H. and Burn, L.S. 2000. Development scenarios for sustainable urban water systems. In O. Sililo et al. (eds), *Groundwater – Past Achievements and Future Challenges: Proc. XXX IAH Congress, Cape Town, South Africa, 26 November to 1 December 2000*: 917–922. Netherlands: Balkema.

Eiswirth, M., Hötzl, H., Burn, L.S., Gray, S. and Mitchell, V.G. 2001. Contaminant loads within the urban water system – scenario analyses and new strategies. In K.P. Seiler and S. Wohnlich (eds), *New Approaches to Characterising Groundwater Flow*: vol. 1, 493–498. Netherlands: Balkema.

Miller, R., Correll, R., Dillon, P. and Kookana, R. 2002. ASRRI: a predictive index of contamination attenuation during aquifer storage and recovery. In P.J. Dillon (ed.), *Management of Aquifer Recharge and Sustainability*. Netherlands: Balkema.

Mitchell, V.G., Diaper, C., Gray, S.R. and Rahilly, M. 2003. UVQ: Modelling the movement of water and contaminants through the total urban water cycle. In *Proc. 28th Hydrology and Water Resources Symposium, Wollongong, NSW, Australia, 10–14 November 2003*. Canberra: Engineers Australia.

Mitchell, V.G. and Diaper, C. 2004. UVQ: A tool for assessing the water and contaminant balance impacts of urban development scenarios. IWA world Water Congress, Sept. 2004, Marrakech, Proceedings on CD-ROM.

Rueedi, J., Cronin, A.A., Moon, B., Wolf, L. and Hötzl, H. 2004. Effect of different water management strategies on water and contaminant fluxes in Doncaster, United Kingdom. In *IWA 4th World Water Congress, Marrakech, Morocco.*

Vanderzalm, J.L., Schiller, T., Dillon, P.J. and Burn, S. 2004. Impact of stormwater recharge on Blue Lake, Mount Gambier's drinking water supply, International Conference on Water Sensitive Urban Design (WSUD2004): Cities as Catchments, Adelaide 21–25 November 2004, Proceedings.

Wolf, L., Held, I., Eiswirth, M. and Hötzl, H. 2004. Impact of leaky sewers on groundwater quality. Acta Hydrochim. Hydrobiol. **32** (4–5), 361–373.

Wolf, L., Eiswirth, M., Held, I., Klinger, J. and Hötzl, H. 2005. Linking urban water infrastructure with numerical groundwater models – case study report Rastatt. *IAH-Special Series*, published in *this volume*.

CHAPTER 4

Urban groundwater problems in Cork city, southwest Ireland

Alistair Allen
Dept of Geology, University College Cork, Cork, Ireland

ABSTRACT: Cork, a coastal city in southern Ireland, with a population of 250,000, is situated at the mouth of the River Lee, which drains into Cork Harbour, an almost enclosed natural harbour. The city centre is located upon a reclaimed marsh and underlain by a deep buried valley, infilled by gravels with a hydraulic conductivity of 5×10^{-3} m/s. It is just above sea level, and within the tidal influence of the River Lee, which fluctuates over a range of 3 m under normal conditions and 4 m during spring tides. Groundwater within the city centre is in hydraulic connection with the River Lee and water table levels vary with the tides over a range of more than 2 m under normal conditions, but with a 30 min–1 hr delay in response from tidal maxima and minima. Groundwater levels are very high throughout the city centre, and at high tide, the water table is only about 30 cm below street level in some parts of the city centre.

Cork is subject to flooding under conditions of heavy rainfall and high tides, and in October 2004 suffered its worst flooding in over 40 years, when the whole of the city centre was flooded to a depth of nearly 1 m. The high water tables contributed to the flooding, and the delayed response of water table rise to the tidal maximum inhibited subsidence of flood water levels. Brackish water intrusion accompanies tidal fluctuations, and has corrosive effects on underground systems including cast iron water pipes and concrete storm drainage and sewer pipes.

Exfiltration from the water supply distribution network, and the sewer and storm water drainage system has contributed significantly to the high groundwater levels in the central city, due to the antiquity of these systems, much of which dates back to the 19th century. However, over the past 5 years, the sewer system has been completely replaced in the city centre and sewage which previously discharged into the River Lee, now goes to a sewage treatment plant located to the east of the city. Nevertheless, leakage of mains water is still in the order of 40% representing a volume loss of 26,000 m^3 per day. Groundwater contamination also results from failure of municipal infrastructure, and in at least one locality has been identified and linked to exfiltration from a sewer.

Building constructions in Cork city centre are also impacted by the high water table, and building sites have to be dewatered for foundation construction and where basements are included in the design. The high hydraulic conductivity of the gravels underlying the city make any drainage operation a costly process, and will restrict any future plans to construct a subway system to relieve traffic congestion in Cork city centre.

1 INTRODUCTION

The interactions of subsurface installations and services with groundwater in the urban environment increasingly represent a major consideration in urban planning decisions.

Pollution of groundwater in urban areas has long been a significant hydrogeological problem, due to leakages from sewer networks, oil tanks and industrial storage systems, leaching of contaminants from landfills and contaminated industrial sites, and runoff from urban transport networks. Shallow water tables are a problem in many cities, particularly those situated in coastal settings, and this problem may be exacerbated by exfiltration from leaking sewers, storm water drains and water supply distribution networks. Also, many low-lying coastal cities may be vulnerable to tidal fluctuations, particularly in areas where the tidal range is considerable, giving rise to both potential flooding of the city and salinisation of underlying groundwater.

High groundwater levels can also have a major impact on urban infrastructure, affecting the foundations of buildings and bridges, and damaging underground telephone, gas and electricity distribution systems, and underground pipeline networks. Furthermore, in coastal regions where the groundwater may be brackish, corrosion of pipelines, particularly older metallic pipes and pipeline joints, may lead to significant damage resulting in exfiltration or infiltration.

Another major problem brought about by high water tables is the need for drainage measures during construction operations, particularly where basements are included in the design. This adds significantly to the complexity and cost of the project.

Cork is a city that due to its geographical and geological setting embodies many of the urban groundwater problems outlined above. Significantly, Cork recently experienced a major flood, the worst episode of flooding in the city for 42 years, which has focused the city council's attention on urban groundwater issues that are discussed in this paper.

2 GEOGRAPHICAL SETTING

Cork is located on the southwest coast of Ireland and with a population approaching 250,000 including the surrounding area, is the second largest city in the Irish Republic. It is situated at the mouth of the River Lee, which drains into Cork Harbour, an almost completely enclosed natural body of water, connected by a narrow channel to the Celtic Sea (Figure 1). Founded by Vikings over one thousand years ago, Cork is located at the lowest bridging point of the River Lee, which historically consisted of a network of channels draining a marsh in what is now the central city. Indeed, the name Cork is the anglicised version of the Irish word for marsh. The city centre is just above sea level, and today the River Lee consists of two distributaries enclosing an elongated island about 3 km long and about 0.5 km in width at its maximum. This part of the city is located within the tidal range of the River Lee, with tidal variation in the order of 3 m under normal conditions and up to 4 m during spring tides. The tidal range extends to the western edge of the island which is enclosed by the distributary channels. Until the 1950s when a dam was built approximately 10 km upstream on the River Lee, the city was subject to frequent flooding during prolonged, intense rainfall events and/or high tides.

3 GEOLOGY AND GEOMORPHOLOGY

The Cork Harbour area lies within the very low-grade Rheno-Hercynian fold-thrust terrane of the late Carboniferous Variscan Orogenic Belt of SW Ireland (Gill, 1962). The area

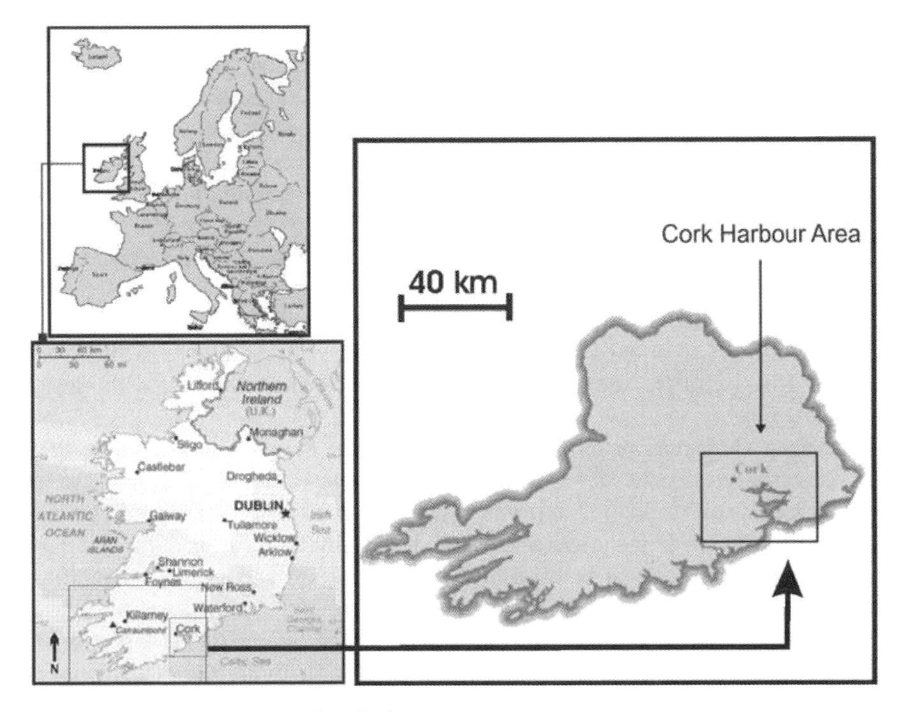

Figure 1. Geographical setting of Cork city.

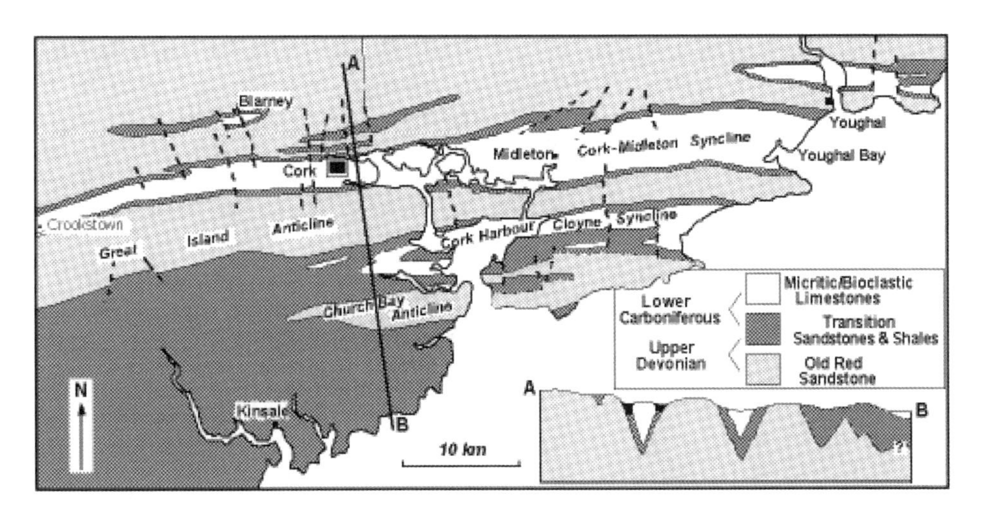

Figure 2. Simplified geological map of the Cork Harbour area.

is characterised by a series of E–W trending anticlines and synclines (Gill, 1962; Cooper et al., 1986), the former cored by Upper Devonian Old Red Sandstone (ORS), consisting of a sandstone/shale sequence, and the latter by massive karstified Lower Carboniferous limestones (Figure 2). The folded sequence is cut by E–W thrusts and steep N–S compartmental faults (Cooper et al., 1986), which slightly offset the various stratigraphic units.

Two major limestone-cored synclines occur in the Cork Harbour area, the Cork-Midleton Syncline and the Cloyne Syncline, separated by the intervening ORS-cored Great Island Anticline. Other ORS-cored anticlines occur to the north and south of the Cork-Midleton and Cloyne Synclines respectively, in particular the Rathpeacon Anticline to the north of the Cork-Midleton Syncline. Overburden of variable thickness, mostly glacial till, overlies bedrock. However, significant sand and gravel deposits, particularly infilling deep buried valleys also occur.

The geomorphology of the Cork Harbour area is strongly controlled by bedrock structure. Alternating E–W trending upland and lowland areas mimic the underlying E–W fold structure. Anticlinal folds cored by ORS sandstones and shales, underlie the high ground, which rises to about 200 metres above sea level (mASL), whilst the synclinal folds, cored by karstified limestones, underlie broad lowland valleys, with elevations of 0–30 mASL.

This geomorphological pattern developed during the Pleistocene glaciation. Prior to this, in the early Tertiary, the Devonian-Carboniferous succession of SW Ireland was overlain by Mesozoic cover, and the Tertiary land surface sloped gently southwards, giving rise to southwards-flowing drainage systems (Nevill, 1963). Intense erosion during the Tertiary stripped away the Mesozoic cover, exposing the Variscan rocks and leading to karstification of the Carboniferous limestones.

At the onset of Pleistocene glaciation (1.8 Ma–10 ka), the Tertiary drainage system of SW Ireland was truncated by glaciers, which advanced outwards from the mountainous region of west Cork and Kerry, preferentially exploiting the weaker shales and karstified limestone coring the synclines, resulting in the development of a number of broad U-shaped lowlands. Superimposed on these E–W lowlands are a number of narrower gravel-infilled buried valleys formed during the late Pleistocene when sea level fell to an estimated 130 m lower than present levels about 18,000 years ago (Mitchell, 1976; Pirazzoli, 1996). In the Cork Harbour area, these valleys represent important local ribbon aquifers. Cork Harbour was also created, probably during the late Pleistocene, by glacial scouring. In post-Pleistocene times, river capture of many of the E–W rivers created during the Pleistocene (e.g. River Lee), by the N–S pre-Pleistocene river systems, has given rise to a trellised drainage pattern in the Cork Harbour area.

Furthermore, the south of Ireland, together with southern England, has been subsiding continuously since the end of the Pleistocene, and the Cork area has been estimated to have sunk 16 m in the last 8,000 years. This has resulted in receding coastlines and the development of ria-type drowned estuaries and mudflats.

4 HYDROGEOLOGY

Major aquifers in the Cork Harbour area occur in both bedrock and overburden deposits. The main bedrock aquifers are the karstified limestones, coring the Cork-Midleton and Cloyne Synclines (Figure 2), which possess significant storage capacity and hydraulic conductivity. In addition, the Upper Devonian Old Red Sandstone coring the anticlinal uplands, and cut by N–S Variscan compartmental faults, can be regarded as a fractured rock aquifer. Overlying these bedrock aquifers, are important overburden ribbon aquifers represented by the gravel-infilled buried valleys, one of which, the Lee Buried Valley (LBV), passes beneath the centre of Cork city (Figures 3 and 4).

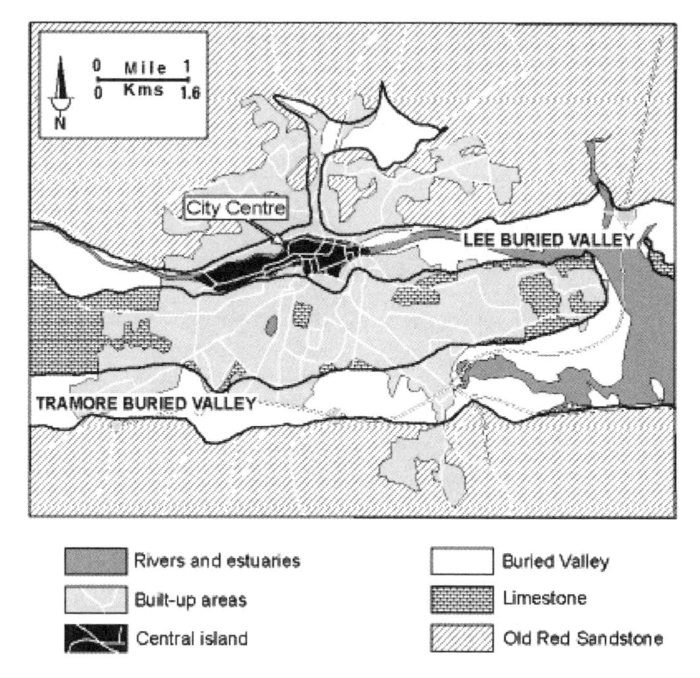

Figure 3. Extent of the Lee and Tramore Buried Valleys in Cork.

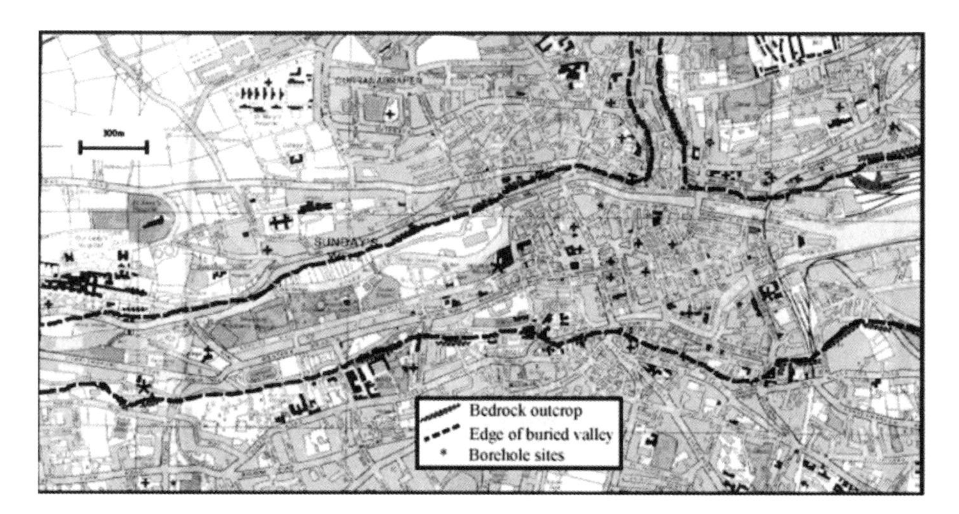

Figure 4. Extent of Lee Buried Valley in central Cork city.

The River Lee in the Cork area is situated at the northern margin of the Cork-Midleton Syncline, where it passes into the southern limb of the Rathpeacon Anticline, and directly overlies the Lee Buried Valley. The floodplain of the River Lee is 0.5–0.75 km in width where it passes through Cork, and is bounded to the north by a series of almost vertical scarps of ORS up to 30 m in height, whilst to the south it is bounded by similar limestone scarps of about 10–15 m height. The Lee Buried Valley is of a similar width to the floodplain, and is

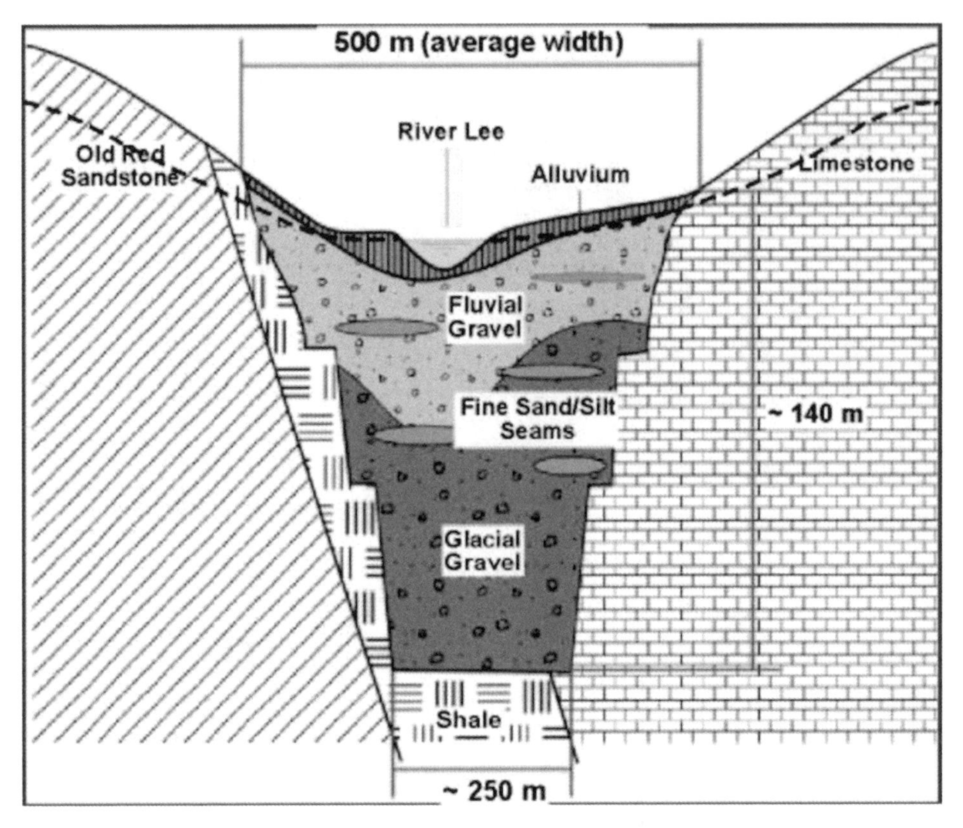

Figure 5. Schematic cross-section of the Lee Buried Valley in Cork (modified after Allen and Milenic, 2003a).

thought to have a stepped cross-section (Figure 5; Allen and Milenic, 2003a), with a proven depth of about 50 m beneath Cork (Allen et al., 1999), but a probable depth in excess of 140 m (Kostic and Milosevic, 2004). It is a significant aquifer with a porosity probably in excess of 20%, hydraulic conductivity of 5×10^{-3} m/s and transmissivity of 1.5×10^{-2} m^2/s (Milenic and Allen, 2005). Gravels occur at relatively shallow depths in the Lee Buried Valley, beneath a layer of alluvium ranging in thickness from 1 to 4 m, although in the centre of Cork, 2–3 m of fill generally overlies the alluvium (Allen et al., 1999). Within the city centre, the water table is very close to the surface (<1 m at high tide) and the aquifer is semi-confined. The aquifer is extremely vulnerable, due to the limited alluvial cover and the shallow depth of the water table.

Cork stretches across the full width of the 3 km wide Cork-Midleton Syncline, and extends into the upland areas forming the northern limb of the Great Island Anticline to the south, and the southern limb of the Rathpeacon Anticline to the north. Another E–W buried valley, the Tramore Buried Valley is located at the southern margin of the Cork-Midleton Syncline (Figure 3). The Tramore Buried Valley has a similar width to the Lee Buried Valley (about 500 m), and probably a similar stepped cross-section but a proven depth of no more than 80 m (Kostic and Milosevic, 2004). Again this buried valley is infilled by fluvioglacial gravels, possessing high storativity, hydraulic conductivity and transmissivity.

The Tramore River, overlying this buried valley, flows into the Douglas Estuary another arm of the inner part of Cork Harbour. Thus Cork straddles all the stratigraphic units indicated in Figure 2, plus the two buried valleys, which underlie the lowest parts of the city. Groundwater in the limestones and the buried valleys is most vulnerable to pollution, the former due to its karstified nature, and the latter due to their high water tables and their location in the areas of lowest relief within the city. The Lee Buried Valley Aquifer, however, is the most vulnerable due to its location beneath the low-lying central city area (Figure 4).

5 URBAN HYDROGEOLOGY OF CORK

Urban groundwater problems in Cork can be summarised as:

- high groundwater levels,
- large tidal fluctuations,
- brackish water intrusion,
- damage to building and bridge foundations by saline reaction with cement and concrete,
- damage to underground cable networks by brackish water,
- corrosion of pipeline networks,
- exfiltration from the water supply distribution system,
- exfiltration from sewers and storm water drains, and
- drainage problems during building foundation and basement construction.

5.1 *High groundwater levels and water table fluctuations*

The whole of the city centre of Cork is situated on the floodplain of the River Lee, at heights of only about 1 metre above sea level, and underlain by the Lee Buried Valley. As static water levels are approximately less than a metre below ground level throughout the extent of the city at high tide, the city is very vulnerable to flooding during periods of heavy and prolonged rain or high tides, a common situation in the south of Ireland. Furthermore because of the enclosed nature of Cork Harbour, under certain weather conditions, particularly south easterly gales, water in Cork Harbour is prevented from escaping out the narrow channel to the open sea and banks up, forcing high tides in the harbour area.

Such weather conditions occurred on 27th October 2004, when a violent storm accompanied by heavy rainfall and gale force south-easterly winds hit the south of Ireland, forcing exceptionally high tides. In the late afternoon of 27th October, high tide reached 2.95 mASL, and the River Lee overflowed its banks, inundating the whole of the central business district. Prior to this however, flooding had already commenced due to back up within storm water drains. The city centre remained flooded to a depth of just under 1 metre for over 4 hours, and at the next high tide early the following morning (28th October), the situation was repeated, with the River Lee again overflowing its banks and flooding the city centre for another 2–3 hours.

The high groundwater levels in Cork city centre may have contributed significantly to the flooding. Groundwater levels in Cork city centre are affected by tidal fluctuations in the River Lee, which vary by 2.5–3.0 m under normal conditions and up to 4 m during spring and neap tides. Groundwater is in hydraulic connection with the River Lee, as illustrated by static water level measurements taken from boreholes in the city centre (Allen and Milenic, 2003b).

These measurements indicate a maximum fluctuation of greater than 2 m under normal conditions (Figure 6). Delay in the response of the groundwater level to the tidal fluctuation ranges from 30 min to 1 hr, so as the river level was dropping after the tidal maximum had passed, static water level was continuing to rise, which would have delayed the withdrawal of flood waters. This strong tidal influence on static water levels suggests that a large-scale engineering flood protection scheme, such as a tidal barrage on the River Lee may not provide a completely successful solution to the flooding problem. Thus, the high groundwater levels in the city centre are of considerable concern.

5.2 *Brackish water intrusion*

The tidal range of the River Lee in Cork extends 3 km from the city centre to the western tip of the central island. Near the eastern tip of the central island, groundwater levels show the greatest fluctuation of over 2 m, and decrease progressively westwards. Electrical conductivity measurements range up to 1085 μS/cm, with maximum values coinciding with highest water levels, and equivalent TDS values range up to 716 mg/l (Milenic, 2004). Furthermore, electrical resistivity investigations of salt water incursion into the gravels of the buried valley near the eastern edge of the central island (Higgs et al., 2002) have revealed the presence of a plume of saline water intrusion exhibiting cyclic variations in penetration, fluctuating concurrently with the tidal cycle of the River Lee, but also with a delay in response.

Thus, there is hydraulic continuity between the River Lee and its estuary and groundwater within adjacent bedrock and overburden. During the waxing phase of the tidal cycle, as river levels rise in advance of groundwater levels, brackish estuarine water infiltrates into the gravels of the Lee Buried Valley. The brackish waters overlie less saline groundwater,

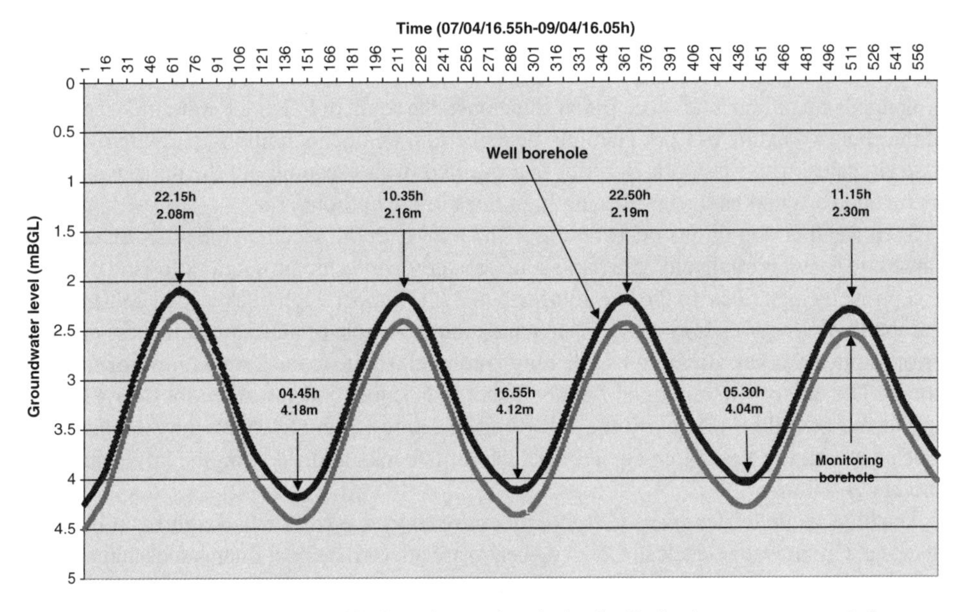

Figure 6. Groundwater level oscillations in two boreholes in Cork city centre, recorded over a monitoring period of 2 days (Allen and Milenic, 2003b).

creating a temporary density inversion, which immediately starts to equilibrate. As the tide reaches its maximum and goes into its waning phase, this partly equilibrated dense surface layer ebbs back into the river, causing fluctuating values of total dissolved solids parallel with fluctuating groundwater levels. The presence of brackish water at shallow depths has major implications for urban infrastructure, as saline water has a corrosive effect on metallic pipes and on concrete and cement structures. Corrosion can lead to weakening of building and bridge foundations, and may also affect underground electrical and telephone cable networks as well as gas pipelines, particularly at joints. This could ultimately lead to electrical shorting with accompanying power cuts and to gas leaks, both of which could have serious consequences.

5.3 *Leakages from sewers and pipeline systems*

Leakages from fractures and breaks in underground pipeline systems significantly impact groundwater levels in the Cork city centre. Exfiltration takes place from water supply distribution pipelines, storm water drains and sewers. Because of the location of the city centre, in the lowest topographical area of the city, all groundwater will flow towards this area. The implication of this flow net pattern is that any increase in groundwater levels in most other parts of the city will ultimately have an effect on the central island of the city.

Cork's current water supply system was first built in the 1860s when a major expansion of the distribution network took place. Treated surface water from the River Lee is pumped from Cork Waterworks on the north side of the River Lee in the west of the city to a number of storage reservoirs on the high ground in the north of the city, from where it is distributed gravitationally all over the city by a 647 km network of water pipes (Figure 7), some of which date back to the creation of the system. Almost 70% of this network consists

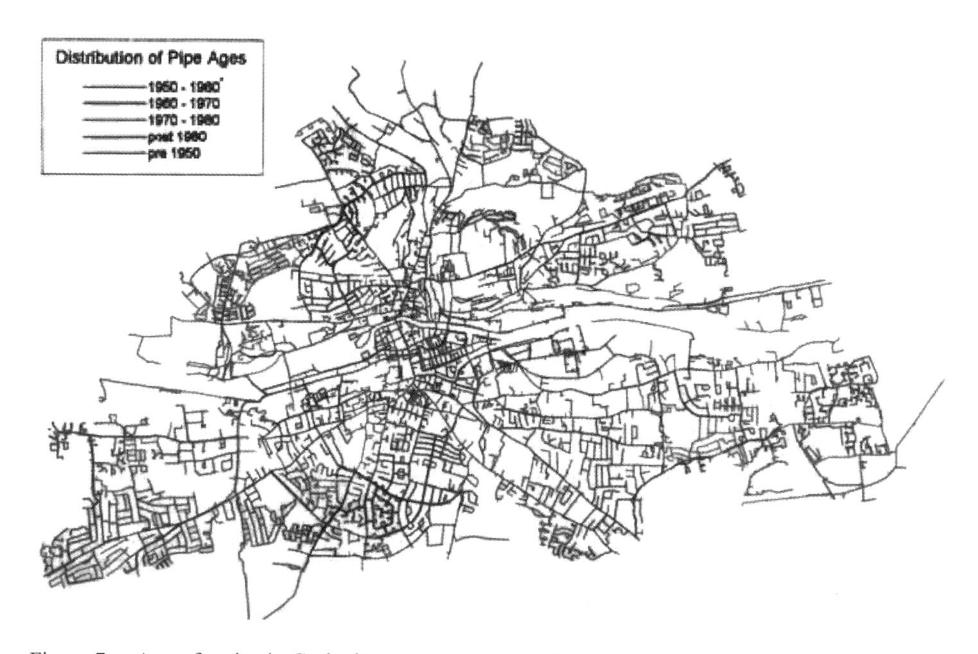

Figure 7. Age of mains in Cork city water supply distribution network (Hanley, 1999).

Table 1. Pipework of Cork water supply distribution network by material (Hanley, 1999).

Pipe material	Total length (m)	%
Cast iron (unlined)	448.3	69.3
Spun iron (unlined)	1.3	0.2
Ductile iron (unlined)	25.2	3.9
Ductile iron (mortar lined)	120.0	18.6
Lead	0.7	0.1
Asbestos cement	8.9	1.4
HDPE	26.7	4.1
MDPE	12.9	2.0
UPVC	2.1	0.3
Other	0.5	0.1
Total	**646.6**	**100.0**

Table 2. Pipework of Cork water supply distribution network by age (Hanley, 1999).

Age range	Total length (m)	%
Pre 1950	149.8	23.0
1950–1960	119.1	18.5
1960–1970	140.8	21.8
1970–1980	86.5	13.4
Post 1980	150.4	23.3
Total	**646.6**	**100.0**

of unlined cast iron pipes (Table 1), laid prior to the mid 1970s (Table 2). These cast iron pipes are concentrated in the older parts of the city, within and immediately surrounding the central island.

Leakage from the water supply system is of the order of 40% out of a daily production of 65,000 m^3, a daily loss of 26,000 m^3. The older cast iron pipes in the central city area are responsible for most of the leakage. Subjected to contact with brackish water twice daily, as groundwater levels rise and fall in response to tidal fluctuations, it is likely that these older cast iron pipes are highly corroded. Leakages in excess of 4.5 $m^3 km^{-1} hr^{-1}$ have been measured in the city centre, representing about 60–70% of the water passing through them.

The central island is bounded to the north and south by steep slopes occupied by the older parts of the city as it spread from the floodplain of the River Lee. Therefore, most of the water leaking from the network, perhaps as much as 20,000 m^3 per day, may ultimately flow to the central island. Taking the area of the central island as 1,500,000 m^2 and the porosity of the gravels/alluvium in the subsurface of the order of 20%, a sheet of water of approximately 7 cm in thickness may be being added daily to groundwater levels in this area by mains water leakage alone.

In addition, exfiltration of storm water and sewage have also until recently occurred from an antiquated sewer system. Within the central island, the old sewer system operated as a combined system, carrying both storm water and sewage together for discharge to the River Lee. Outside the central island, at a height of approximately 6 m OD (Ordnance

Datum) on the slopes to the north and south of the River Lee, a back-up system of storm water drains was installed for use during extremely wet conditions when overflow of the sewers took place, usually at a discharge rate of about six times the dry weather flow (DWF). However, no such system has been installed in the central island, so during flood events, foul water overflows into the city centre streets.

The old sewer network consisted mainly of brick-lined drains, some several centuries old, but also some newer concrete and clayware pipes. Many of the brick-lined drains had undergone partial collapse, so significant exfiltration was occurring. The rigid concrete pipes were prone to breakage when subjected to stresses caused by subsurface movements, perhaps resulting from piping. The concrete pipes were also susceptible to corrosion by H_2S released from the sewage. Failure at joints sealed by mortar also occurred due to the susceptibility to corrosion by H_2S and by brackish groundwater. However, no information on the extent of leakage from the old sewer system or from the back up storm water drains is available, as no monitoring was ever undertaken by the city council.

Evidence for failure of one of the old sewers has been identified. Analyses of water from a test borehole in the central island, indicated significant contamination (Allen et al., 1999). Subsequent analyses four years later (Milenic, 2004) confirmed the original analyses, recording electrical conductivity values in excess of 16,000 μS/cm and total dissolved solids values of 11,000 mg/l, indicating that this problem was continuous and had worsened.

There is the potential for health problems to arise if groundwater contaminated by sewage exfiltration came into contact with water supply pipes and was able to infiltrate into them. However, high-pressure flows are continuously maintained in the water pipes to prevent such a situation, and also to prevent brackish water infiltration at high tides.

In a major construction project over the past 4–5 years, a new sewer system, the Cork Main Drainage Scheme has been installed in the central island. The old sewer system in the central city area has been replaced by a new sewer system, which now carries sewage to a new sewage treatment plant constructed at Little Island, to the east of the city across the River Lee. New concrete and PVC pipes have replaced the old sewers, which are now disconnected from all sewage discharges from buildings. The old sewers have been left in place as storm water drains, except where they are partially collapsed or on the point of collapse, where they have been replaced by new concrete pipes, and storm water alone is now transported for discharge into the River Lee. Some of the old cast iron water supply pipes in the city centre area were replaced at the same time by ductile iron pipes, but unfortunately this has failed to reduce the total leakage from the water supply system.

5.4 *Drainage problems during building constructions*

High water tables lead to problems of drainage during construction operations. This is particularly true in the case of Cork, due to the presence of the gravel-infilled Lee Buried Valley underlying all of the central city area. Most of the newer constructions in the city centre area are piled into the gravels, because of the depth to bedrock, but older constructions have only normal foundations and could be vulnerable to settlement where founded within the alluvium overlying the gravels. Furthermore, where foundations or basements in excess of a metre or so are part of the design, drainage measures are required, which due to the high conductivity of the gravels can be extremely complex and expensive.

An art gallery, with a basement incorporated within the design, situated on the south bank of the South Channel of the River Lee, two kilometres to the west of Cork city centre has recently been completed. The basement has a depth of 3 m with lift shafts and a sump to a further 2 m depth, has dimensions of 23.3 × 17.5 m, giving a footprint of approximately 400 m^2. The total volume of the hollow basement below the static water level is 1080 m^3. In order for this phase of the construction to be completed, excavation to a depth of 5 m was required, necessitating drainage measures during construction of the basement (Allen and Milenic, 2003b). This involved continuous drainage of the site over a period of approximately 4½ months from the beginning of December 2002 to late April 2003 during relatively wet winter conditions. Mean rainfall for such a period is of the order of 50% of an annual mean rainfall total of 1200 mm, based on rainfall recordings at Cork Airport over 40 years (Milenic, 2004). Due to the high hydraulic conductivity of the gravels of the buried valley at the site (5×10^{-3} m/s), severe drainage problems were encountered. These problems were exacerbated by high rainfall over the period leading to significant recharge.

A total of fifteen 100 mm and 150 mm pumps operating at full capacity were required initially to drain the area for excavation. Eleven of these were submersible pumps set in drainage wells drilled at intervals around the margins of the excavation in order to stabilise its slopes. In addition, some parts of the excavation also required sheet piling to ensure stability. The core of the excavation was drained by means of four 150 mm sump pumps placed in two lift shafts and a sump. To maintain drainage over the construction period, twelve of the pumps operated continuously for 4½ months.

No measurements of flow rate or discharge were taken by the dewatering contractor, but on the basis of the pump manufacturers stated maximum flow rates for the various pumps, estimates of discharge from the site suggest that at the stage when the water table was being drawn down, perhaps as much as 400,000 gals/hr (500 l/sec) were being pumped from the site.

A further problem encountered in construction of the basement structure was the buoyancy of the hollow concrete basement, which prevented the pumps being turned off until the concrete ground floor was poured and walls were constructed to first-floor level. Failure to do so could have resulted in the buoyant basement rising with the rapidly recovering water table when the pumping system was shut down, which could have destabilised the foundation piles, and significantly weakened them. This prolonged pumping operations and further increased construction costs.

At this site, towards the west of the island, water table fluctuations due to tidal effects are low (approximately 40 cm range), but at a recently completed multi-storey hotel complex in the central city area near the east end of the island (Allen and Milenic, 2003b), water table fluctuations in excess of 2 m are an added problem. Here, the 8 m deep basement excavation for the hotels' underground car park required drainage measures to be implemented for over a year.

Work on the Cork Main Drainage Scheme was also subject to interference by high water levels, particularly fluctuations in water level due to the tidal influence. Thus, it was necessary to organise pipe-laying operations round the low water level intervals, with changes in shifts coinciding with water level maxima.

6 DISCUSSION AND CONCLUSIONS

As is evident from the foregoing, coastal cities in general are subject to special urban groundwater problems, particularly where situated on estuaries with large tidal ranges.

Cork is an example of such a coastal city, but is also more complex due to the unique hydrogeological conditions that are a result of the interaction between its geographical location and the underlying geology. The presence of a deep buried valley infilled with high permeability gravels directly beneath the low-lying central city area, together with a high water table and a large tidal range, create exceptional urban groundwater problems for Cork, as has been described above. Flooding is a major problem associated with high tides, generated by forcing conditions of south-easterly gales and heavy rainfall. A receding coastline in the south of Ireland and predicted rise in sea levels due to global warming does not bode well for the future of Cork, and severe flooding may become a more frequent occurrence. In recognition of this scenario, engineering solutions are being considered, most of which are extremely costly.

High water tables and fluctuating groundwater levels are a problem for all construction operations in the central island of Cork. Such conditions are characteristic of many coastal cities, but in Cork they are exacerbated by the extremely high hydraulic conductivity of the gravels underlying the city. Furthermore, where water tables are high, construction of subsurface transport infrastructure such as subways and underground car parks may not be feasible or certainly exorbitantly expensive. Furthermore, this type of infrastructure will always be vulnerable to flooding, particularly where tidal fluctuations are large. Therefore, the development of a public transport subway system beneath Cork to relieve traffic congestion in the city centre may not be a viable option, given subsurface geological and hydrogeological conditions in the central city area.

Salinisation of groundwater in coastal cities due to brackish water infiltration can lead to corrosion of pipeline systems and underground cable networks. In Cork this has led to extensive corrosion damage to the cast iron and concrete public water supply system as well as the sewer/storm water drainage network, which has resulted in exfiltration from these systems. Leakage from the public water supply system is a major problem in Cork, costing the local authority over €660,000 per year. In addition, apart from the legal requirement to cease sewage discharge to the River Lee, the potential for major exfiltration from the sewer system and subsequent contamination of groundwater, was also a justification for Cork City Council to undertake a complete overhaul of its antiquated sewer system in the central city over the past 5 years. Such exfiltration not only affects the quality of groundwater in the gravel aquifer underlying the city, but also has a negative impact on already high groundwater levels in the city centre area.

Solutions to Cork's urban groundwater problems are not readily obvious. The Cork Main Drainage Scheme has undoubtedly brought about a reduction in exfiltration from sewage disposal pipes with an improvement to the quality of groundwater in the Lee Buried Valley Aquifer, a major groundwater resource in the Cork area. Reduction in exfiltration from storm water drains is another obvious benefit of the scheme. However, no reduction in public water supply losses has resulted from replacement of water pipes during the scheme. This leakage represents an enormous cost to the city council and the taxpayer and has a major impact on groundwater levels in the city centre area, with implications for future potential flooding. Clearly this is a problem that needs addressing, but after 5 years of disruption to traffic and businesses due to the excavations involved in the Cork Main Drainage Scheme, Cork citizens would be less responsive to a new round of excavations.

The mitigation of Cork's flooding problem is another area under consideration. A tidal barrage scheme on the River Lee has been mooted, but where to site it and potential impacts require a major investigation. Also, given the enormous costs involved, government funding would be required and such a project, if adopted, would necessarily be a long-term solution. In the short term, wide-ranging hydrogeological investigations of groundwater behaviour within the whole of the Lee floodplain area can give vital information, which may allow less expensive engineering options to be adopted in the short term. These, in themselves may provide satisfactory flood prevention and control measures for the city.

ACKNOWLEDGEMENTS

Sincere thanks are due to a number of senior Cork City Council engineers including Dennis Duggan, Brendan Goggin, Jack O'Leary, Pat O'Sullivan and Eamonn Walsh, who have provided much data and information on various aspects of the water supply, sewer and storm water networks serving Cork, the Cork Main Drainage Scheme, and the October 2004 flood. Simona Kralickova is thanked for help with drafting the maps. Reviews by Beatriz Garcia-Fresca and an anonymous reviewer have helped to improve the manuscript.

REFERENCES

Allen, A.R., McGovern, C., O'Brien, M., Leahy, K.L. and Connor, B.P. 1999. Low enthalpy geothermal energy for space heating/cooling from shallow groundwater in glaciofluvial gravels, Cork, Ireland. In: Fendekova, M. and Fendek, M. (Eds.) *Hydrogeology and Land Use Management.* XXIX IAH Congress, Bratislava, Slovak Republic, IAH, Bratislava, pp. 655–664.

Allen, A.R. and Milenic, D. 2003a. Low enthalpy geothermal energy resources from groundwater in fluvioglacial gravels of buried valleys. *Applied Energy*, 74, 9–19.

Allen, A.R. and Milenic, D. 2003b. Drainage Problems during Construction Operations within a Buried Valley Gravel Aquifer. In: Petrič, M., Pezdič, J., Pirc, S. and Trček, B. (Eds.) *Proceedings 1st International Conference on Groundwater in Geological Engineering*, Bled, Slovenia, September 2003. 11 pp.

Cooper, M.A., Collins, D.A., Ford, M., Murphy, F.X., Trayner, P.M. and O'Sullivan, M. 1986. Structural evolution of the Irish Variscides. *J Geol Soc Lond*, 143, 53–61.

Gill, W.D. 1962. The Variscan Fold Belt in Ireland. In: Coe, K. (Ed.) *Some Aspects of the Variscan Fold Belt*, Manchester Univ Press, pp. 49–64.

Hanley, R. 1999. Strategic operational management and investment plan. Cork Water Network Management Project. Carl Bro Ireland.

Higgs, B., Unitt, R., Walsh, B. and George, D. 2002. Applications in environmental geophysics in the Cork area. In: *Environ 2002, 12 Irish Environmental Researchers Colloquium*, Abstr vol. p. 57.

Kostic, S. and Milosevic, D. 2004. Geophysical Investigation of the Lee and Tramore Buried Valleys of the Cork-Midleton Syncline, SW Ireland. Faculty of Mining and Geology, University of Belgrade, Belgrade, Serbia.

Milenic, D. 2004. Evaluation of groundwater resources of the Cork City and Harbour area. Unpublished PhD thesis, University College Cork, Ireland. 485 pp.

Milenic, D. and Allen, A.R. 2005. Buried valley ribbon aquifers – a significant groundwater resource of SW Ireland. In: Bocanegra, E.M., Hernández, M.A. and Usunoff, E.J. (Eds.) *Groundwater and Human Development*, A.A. Balkema, Amsterdam, Ch. 14, pp. 171–184.

Mitchell, G.F. 1976. The Irish Landscape. Collins, London, p. 68.

Nevill, W.E. 1963. *Geology and Ireland: with physical geography and its geological background.* Allen Figgis, Dublin.

Pirazzoli, P.A. 1996. *Sea Level Changes – the last 20,000 years.* John Wiley & Sons.

Sewer exfiltration

CHAPTER 5

Impact on urban groundwater by wastewater infiltration into soils

Ulf Mohrlok, Cristina Cata and Meike Bücker-Gittel
Institute for Hydromechanics, University of Karlsruhe, Kaiserstrasse 12, D-76128 Karlsruhe, Germany

ABSTRACT: Damaged sewers allow wastewater to infiltrate into the unsaturated zone, which increases the risk for groundwater contamination. This study examines different scenarios of wastewater infiltration into soils pertaining to the impact on urban groundwater resources. A numerical model WTM (Water Transport Model), is applied to simulate various infiltration conditions and hydraulic soil parameters, using soil types assumed to follow the van Genuchten model of soil characteristic functions. The results show that the hydraulic soil parameters together with the high infiltration rate have a significant influence on the seepage flow in relation to plume development and arrival times of contaminants to the water table.

1 INTRODUCTION

Urban wastewater is collected in sewers and transported to wastewater treatment plants to prevent contamination. However much of the sewer infrastructure in cities was built several decades ago and it has not been properly maintained due to lack of finances and technical expertise. There is the risk of wastewater leaking from these damaged sewers and then infiltrating down to groundwater, affecting its quality (Härig, 1991; LfU, 1999). The risk of leakage from sewers is difficult to quantify and in addition, there are uncertainties in estimating infiltration rates (Dohmann and Hausmann, 1996), making it difficult to take accurate measurements on a large urban scale. Furthermore, the possible transformation processes during subsurface migration are not well understood in real field conditions. In several experiments and field studies, important observations were reported with respect to the exfiltration of wastewater from sewers (Rauch and Stegner, 1994) as well as transformations and related clogging effects in the soils (Vollertsen and Hvitved-Jacobsen, 2002).

In order to predict the fate of wastewater contaminants in the subsurface, the simultaneous movement of water and the solutes in it must be well understood. This paper focuses on the impact water fluxes from single sewers have on soils and groundwater.

Flow in the unsaturated zone is determined by capillary effects, which can be explained by well-known retention relationships (van Genuchten, 1980; Brooks and Corey, 1966). Based on the finite element approach, several numerical codes, such as Hydrus2D (Simunek et al., 1998) and FEFLOW (Diersch, 1999), have been developed to simulate water flow and the associated solute transport in the unsaturated zone. In this paper, the recently developed numerical code WTM (Water Transport Model) (Bücker-Gittel et al., 2003), which is based on the random-walk approach has been used to simulate infiltration from sewer leaks.

The results of this study focus on the influence of soil parameters, infiltration rate and the thickness of the unsaturated zone on water content and contaminant dispersal, in relation to the overall impact to groundwater resources.

2 NUMERICAL APPROACH

2.1 *Unsaturated water flow*

Transport of dissolved contaminants in the subsurface depends on the water flow in the unsaturated and saturated pore space of geological formations. Wastewater from leaking sewers infiltrates into the unsaturated zone, where the water flow is commonly described by the Richards equation:

$$\frac{\partial \theta}{\partial t} = \vec{\nabla} \cdot [K(\theta)(\vec{\nabla}\psi(\theta) + 1)] \tag{1}$$

This equation relates the temporal change in the water content $\partial\theta/\partial t$ (1/s) to the gradient in water suction $\vec{\nabla}\psi(\theta)$ and to the hydraulic conductivity $K(\theta)$ (cm/s).

Several authors (van Genuchten, 1980; Brooks and Corey, 1966; Mualem, 1976) have derived retention relationships with different parameterizations to describe these associations. The variability of the water content is limited by the maximum value of the saturated water content θ_S (theoretically the porosity) and the residual water content θ_r, which cannot be changed by increasing the capillary forces. The unsaturated hydraulic conductivity, $K(\theta)$ (cm/s), is always less than the saturated hydraulic conductivity K_S (cm/s). These parameters together with the parameters from the retention relationships characterize the hydraulic properties of specific soils.

With regard to the development of a numerical solution technique, Equation (1) can be formulated as a Fokker-Planck equation for the water content θ (Bear, 1979):

$$\frac{\partial \theta}{\partial t} - \vec{\nabla} \cdot (D\vec{\nabla}\theta) - \left(\frac{\mathrm{d}K(\theta)}{\mathrm{d}\theta}\right)\frac{\partial \theta}{\partial z} = 0 \tag{2}$$

with the capillary diffusivity $D(\theta)$ (cm^2/s)

$$D(\theta) = -K(\theta)\frac{\mathrm{d}\psi}{\mathrm{d}\theta} \tag{3}$$

The second term in Equation (2), the diffusion term, represents the spreading of seeping water due to capillary forces, and the third term describes the vertical movement due to gravity. This Fokker-Planck equation defines a water transport problem described by the volumetric water content θ, which could be solved by the random walk approach (Kinzelbach, 1986).

2.2 *Water transport model*

Applying this approach to the unsaturated water flow, the total pore water is described by a certain number of particles. Each particle, representing the same fixed water volume, is moved by an advective step according to the first term of Equation (4), and an additional random step, representing the diffusive-dispersive transport (second term of Eq. 4). As an example, the vertical step Δz of a water particle can be calculated using the following equation:

$$\Delta z = \left(\frac{\partial D}{\partial z} - \frac{K(\theta)}{\theta} \right) \Delta t + Z\sqrt{2D\Delta t} \tag{4}$$

where Z is a normally distributed random variable with an average of 0 and a standard deviation of 1. In order to account for non-uniform flow fields, the advective step includes a correction term $\partial D/\partial z$ (Kinzelbach, 1986; Uffink, 1990). The superimposed particle steps generate a particle distribution representing the volumetric water content, which fulfils the Fokker-Planck equation (Eq. 2). By applying mass loads to the particles, the unsaturated water flow and the advective-dispersive transport of dissolved contaminants can be computed simultaneously. In order to balance the water content and the concentrations, the model domain is subdivided into uniform cells, where the amount of particles is equivalent to the total pore volume. This random walk approach is implemented in the recently developed code WTM (Bücker-Gittel et al., 2003).

3 NUMERICAL INVESTIGATIONS

3.1 *Model setup and parameters*

This numerical code was used for running different simulations, calculating three-dimensional flow and transport from a sewer leak towards the groundwater table within homogeneous soils. The sewer leak is placed at different heights above the bottom (i.e. $z = 0\,\text{m}$) where the groundwater table is represented by saturated conditions. The lateral boundaries were defined as no-flow boundaries. The model domain dimensions and the boundary conditions are given in Table 1.

Three different soil types were considered and their hydraulic properties described by the van Genuchten model (1980) of soil characteristic functions (Table 2). All model runs refer to a continuous infiltration of pure water with different infiltration rates; this was done for 12.5 h (sim. 1) simulation time and 2 h (sims 2–4), respectively (Table 1).

Furthermore, a solute pulse injection of one-minute duration with an input concentration $C_{in} = 1\,\text{mg/L}$ into an initially clean environment (i.e. $C_0 = 0$) is applied 30 minutes after the water infiltration has started (sims 2–4). The solute is considered non-reactive.

3.2 *Results and discussions*

The water content distributions during simulation 1 computed at 5, 60, and 750 minutes after the clean water infiltration started, are presented in Figure 1. The initial water content was calculated using the Brooks and Corey retention relationship (1966), with the groundwater table at the bottom of the model (i.e. $\psi = 0$ at $z = 0$). At the beginning of the infiltration, the wetting front is almost radial, while with increasing time it shows more downward,

Table 1. Model domain dimensions and boundary conditions.

Simulation number	x-length [m]	y-length [m]	z-length [m]	Leak elevation [m]	Q [mL/min]	T_{max} [h]
1	0.8	0.8	2.2	1.5	246	12.5
2	0.6	0.6	0.9	0.75	50	2
3	0.6	0.6	0.9	0.75	200	2
4	0.6	0.6	0.9	0.75	200	2

Table 2. Hydraulic soil parameters (model input data).

Simulation number	K_{sat} [m/s]	θ_S [cm³/cm³]	θ_r [cm³/cm³]	α_{vG} [1/m]	n_{vG} [–]
1	$1.8 \cdot 10^{-4}$	0.41	0.060	6.00	1.700
2	$1 \cdot 10^{-4}$	0.40	0.060	6.00	1.700
3	$1 \cdot 10^{-4}$	0.40	0.060	6.00	1.700
4	$1.6 \cdot 10^{-4}$	0.41	0.065	5.41	1.838

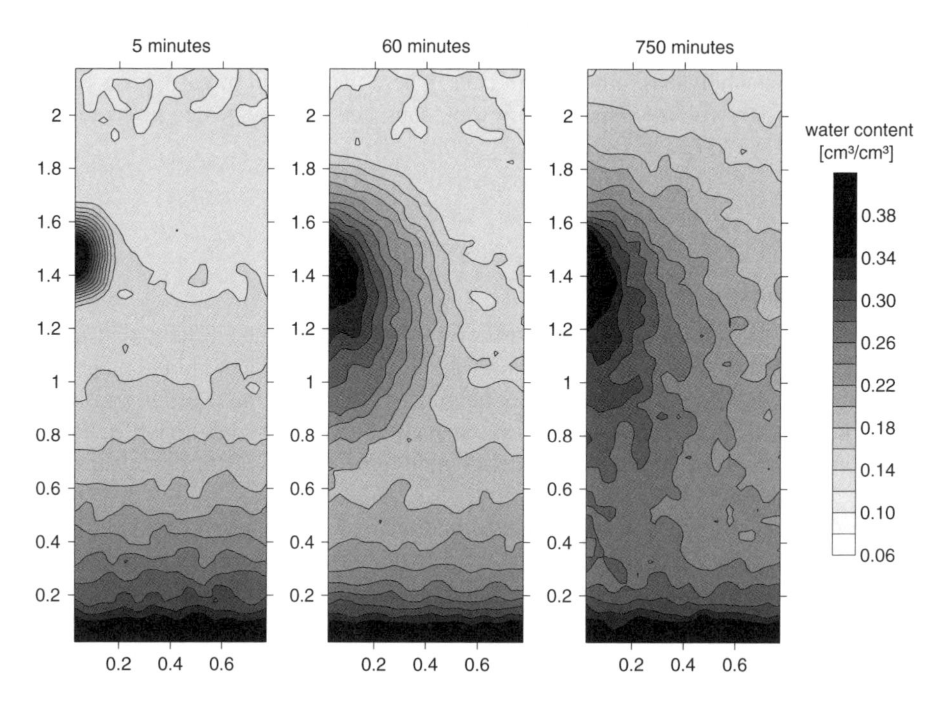

Figure 1. Water content distribution at 5, 60, 750 minutes after starting the water infiltration (sim. 1).

rather than lateral spreading behaviour. Even for the high infiltration rate, only the soil near the leak was saturated, whereas the lower portions of the soil, down to the groundwater table remained unsaturated.

The low number of particles representing the water content in a cell caused irregularities in the water content distribution. This could be solved by using more particles, but

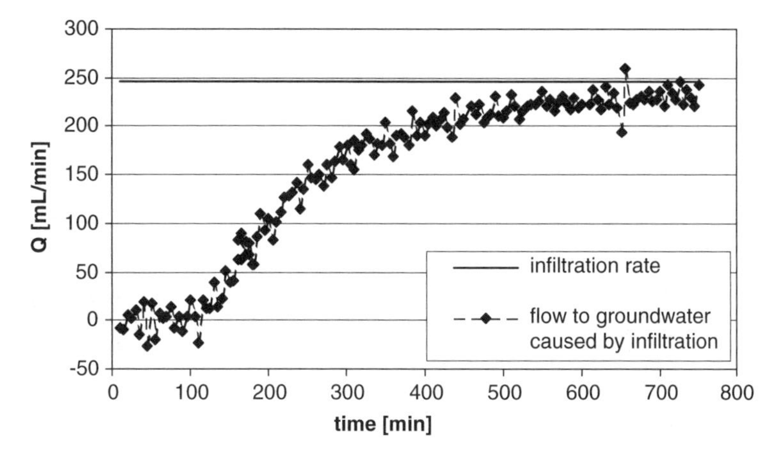

Figure 2. Developing groundwater recharge rate for 246 mL/min infiltration (sim. 1).

that would require more CPU time for the calculations, although it could decrease these irregularities.

After approximately 2 hours, the infiltration reached the groundwater table (Figure 2). The recharge rate increased continuously up to 95% of the infiltration rate. A quasi steady-state condition was reached after about 10 hours. The missing flow had spread laterally due to the remaining gradient in water suction.

The infiltration rates and the soil properties influenced the water content and solute distribution as demonstrated by simulations 2–4 (Figure 3). Lateral development of the water content distribution was observed for $Q = 50$ mL/min (Figure 3a), while the waterfront moved mainly downward for $Q = 200$ mL/min (Figures 3b–c). Depending on the soil properties and the infiltration rates, different water content averages were obtained between the leak elevation and the groundwater table.

A solute pulse injection of one-minute duration was applied into an initially clean environment 30 minutes after the continuous water infiltration had started. Large amounts of lateral spreading which caused a strong dilution effect in horizontal and vertical directions, was observed in all simulations. As expected, the solute moved faster in the simulations with $Q = 200$ mL/min than $Q = 50$ mL/min. As a result of continuous water infiltration, the solute concentration that reached the groundwater table (the bottom layer) was about 0.001 mg/L, taking up to 1.5 hours for $Q = 50$ mL/min, and only 1 hour for $Q = 200$ mL/min.

4 CONCLUSIONS

The results in this study indicate that hydraulic parameters, infiltration rates and the thickness of the unsaturated zone have a significant influence on water content, solute transport distribution development and arrival times at the groundwater table.

The lateral spreading of water and contaminant has two different impacts. First, a significant part of the contaminant flux is stored in the soil for a relatively long time and causes a risk potential for the groundwater resources. Second, the contaminant flux reaching the groundwater is highly diluted and retarded.

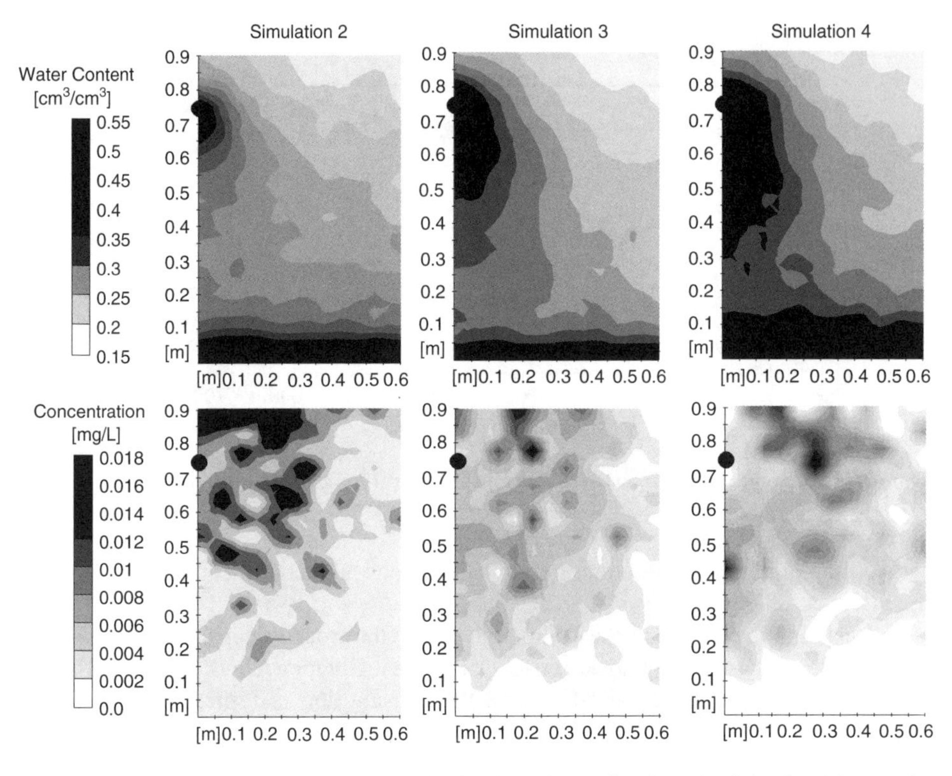

Figure 3. Water content and contaminant distribution 1 hour after the pulse injection (sims 2–4).

Furthermore, water content together with infiltration conditions define the residence time in the unsaturated zone, and therefore are important for the transformation processes. In the above cases, for high infiltration rate, the solute was washed out from the domain faster than for the low infiltration rate. The significance of this is that in the low infiltration rate, if reactive transport is considered, chemical and microbially mediated reactions could take place.

ACKNOWLEDGMENT

The German Research Foundation (DFG) within the research group unit FOR 350 "Gefährdungspotenzial von Abwasser aus undichten Kanälen für Boden und Grundwasser" (Risk potential of wastewater from damaged sewers for soil and groundwater) is funding the presented work. This paper is dedicated to our colleague Dr. Matthias Eiswirth, an active promoter of urban groundwater hydrogeology research, who has been taken from our community far too soon.

REFERENCES

Bear, J. 1979. *Hydraulics of Groundwater*. McGraw-Hill Inc., New York.

Brooks, R.H. and Corey, A.T. 1966. Properties of porous media affecting fluid flow. *J. Irrig. Drainage Div., 92, Proc. Am. Soc. Civil Eng. (IR2)*, 61–88.

Bücker-Gittel, M., Mohrlok, U. and Jirka, G.H. 2003. Modelling unsaturated water transport using a random walk approach. In K. Kovar and Z. Hrkal (eds.), *Calibration and reliability in groundwater modelling – a few steps closer to reality*, IAHS publication no. 277, Wallingford, UK, pp. 17–21.
Diersch, H.-J.G. 1999. *User Manual FEFLOW*, Version 4.8. WASY GmbH, Berlin.
Dohmann, M. and Hausmann, R. 1996. Belastung von Boden und Grundwasser durch undichte Kanäle. *Abwasser Spezial II*, GWF 137 (15), S2–S6.
Härig, F. 1991. Auswirkungen des Wasseraustausches zwischen undichten Kanalisationssystemen und dem Aquifer auf das Grundwasser. Inst. f. Wasserwirtschaft, Hydrologie und landwirtschaftlichen Wasserbau (Universität Hannover), 76, 155–258.
Kinzelbach, W. 1986. *Groundwater Modelling*. Elsevier Science Publishers B.V., New York.
LfU (Landesanstalt für Umweltschutz Baden-Württemberg, Karlsruhe): Pilotprojekt Karlsruhe: Änderung der Grundwasserbeschaffenheit auf dem Fließweg unter der Stadt, Grundwasserschutz 7, 1999.
Mualem, Y. 1976. A new model for predicting the hydraulic conductivity of unsaturated porous media. *Water Resources Research* 12, 513–522.
Rauch, W. and Stegner, T.H. 1994. The colmation of leaks in sewer systems during dry weather flow. *Groundwater* 31, 556–565.
Simunek, J., Sejna, M. and van Genuchten, M.Th. 1998. The HYDRUS-1D Software Package for simulation water flow and solute transport in two-dimensional variably saturated media, Version 2.0 IGWMD-TPS-70, International Ground Water Modeling Centre, Colorado School of Mines, Golden, CO.
Uffink, G.J.M. 1990. Analysis of dispersion by the random walk method. PhD Thesis, Geotechnical Laboratory, University Delft, Netherlands.
Van Genuchten, M.T. 1980. A closed-form equation for predicting the hydraulic conductivity of unsaturated soils. *Soil Sci. Soc. Am. J.* 44, 892–898.
Vollertsen, J. and Hvitved-Jacobsen, T. 2002. Exfiltration from gravity sewers – a pilot scale study. *Water Sci. Technol.* 47 (4), 69–76.

CHAPTER 6

Direct measurements of exfiltration in a sewer test site in a medium-sized city in southwest Germany

I. Held, J. Klinger, L. Wolf and H. Hötzl

Department of Applied Geology, University of Karlsruhe, Kaiserstrasse 12, 76133 Karlsruhe, Germany

ABSTRACT: Severe deterioration in the sewer systems of many European cities as a result of age and structural deficiencies leads to groundwater ingress and wastewater exfiltration. Groundwater ingress is generally determined through measurements of night-time sewage flow. Exfiltration measurements in pipes provide estimates of the flow of exfiltrating water. Extrapolation of these data on the total defect areas determined by closed-circuit television (CCTV) inspection suggests that the amount of exfiltrating sewage is sufficient to account for a significant fraction of urban groundwater recharge. Laboratory tests have been conducted to investigate clogging (Dohmann et al., 1999; Vollertsen and Hvitved-Jacobsen, 2003) but detailed knowledge about the variability of the colmation layer is missing. In Rastatt, a medium-sized city in SW-Germany, detailed investigations have been undertaken to assess the impacts of sewage exfiltration on groundwater. A purpose-built exfiltration test site has enabled investigation of exfiltration from an active sewer under normal operating conditions in a qualitative and quantitative manner. Sewage seeps from a crack in the pipe into a steel tank underneath and the exfiltration rate is recorded by a pluviometer. Soil moisture is measured at different depths underneath a second identical trench. Depth specific soil water is sampled over intervals of 7 or 14 days. The waste water quality is monitored through regular and online measurements of major ions. Waste water discharge is permanently recorded by a flow meter accounting for the fill level, the velocity and the pipe geometry.

The results show that the exfiltration rate is ruled by the existence of a colmation layer which varies over time. The clogging layer is removed during strong rain events and, aided by high fill levels and dilution of wastewater, leads to exfiltration rates (>200 L/d) much higher then normal base flow (6–10 L/d).

1 INTRODUCTION

In Germany, groundwater is the source of drinking water supply for over 80% of the population (Eiswirth and Hötzl, 1996). Matthias Eiswirth was one of the first scientists to recognize the importance of urban hydrogeology. In 1992 he started detailed investigations of urban groundwater in Rastatt, a medium-sized city (50,000 inhabitants) in southwestern Germany. Within the broad spectrum of anthropogenic contaminant sources, his major interest soon became directed towards leaky sewers and their impact on groundwater. These studies result from two projects Matthias initiated and significantly influenced.

The groundwater chemistry in the urban area of Rastatt reflects an area-wide anthropogenic influence, with elevated concentrations of sodium, potassium, phosphorous and boron, a generally increased specific electrical conductivity (SEC) and decreased oxygen contents (Wolf et al., 2004). By using x-ray contrast media and microbiological indicators (*E.coli*, Enterococci, Coliform bacteria) to analyze the areas in close proximity to leaky sewers, the impacts of sewage exfiltration can be established (Paul et al., 2004). Determination of the magnitude of the exfiltration is however, problematic. Mass balances are required due to the strong variability of wastewater composition and temporal variations in the groundwater. The basis for any mass balance is knowledge of the exfiltration rate. Therefore, direct exfiltration measurements on specific leaks and on one asset have been undertaken. An exfiltration test site has been built where the temporal characteristics of the exfiltration procedure could be observed under normal operating conditions.

2 INVESTIGATIONS ON SEWAGE EXFILTRATION IN A CASE STUDY CITY

2.1 *Direct measurements under test conditions*

Two different types of direct exfiltration measurements have been undertaken. Both assess exfiltration under variable head pressures with pressure probes through the lowering of the water level over time. For the exfiltration from single leaks, a small section (~0.8 m) of a pipe was blocked with a double-packer system containing two packer-discs, a camera and a pressure transducer. This device was used to measure the exfiltration of typical defects like leaky joints, broken shards and cracks in different pipe materials.

One of the main parameters that regulate the processes of exfiltration is an organic mat which builds on the defect and the soil which reduces free drainage through sewer cracks and joints. This microbial sealing is commonly referred to as a colmation layer (Burn et al., 2005).

Several interesting effects could be observed during these measurements:

- Cleaning the pipe, prior to measurement, significantly affected exfiltration rates (Figure 1a). The high pressure flushing of the pipe destroyed the sealing colmation layer resulting in much higher exfiltration rates.
- With the colmation layer damaged, the exfiltration rate was mainly dependent on the fill level in the pipe (Figure 2).
- The first measurement at a leaky joint showed an average exfiltration rate of approximately 0.4 L/min (Figure 1b). Subsequent measurements at the same leak at the same day produced much lower exfiltration rates due to the saturation of the surrounding bedding material resulting from the enhanced exfiltration during the first experiment.

Besides the single-leak tests, total exfiltration was measured along a 50 m pipe section by blocking a whole asset of a 600 mm sewer (DN600) with a variety of leaks known from CCTV inspections (Figure 3). The sewer was filled with water up to a pressure level of 1.4 m in the manhole. This process took about one hour. During that time filling and exfiltration occurred at the same time thus the exfiltrating water volume was not determined. In the following 30 minutes exfiltration rates were measured. This 1.5-hour exfiltration of fresh water resulted in a change in the SEC, as well as the water level in a groundwater monitoring well, 2 m from the pipe. Although the fresh water was less mineralised than the groundwater, there was a rise in SEC which can be ascribed to the washing out of

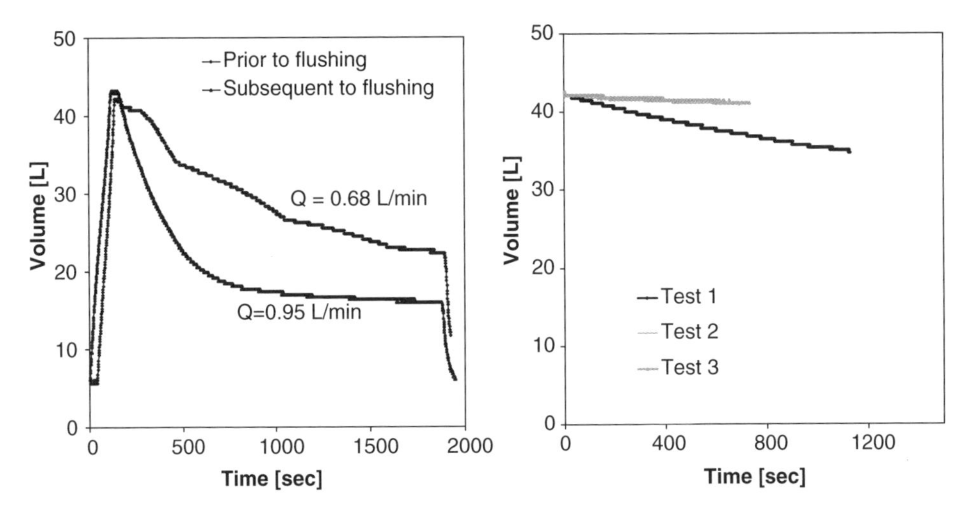

Figure 1. (a) Increase in the exfiltration rate through destruction of the colmation layer on a broken shard (left); (b) decrease in the exfiltration rate through saturation of the soil on a leaky joint (pipe diameter 300 mm) (right). Volume [L] indicates the volume of the water in the blocked compartment of the pipe.

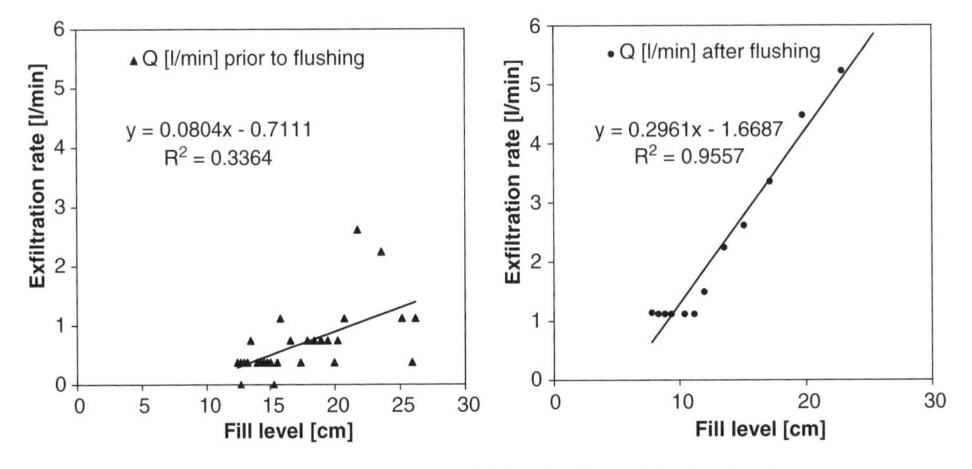

Figure 2. Exfiltration rate versus fill level with (a) and without (b) colmation layer.

small solid particulates and extruding of the moisture front in the unsaturated zone. During the pure water exfiltration procedure (30 minutes) about 600 L of water was lost to the underground.

It proved very difficult to quantify exfiltration through direct measurement. Observed exfiltration rates are considered to be unrealistically high as the test conditions (usage of fresh water, test fill levels higher than operation fill levels) promote exfiltration. A reproduction of the measured values was very difficult. Installation and removal of the packer system disrupted clogging layers and led to progressively higher exfiltration rates during the course of the experiments. Results are thus considered to represent worst-case scenarios.

Exfiltration under normal operating conditions is a very dynamic process due to variations in the fill level of the pipe, flow and composition of the wastewater, especially in combined

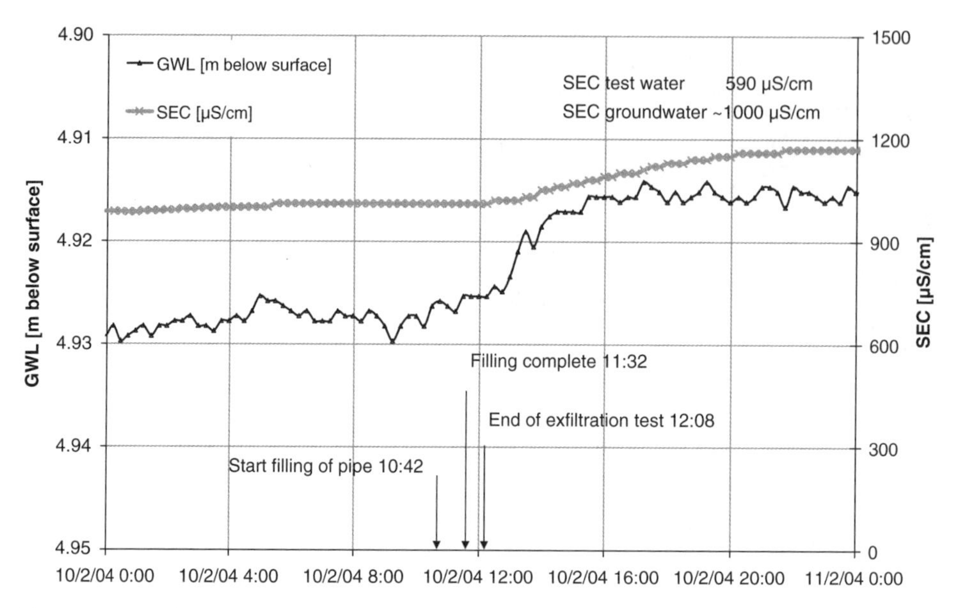

Figure 3. Reaction of SEC and groundwater level during an exfiltration test along a 50 m asset in a DN600 sewer.

sewers. The dilution of sewage with rainwater during storm events in combined sewers, along with high water levels can lead to a pronounced increase in exfiltration. However, the qualitative impact on groundwater in such a case remains very small due to the low contaminant load of the exfiltrating water.

To investigate undisturbed exfiltration over time, a test site was set-up in July 2004 where exfiltration and its effects on the unsaturated zone were monitored.

2.2 *Exfiltration test site Kehler Strasse*

The test site lies at the outflow of a catchment area featuring mainly domestic wastewater (Figure 4).

For the test site, a concrete foul sewer (diameter 500 mm) was excavated and two artificial leaks were cut in the bottom of the pipe (Figure 5). In the style of the damage symptoms of leaky joints, the leaks are designed as two transversal trenches with an opening width slightly over 1 cm and a total defect area of approximately 20 cm^2 each. As a fortunate coincidence, the sewer had been rehabilitated with a PVC liner which ensures that the two milled trenches are the only leaks along the test section and all measured effects derive from the exfiltration through the artificial leaks. To ensure that the water outflow does not enter into the space between pipe and liner and flow along the slope of the pipe, the pipe concrete was removed around the trenches so that all water runs out of the leak, past the liner and seeps downwards due to gravity. As the natural sediments surrounding the pipe (medium to coarse sand and gravel) was not suited for the intended instrumentation, the sections directly underneath the leaks were filled with homogeneous, narrow medium grained sand which is often used during the construction of sewers as pipe bedding. The exfiltrating sewage from one leak is collected in a steel tank underneath the pipe and its volume is permanently recorded with a pluviometer. Underneath the second leak, the soil

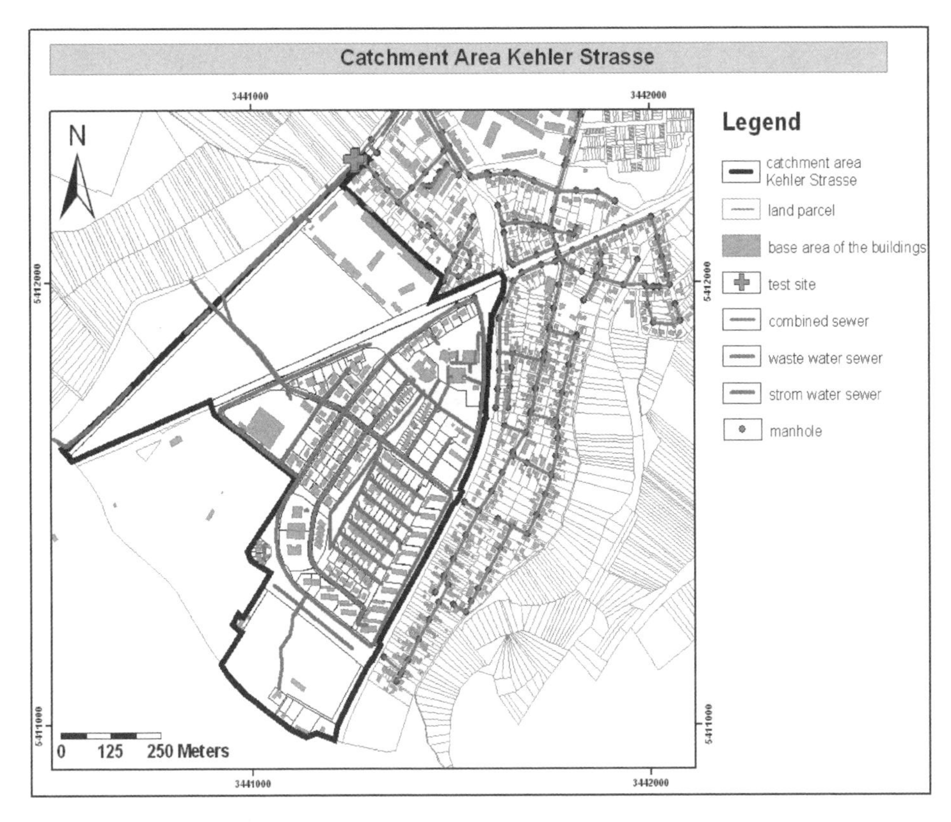

Figure 4. Catchment area of the exfiltration test site Kehler Strasse (cross in northern section of map) in Rastatt, SW Germany.

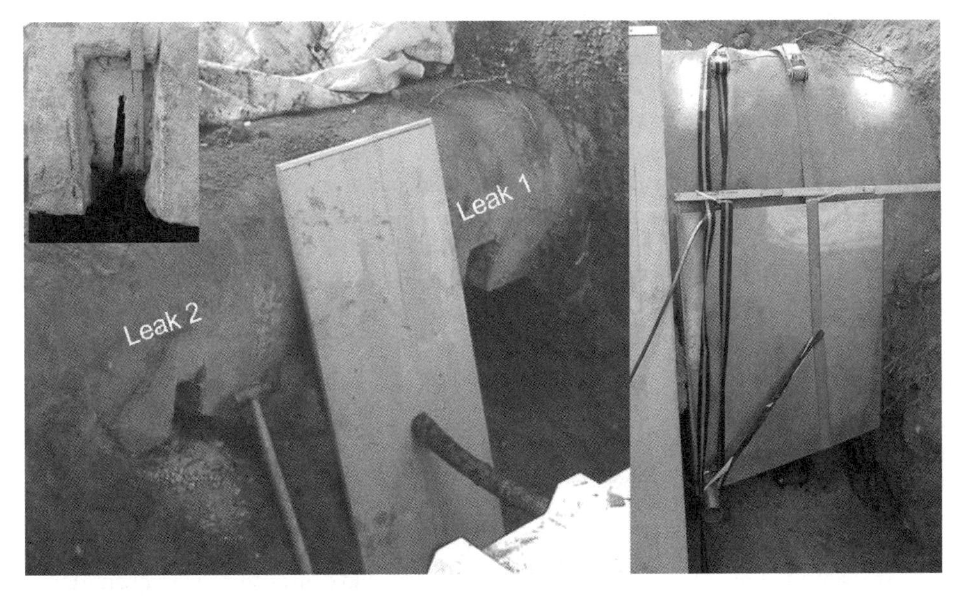

Figure 5. Excavated sewer during set-up of test site Kehler Strasse. Top left: detail of leak design, right: collecting tank underneath leak 1.

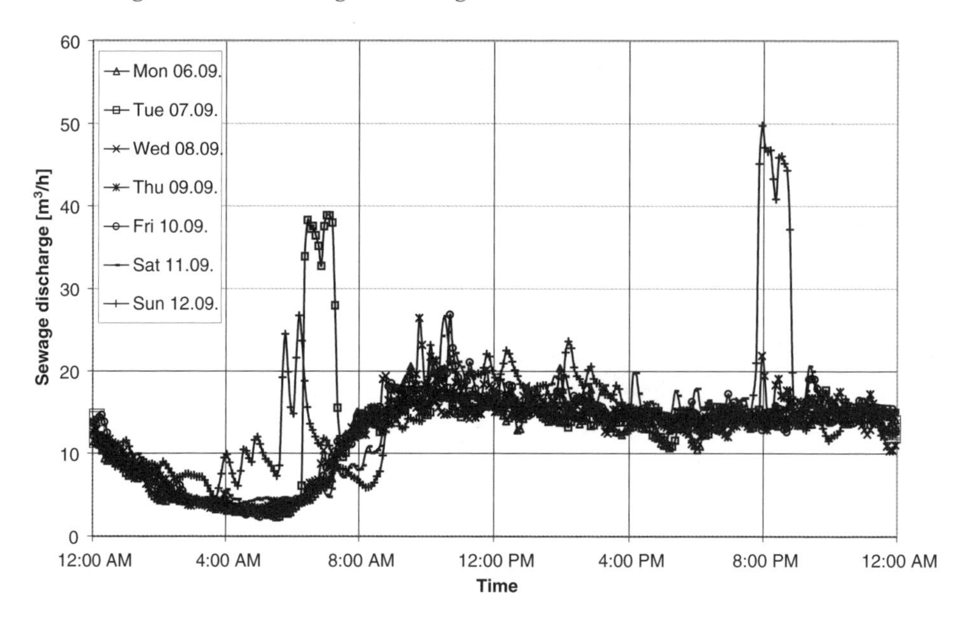

Figure 6. Characteristic diurnal sewage discharge with three rain events disturbing the regular pattern.

moisture and the matrix potential are recorded at different depths. Soil water is sampled in 7 or 14 day periods at various depths from 10 cm to 60 cm. The waste water quality is monitored through regular measurements of major ions. Waste water discharge is permanently recorded by a flow meter accounting for the fill level, the velocity and the pipe geometry.

2.2.1 *Sewage flow characteristics at the test site*

Minimum night-time sewage flows usually occur between 2 am and 6 am. The baseflow of 3 m^3/h results from permanently running water taps and toilet flushing, rather than groundwater ingress. Highest discharge usually occurs around 10 am (20 m^3/h), slowly decreasing over the rest of the day (Figure 6). Even though the sewer is designed as a separate system from the wastewater system, discharge shows a distinctive influence from storm events during which time baseflow can drastically increase. The highest measured discharge was 500 m^3/h. This led to a system overload and sewage overflow from manholes. Rain events clearly influence the sewage water composition causing dilution to less then 300 μS/cm. Average daily variations of the specific electrical conductivity typically range from 1200 μS/cm to 1800 μS/cm with short-time rises to 3500–4000 μS/cm especially during week days (Figure 7).

2.2.2 *Soil water*

Suction cups have been installed to sample soil water at depths of 10 cm, 30 cm, 50 cm and 60 cm underneath one of the leaks. Suction cups were positioned centrally, upstream and downstream as well as on both sides of the leak in horizontal planes. The soil water is collected over a period of seven days, volume and SEC are determined and depending on the volume, mixed samples of 7 or 14 days were analysed for major ions. Analyses and interpretations are on-going.

First analyses of boron as typical waste water content and conservative marker substance suggest that the seepage water concentration reflects the impact of rain on the sewage

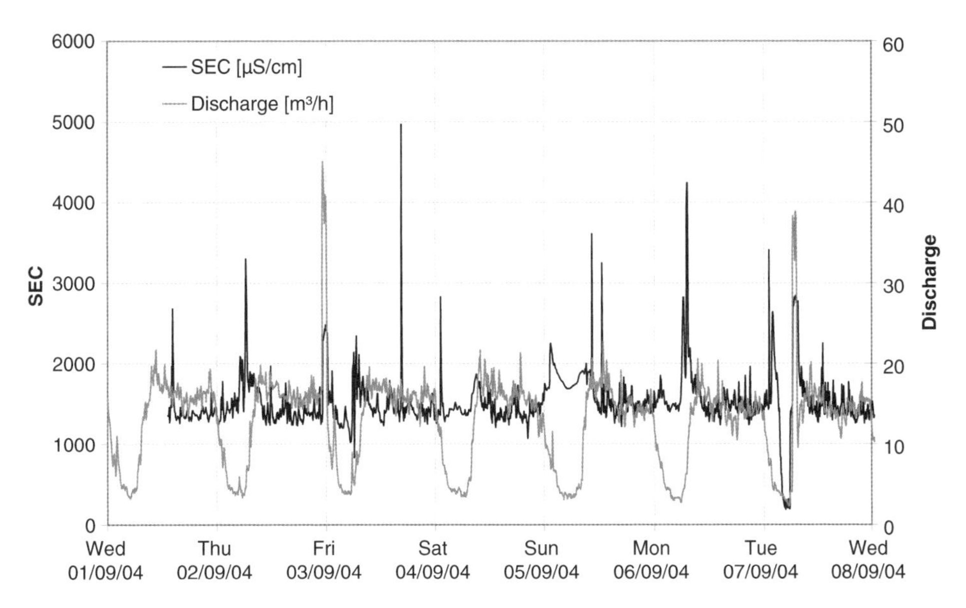

Figure 7. Typical pattern of sewage discharge (grey) and SEC (black) in the wastewater at Kehler Strasse.

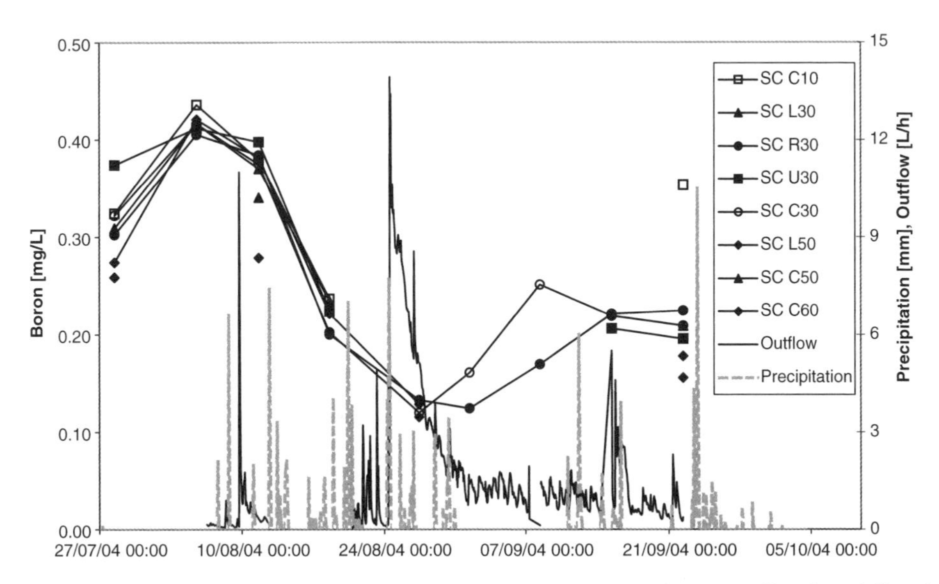

Figure 8. Decrease of boron concentration in seepage water as result of strong exfiltration of diluted sewage during rain events.

composition. Lowest boron concentrations can be found in samples with rain events in the seven-day collecting period (0.1 mg/L compared to the highest measured concentration of 0.4 mg/L). During periods of continuous rainfall, the concentration in the suction cups decreased gradually (Figure 8). Boron concentrations also show a decrease with depth. The highest concentrations within one horizontal plane are not bound to the central position indicating the existence of preferential flow paths (Figure 9).

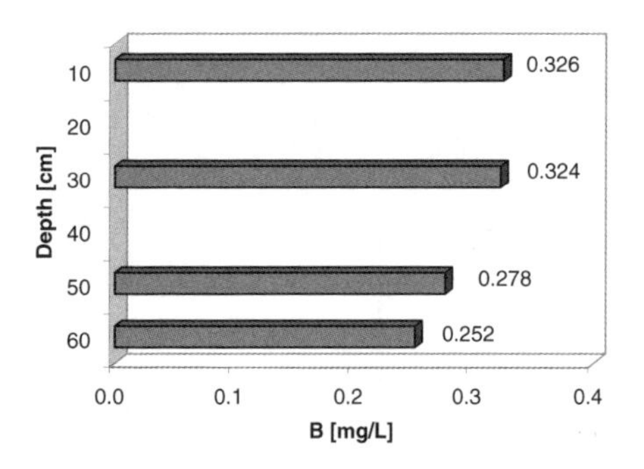

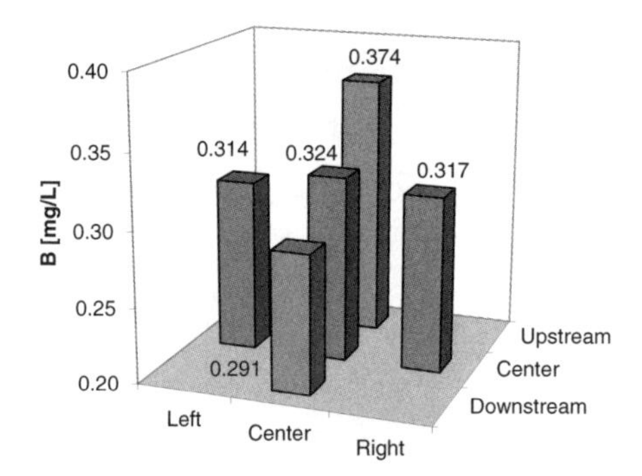

Figure 9. Decrease of boron concentration with depth centrally underneath leak 2 (left) and variations of boron concentration in 30 cm layer (right).

However, the results shown in this paper represent only the initial phase of the investigation. Detailed results and interpretations are to be presented in future publications.

2.2.3 *Soil moisture underneath the artificial leak*

Between the measuring planes of the suction cups, three layers of TDR (Time Domain Reflectometry) probes monitor the soil moisture underneath leak 2 (20 cm, 40 cm, 60 cm depth). A reference TDR probe measures the natural soil moisture without artificial influence. TDR records show a rapid decrease of the soil moisture, starting at over 30% at the time of exfiltration, to only ~17% three hours later. In the following four days, the soil moisture levelled off to average values between 5% and 10%. The soil moisture follows in diminished form the pattern of the fill level of sewage in the pipe indicating higher exfiltration with higher fill levels. According to their depth the TDR probes show a fast reaction to prominent increase of sewage discharge especially during rain events which can go up short-term to ~30%. Figure 10 illustrates the soil moisture front during enhanced exfiltration with a velocity of ~2.9 cm/min.

2.2.4 *Exfiltration underneath leak 1: outflow rates and variability of colmation*

Unfortunately due to technical problems, there are no observations from the very beginning of exfiltration from the sewage collecting box. First available data of the rain gauge two weeks after the start of the test site indicate an outflow of 100 L in 6 days. The general outflow then was between 0.10 and 0.19 L/h (~4.5 L/d). This equals the outflow rate after 4 months of exfiltration (November 2004) implying that it took less than 14 days to completely build up the self-sealing colmation layer after the creation of the leaks.

These outflow rates show significant variations over time. Storm events coincide with a strong rise in sewer discharge followed by a slow fall over several days (Figure 11). Enhanced exfiltration is a combined effect of the high fill level and the turbulent discharge flush resulting in the destruction of the clogging layer and dilution by rainwater. Thick waste water without dilution of rainwater leads to stronger clogging.

During a series of typical summer storms (e.g. 24.08.2004) the outflow increased from less than 0.1 L/h to 4.5 L/h during a first short rain event and went up to as high as 14 L/h, four hours later during a heavy rain event of 1.5 hours. This is equivalent to a daily flow of 3.4 L/d at the start of the rainfall event and 334.3 L/d for the peak outflow! The tailing of the outflow lasted for 8 days (with some minor rain events in this period of time only slightly affecting the overall trend) before reaching steady state flow which clearly reflects the

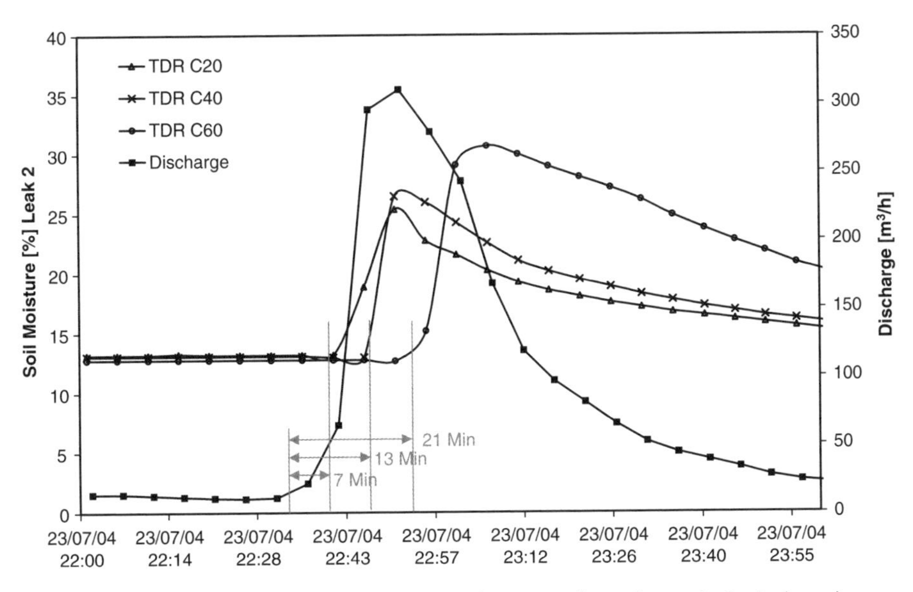

Figure 10. Reaction of soil moisture in TDR probes centrally underneath the leak on increased exfiltration according to depth.

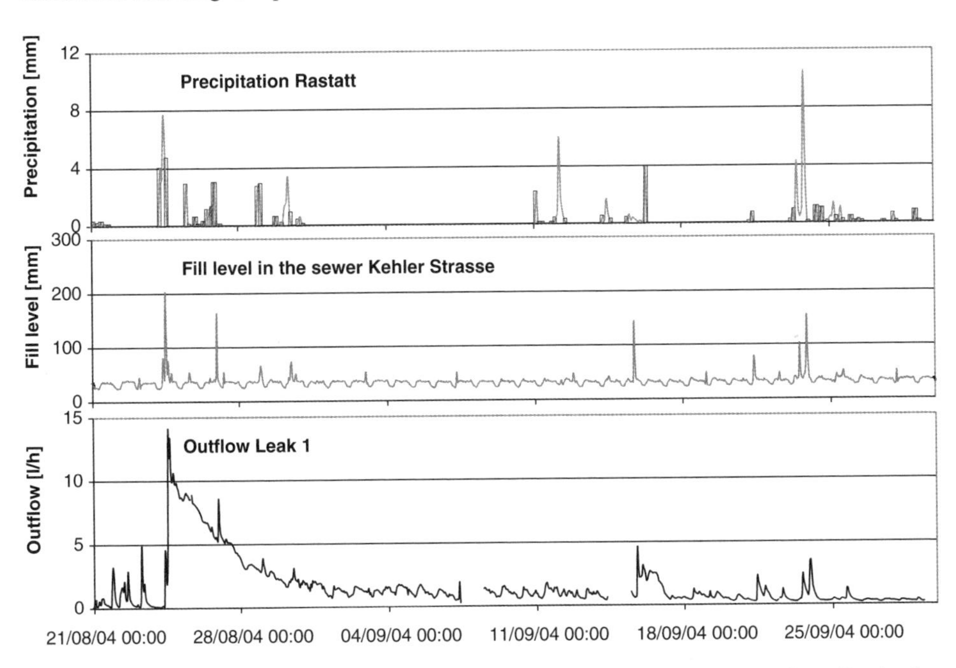

Figure 11. Correlation of precipitation, waste water fill level and outflow from the collecting box underneath leak 1.

pattern of fill level. However, the average daily discharge volume of over 20 L/d was higher than before the storm event (~16 L/d). After two more extreme events, with short time maximum of the outflow of 4.60 L/h (110.4 L/d) and 3.54 L/h (85.0 L/d) respectively, the outflow rates reverted back to the original low values of less then 1 L/h (~15 L/d) and below 0.5 L/h

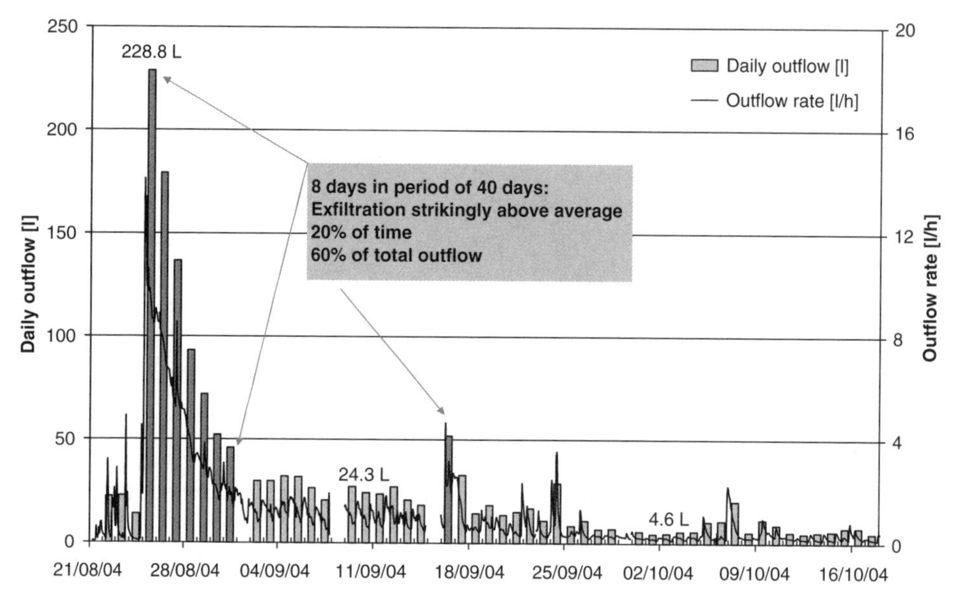

Figure 12. Variation of daily outflow volumes over a period of about 40 days.

(~7 L/d) one week later. Surprisingly, although the wastewater discharge during the third storm event was higher than during the previous event, the formation of a particulate clogging layer did not seem to be completely stopped by full destruction of the colmation layer. The sealing process was only slightly disturbed and the overall clogging continued to be a very effective sealant, resulting in a low and quite steady outflow with only diminished influence of the fill level in the pipe.

These results highlight the dynamic nature of exfiltration and clearly show the importance of the colmation layer. It further demonstrates the need to consider the colmation and its failures in any kind of exfiltration assessment. For a quantitative perspective, approaches assuming a constant colmation layer are very likely to be inaccurate. Even if the events leading to a destruction of the colmation layer are only of short duration they can account for a significant proportion of the total exfiltrating volume. Over a period of 40 days, the exfiltration at the test site amounted to 1500 L with variations in volumes ranging from 4.6 L to 228.8 L per day. For 8 days, daily outflow was clearly above average. As a consequence, 60% of the total outflow occurred in just 20% of the period under consideration (Figure 12).

3 IMPACT OF WASTEWATER EXFILTRATION ON URBAN GROUNDWATER RECHARGE

The pipe network in the test city compared to other cities in Germany is considered to be very well maintained and in above average condition. CCTV inspections of the sewer network showed, nevertheless, a large number and variety of defect types including cracks, broken shards, root intrusions, joint displacements and even areas on the edge of collapse. Along 208 km of inspected pipes 31,006 defects were recorded (Wolf et al., 2004). The

Table 1. Hypothetical calculation of exfiltrating volumes during dry weather flow on 189 km of pipes.

Exfiltrating volumes		1 leaky joint every 10 m		1 leaky joint every 5 m		1 leaky joint every 2 m		1 leaky joint every 1 m	
	[%] exfiltration	Monthly [m³]	Yearly [m³]	Monthly [m³]	Yearly [m³]	Monthly [m³]	Yearly [m³]	Monthly [m³]	Yearly [m³]
Pipes <DN500	50	1216	14,794	2432	29,587	6080	73,968	12,159	147,936
Pipes ⩾DN500	100	975	11,865	1950	23,730	4876	59,324	9752	118,648

Table 2. Hypothetical calculation of groundwater recharge arising from leaky joints.

Groundwater recharge		1 leaky joint every 10 m		1 leaky joint every 5 m		1 leaky joint every 2 m		1 leaky joint every 1 m	
	[%] exfiltration	Monthly [mm]	Yearly [mm]	Monthly [mm]	Yearly [mm]	Monthly [mm]	Yearly [mm]	Monthly [mm]	Yearly [mm]
Pipes <DN500	50	0.09	1.13	0.19	2.25	0.46	5.64	0.93	11.27
Pipes ⩾500	100	0.07	0.90	0.15	1.81	0.37	4.52	0.74	9.04

design of the artificial leaks in the test site was created to replicate damaged joints, which is one of the most common defect types. In Rastatt, leaky joints made up as much as 24% of the total defects. Tables 1 and 2 show rough calculations of hypothetic exfiltration from leaky joints for 189 km of inspected network based upon low outflow conditions of the test site Kehler Strasse (~6 L/d) and the resulting groundwater recharge for the city area of 13.2 km^2.

The calculations were performed using the measured exfiltration rate from the test site for pipes with diameter of 500 mm or larger. For pipes with smaller diameter only 50% of the measured exfiltrated volume was assumed. Exfiltration through other types of defects was not considered.

With a natural groundwater recharge of ~350 mm (Eiswirth, 2002), the recharge through sewage exfiltration could account for a significant portion of the total groundwater recharge in urban areas. Compared with other investigations, these assumptions seem to be in reasonable range. For a medium sized city in the U.K., Rueedi et al. (2005) determined the fraction of groundwater recharge attributed to sewage exfiltration to be 20–35 mm/a. This result was obtained by means of a mass balance approach using typical waste water constituents.

Using the water balance, Yang et al. (1999) calculated a groundwater recharge through leaky sewers for the Nottingham aquifer of 10 mm/a, estimating that about 1% to 2% of the average sewer flow is lost through exfiltration.

The extrapolation of the results to the city scale is still associated with large uncertainties (defect distribution, fill levels, etc.). An attempt to quantify the range of possible contributions of leaky sewers to the groundwater recharge is documented by Wolf and Hötzl (this volume).

4 CONCLUSIONS AND OUTLOOK

To obtain realistic exfiltration rates from leaky sewers, a sewer test site was installed to monitor exfiltration under normal operating conditions, including temporal variations. The investigations at the exfiltration test site, Kehler Strasse (DN500), developed a leak with the exfiltrating volume of 20 cm^2 and exfiltration rates between 0.25 and 0.40 L/h for average conditions. Storm events caused the leakage rate to increase to up to 13.9 L/h. The collection of the outflow from one of the leaks in a collecting tank underneath shows that on a daily basis, large differences of base outflow (~6–10 L/d) and extreme events (max. ~230 L/d) occur. Maximum exfiltration rates are observed in combination with rain events, resulting in high fill levels and strong dilution of the sewage. Short-term variations in the outflow follow the pattern of sewage discharge and fill level in the pipe. The exfiltration process is further regulated by an efficient sealing of the leak through clogging. The build-up of an initial colmation layer took less than two weeks after the start of the exfiltration. Outflow tailing after maximum exfiltration takes several days, reflecting the recovery of the colmation layer after damage or destruction through massive flushing. Observed exfiltration characteristics in active pipes show the potential of leaky sewers to account for significant groundwater recharge. Exfiltration is most predominant under the influence of rain events when exfiltrating volumes can exceed the base outflow by several magnitudes and can also dominate the total outflow over a certain period of time.

The focus in this paper has been on a quantitative assessment of the exfiltration through sewer leaks. Beside the volumes, the qualitative impact on groundwater is of great importance. Waste water can contain a variety of substances hazardous for human health with the potential for severe groundwater contamination (halogenated hydrocarbons, PAH, microbiological pollutants). Furthermore, the effects (especially long-term effects) of other sewage specific pollutants like pharmaceutical residues remain poorly understood.

These data from the test site will provide a more detailed understanding of the exfiltration process especially regarding the difference between dry weather flow and storm events. Further investigation, especially about seepage water chemistry, is underway or in the planning stages. This includes a detailed mass balance model which will help in a risk assessment of waste water exfiltration on groundwater. The measurements will be used to validate and test models that are currently being developed by the AISUWRS project for pipe leakage and unsaturated zone processes (Burn et al., 2005).

REFERENCES

Burn, S., Desilva, D., Gould, S., Meddings, S., Moglia, M., Sadler, P. and Tjandraatmadja, G. 2005. Pipeline Leakage Model (PLM) Manual, CMIT Report CMIT (C)-2005-219.

Dohmann, M., Decker, J. and Menzenbach, B. 1999. Untersuchungen zur quantitativen und qualitativen Belastung von Boden-, Grund- und Oberflächenwasser durch undichte Kanäle. In: Dohmann, M. (Hrsg.): *Wassergefährdung durch undichte Kanäle – Erfassung und Bewertung*. Springer Verlag, Berlin, Heidelberg.

Eiswirth, M. and Hötzl, H. 1996. Anthropogene Grundwasserbeeinflussung in urbanen Räumen. In: Thein, J. and Schäfer, A. (Hrsg.): *Geologische Stoffkreisläufe und ihre Veränderungen durch den Menschen. – Schriftenreihe DGG*, 1: 25–27.

Eiswirth, M. 2002. Bilanzierung der Stoffflüsse im urbanen Wasserkreislauf – Wege zur Nachhaltigkeit urbaner Wasserressourcen. Habilitation. University of Karlsruhe.

Paul, M., Wolf, L., Fund, K., Eiswirth, M., Held, I., Winter, J. and Hötzl, H. 2004. Microbiological condition of urban groundwater close to leaky sewer systems. Paper submitted for *Acta hydrochimica et hydrobiologica*.

Rueedi, J., Cronin, A.A. and Morris, B.L. 2005. Estimating sewer leakage using hydrochemical sampling of multilevel piezometers. Submitted at *Water Resources Research*.

Vollertsen, J. and Hvitved-Jacobsen, T. 2003. Exfiltration from leaky sewers: a pilot scale study. *Water Sci. Technol.* 47 (4), 69–76.

Wolf, L., Held, I., Eiswirth, M. and Hötzl, H. 2004. Impact of leaky sewers on groundwater quality beneath a medium sized city. *Acta hydrochimica et hydrobiologica*, in print.

Yang, Y., Lerner, D.N., Barrett, M.H. and Tellham, J.H. 1999. Quantification of groundwater recharge in the city of Nottingham. *Env. Geology* 38 (3).

CHAPTER 7

Upscaling of laboratory results on sewer leakage and the associated uncertainty

Leif Wolf and Heinz Hötzl

Department of Applied Geology, University of Karlsruhe, Kaiserstrasse 12, 76133 Karlsruhe, Germany

ABSTRACT: Leakage from defective sewer systems has been the subject of several investigations during the last two decades. This paper compares the published results at the laboratory scale and investigates the possibilities of applying this small scale process knowledge to the city scale. All laboratory experiments show the importance of a colmation layer and its strongly varying hydraulic conductivity. Upscaling the small scale models to a real sewer system requires the knowledge of several parameters with high spatial and temporal variability which results in a significant uncertainty. A Monte Carlo simulation was applied to the data set of a real sewer network and resulted in most probable exfiltration rates of 10 l/day/detected leak, equal to 4 mm/a artificial groundwater recharge from sewers. However, the uncertainty involved is very significant. With a probability of 95% the resulting groundwater recharge will be below 65 mm/a. This large range of possible exfiltration rates underscores the necessity for groundwater quality monitoring and the use of marker species approaches.

1 INTRODUCTION

Leaky sewers are suggested as a potential source for contaminants in urban aquifers (e.g. Barret et al., 1999; Sacher et al., 2002; Dohmann et al., 1999; Härig and Mull, 1992; Ellis, 2001; Eiswirth and Hötzl, 1997). The European Union standard EN 752-2 recognises this problem and therefore demands the structural integrity of urban sewer systems including their watertightness (Keitz, 2002). However, major difficulties still exist in the quantification of the volumes exfiltrating from leaky sewers and sometimes contradictory statements can be found. This paper compares the results of different studies and tries to apply them to a demonstration sewer catchment in Rastatt (SW-Germany).

Härig and Mull (1992) attempted to quantify the effects of sewage exfiltration for the City of Hannover by balancing the water quality data. However, a significant uncertainty remains as the substances considered (e.g. sulphate) could derive from a number of different urban contamination sources. In consideration of the multitude of sources for typical urban contaminants, Barret et al. (1999) tried to use sewage specific marker species to detect the influence of wastewater on groundwater underlying the City of Nottingham. Dohmann et al. (1999) attempted to quantify exfiltration rates based on laboratory and field studies. However, the extrapolation to larger scales was based on very general statistical information on the structural condition of the sewer system. Extrapolations, which are based on CCTV-inspection data have been attempted by Wolf et al. (2005). An approach to validate the assumptions on sewer leakage by the measurement of marker species distributions in the urban aquifer was described in Wolf et al. (2004).

2 QUANTIFICATION OF SEWER LEAKAGE AT A SINGLE LEAK

2.1 *Existing studies*

Rauch and Stegner (1994) set up a test site at a wastewater treatment plant and measured exfiltration rates using various pipes, different leak sizes, and different materials surrounding the pipes. They observed a strong decrease in exfiltration rates after the experiment started and all experiments were stopped after a maximum duration of 1 hour. They applied Darcy's law to their set up and calculated leakage factors between 86.4 and 864 l/d (i.e. 0.01 and 0.001 l/s). A leakage factor is also the ratio between thickness of the colmation layer and the hydraulic conductivity of the layer (1).

$$L = \frac{h_{colmation}}{K} \tag{1}$$

L: Leakage factor [1/T]
$h_{colmation}$: Thickness of the colmation layer [L]
K: Hydraulic conductivity of the clogging layer [L/T]

Assuming a thickness of 0.01 m for the colmation layer this corresponds to hydraulic conductivities between 8.64 m/d and 0.864 m/d (i.e. 1.0×10^{-4} m/s and 1.0×10^{-5} m/s). From depth-specific sampling of organic matter content in the bedding material they concluded that the thickness of the colmation layer is between 0.01 and 0.05 m.

Dohmann et al. (1999) also set up a test site at a wastewater treatment plant and used a concrete pipe with a 4 mm wide longitudinal crack in the bottom section. The surrounding material was medium to coarse sand. After 10 days of operation they measured exfiltration rates of 1.2 l/d per m length of the crack. This would correspond to an extremely low hydraulic conductivity of the clogging layer of 2.68×10^{-3} m/d (i.e. 3.1×10^{-8} m/s). Based on a large number of additional experiments, Dohmann et al. (1999) calculated a best case (31 million m^3 wastewater exfiltration/year) and a worst case scenario (445 million m^3 wastewater exfiltration/year) for the western part of the German Federal Republic with a total length of 193.156 km combined and 87.221 km separate wastewater sewers. The best case calculation is based on the assumption of a dry weather exfiltration rate of 4.08 l/d per crack, 19.92 l/d per crack during rain events and 6697.2 l/d per crack for storm events (the author assumes a hydraulic pressure of 1.5 m during these events; only applied for 26 h in one year). The worst case calculation is based on the assumption of a dry weather exfiltration rate of 7.68 l/d per crack, 37.68 l/d per crack during rain events and 24744 l/d per crack for storm events.

In a similar experimental set up, Dohmann et al. (1999) tested the influence of different soil types on the exfiltration behaviour and found significant lower flow rates for silty soils as discussed in the following chapter.

Rott and Zacher (1999) described experiments in which they also set up a sewer test site and operated it for a total of 405 days. Both longitudinal as well as transversal cracks were simulated. Wastewater was applied to the system each day, but only for 10 hours. Typical exfiltration rates ranged between 0.012 l/d/cm^2 to 0.68 l/d/cm^2. However, the discontinuous application prevents comparison with the other experiments.

Vollertsen and Hvitved-Jacobsen (2003) described pilot scale experiments conducted with raw sewage water in Denmark. They simulated the effect of storm events, flushing of pipes, and alternating infiltration/exfiltration conditions. Furthermore, they studied different leak types and the influence of different soils. Based on their experimental data, these researchers calculated leakage factors and hydraulic conductivities for different set ups. For holes and cracks they measured typical exfiltration rates of 0.06 l/d/cm^2. Assuming a thickness of the colmation layer of 0.01–0.02 m they estimated the hydraulic conductivity of the clogging layer to be 0.17 m/d (i.e. 2×10^{-6} m/s). For open joints they measured significantly lower exfiltration rates of 0.02 l/d/cm^2. During the simulation of storm events, 20 to up to 56 times higher exfiltration rates were measured.

As part of a multidisciplinary research project at the University of Karlsruhe, a long-term experiment on sewer leakage was conducted at the local wastewater treatment plant (Forschergruppe Kanalleckagen, 2002). A leak (0.4 cm $\times$ 14 cm) was cut into a 0.2 m diameter sewer running through a 1.5 m $\times$ 1.5 m $\times$ 3 m box filled with medium sand. Hydraulic and chemical processes were monitored using TDR-probes, tensiometers and suction cups. All effluent was collected at the bottom. Typical exfiltration rates of the first 100 days were around 2 l/d. However, 10 times higher exfiltration rates were observed following minor manual damage to the colmation layer. After one year of operation, exfiltration rates amounted to only 0.6–1 l/d.

Held et al. (2005) describe the results of a new sewer test site which was constructed beneath a real sewer at the city of Rastatt. Differing from the other cited experimental set ups, this test site is subject to the natural variations of water level and sewage composition induced by rain events. The daily summation of the outflow of the collecting tank underneath one leak shows large differences of base outflow ($\sim$6–10 l/d) and extreme events (max. $\sim$230 l/d). From the set up of the experiment it may be possible that a small void was present between the leak and the surrounding soil. This situation is also frequently occurring at non-artificial leaks where intermittent infiltration has washed away the surrounding soil.

2.2 *Mathematical description of the exfiltration process*

The results of laboratory experiments on sewage exfiltration performed in recent years under various conditions and exfiltration rates have been described in the previous section. Most of the experiments revealed no clear correlation between leak size and exfiltration rates (esp. Dohmann et al., 1999). This leads to a broad band of uncertainties in defining hydraulic conductivities for the clogging layer and consequently only few attempts were undertaken by the previous authors to derive such values.

However, without calculating the coefficients, comparison between the different exfiltration rates measured in different experiments is not possible. As different results are available, this paper tries to compare the individual findings and to analyse the variability in a second step. The concept of calculating hydraulic conductivities (K-values) in this case is given in Figure 1. Unlike the situation at a leak with free drainage conditions which could be calculated using Torricelli's law, the exfiltration process is predominantly influenced by the surrounding sediments, which calls for Darcy's law.

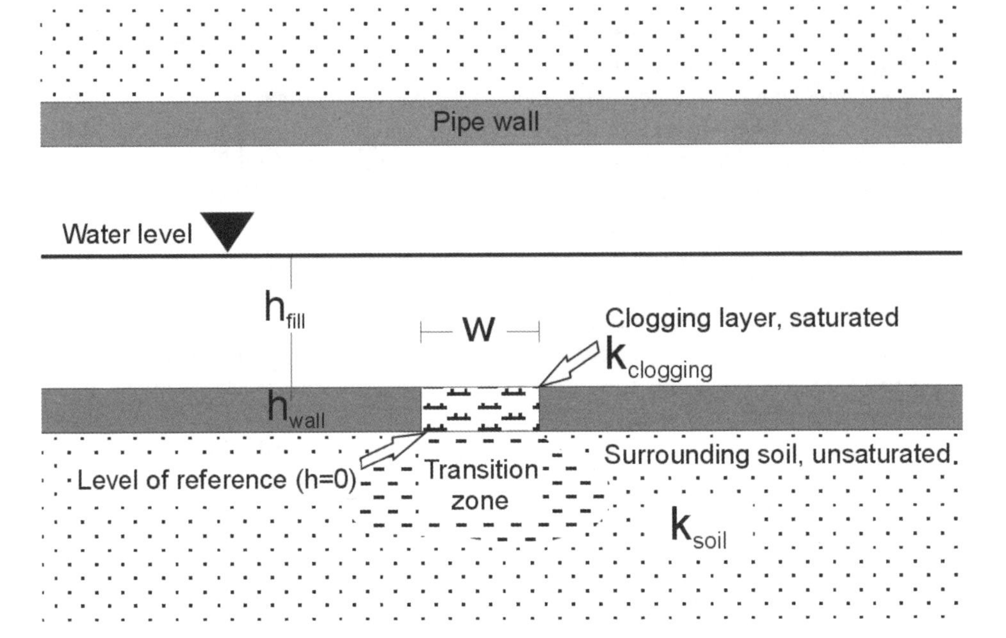

Figure 1. Schematic representation of the clogged leak.

For the following calculations it is assumed that the exfiltration behaviour is mainly controlled by the filling of the crack with a mixture of sediment and bio-film (i.e. colmation layer) and that the pipes are bedded in a clean medium sand. Medium sand is both the standard test environment for most of the available experiments and the most commonly used fill material for the utility trenches around the pipes.

Calculating K-values for the clogging layer following Darcy's law (2;3):

$$Q = K \cdot I \cdot A \tag{2}$$

equivalent to

$$K = \frac{Q}{I \cdot A} \tag{3}$$

with:

$$I = \frac{h_{fill} + h_{wall}}{h_{wall}} \tag{4}$$

A: Size of the leak [L^2]
Q: Exfiltrating flow [L^3/T]
K: Hydraulic conductivity of the clogging layer [L/T]
I: Hydraulic gradient [-]

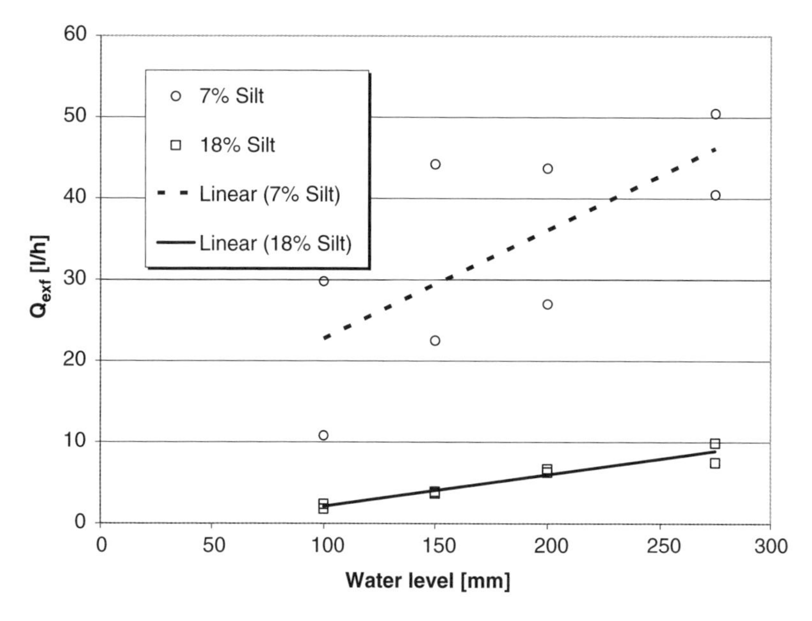

Figure 2. Exfiltration from a sewer leak in relation to water level and soil type (redrawn from Dohmann et al., 1999).

h_{fill}: Water level inside the sewer [L]
h_{wall}: Pipe wall thickness [L]

Equation 4 defines the gradient which is the driving force for the water movement. It is assumed that all exfiltrating water has to pass the slot (i.e. leak) which is filled with low hydraulic conductivity sediments. The level of reference for the calculation of this height difference is set at the outer border of the pipe. This includes the assumption that only a section equivalent to the pipe wall thickness is responsible for the exfiltration process. The unsaturated soil outside the pipe is assumed to ensure free drainage.

The given equation is only valid if the hydraulic conductivity of the clogging layer is indeed the control factor for wastewater exfiltration.

Under field conditions, two cases can be distinguished:

- $K_{\text{soil}} > K_{\text{clogging layer}}$: Clogging layer is determining the exfiltration rate
- $K_{\text{soil}} < K_{\text{clogging layer}}$: Hydraulic conductivity of the soil is the limiting factor.

The influence of the bedding material on the exfiltration rate was observed by Dohmann et al. (1999) and is demonstrated in Figure 2. A silt content of 18% leads to a significantly lower flow rate compared to sand with just 7% silt.

2.3 *Comparing hydraulic conductivities of the colmation layer*

The *K*-values were calculated from the laboratory data from the various sources. They range between 1.15×10^{-4} m/s and 3.1×10^{-8} m/s. This not only reflects the variability of different leak sizes but also the different set ups which lead to different ages and conditions of the clogging layer. For example, the Dohmann experiments were performed with a very young or almost non-existing colmation layer and consequently show high *K*-values. As the clogging layer had not yet built up to its full extent, Dohmann et al. (1999) also noted a strong

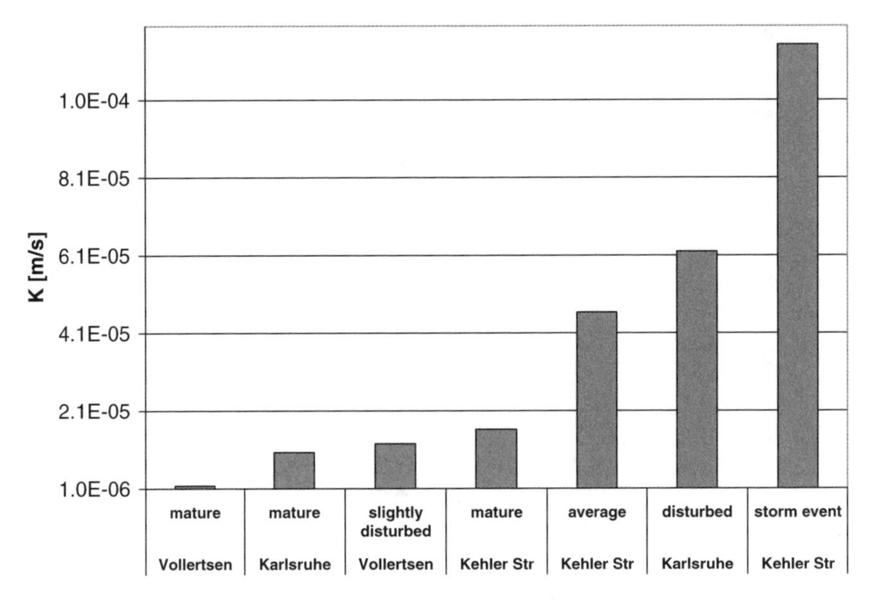

Figure 3. Hydraulic conductivities in comparison with the condition of the clogging layer derived from experiments in Aalborg (Vollertsen and Hvitved-Jacobson, 2002), Karlsruhe (Forschergruppe Kanalleckagen, 2002), and Rastatt-Kehler Strasse (Held et al., 2005).

influence of the grain size of the surrounding soil (Figure 2). However, given the information gathered through the Karlsruhe (Forschergruppe Kanalleckagen, 2002) and Aalborg (Vollertsen and Hvitved-Jacobson, 2002) experiments, it is likely that the exfiltration rates would have also continued to decrease in the Dohmann et al. (1999) experiments because the conductivity of the clogging layer would have fallen below the conductivity of the surrounding soil.

Emphasizing the importance of the age of the undisturbed clogging layers is Figure 3, which compares only hydraulic conductivities of clogging layers built up on medium sand.

3 EXTRAPOLATION TO AN EXISTING CATCHMENT

3.1 *Condition monitoring with CCTV*

Obeying recent modifications in German law, the City of Rastatt has surveyed more than 90% of its sewer network using CCTV (Closed-Circuit Television, i.e. a video recorder mounted on a vehicle) inspections during the last five years. Each observed damage was classified according to ATV-advisory leaflet M143 (ATV, 1999) and stored in a database. This sewer defect database was given spatial reference in the GIS system (Figure 4). One has to consider that there is a considerable underestimation of defects on the sewer bottom, as during the inspection the latter was frequently obscured by the remaining water. This is of special importance as the bottom section of the sewer during operation is always filled with base flow sewage and therefore prone to constant exfiltration.

For the further calculations a smaller catchment area has been selected upstream of the sewer test site Rastatt-Danziger Strasse (Figure 4). The demonstration catchment is drained by a combined sewer system and covers an area of 220 hectares with a mixed land use of

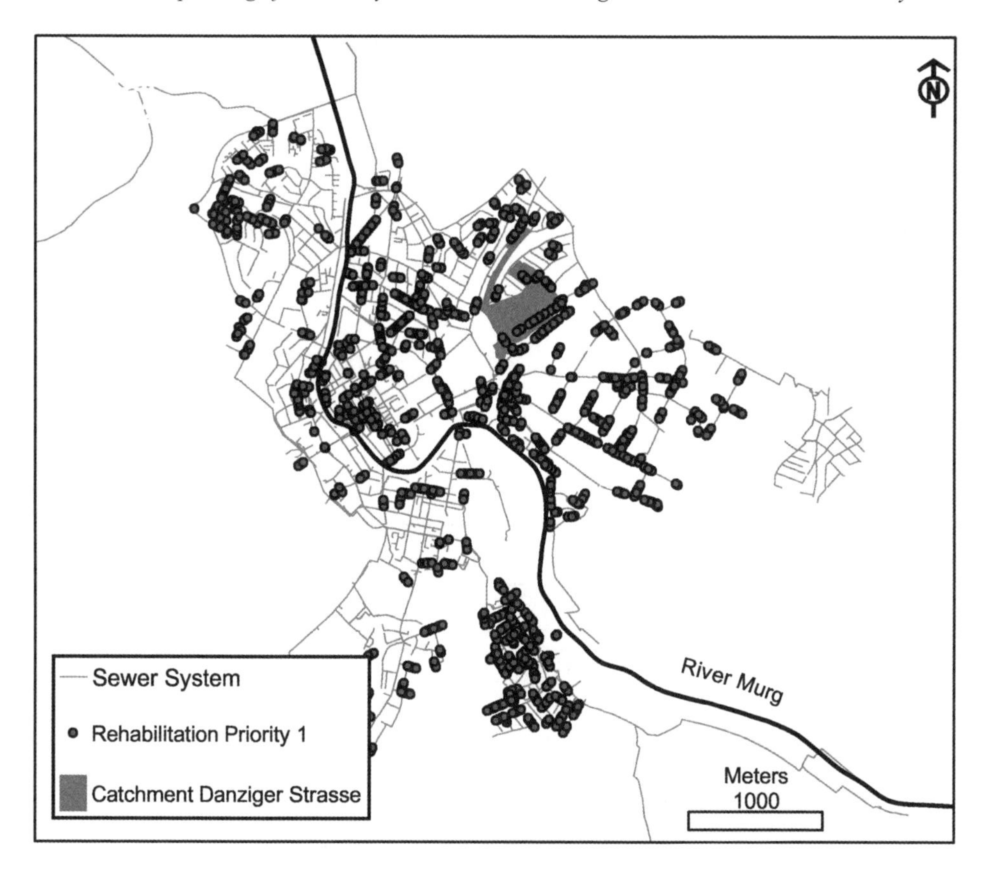

Figure 4. Spatial distribution of sewers with top rehabilitation priority in Rastatt.

residential and commercial buildings. More details, including water balance calculations, have been described in (Eiswirth et al., 2004) and (Klinger and Wolf, 2004). Table 1 shows an analysis of the existing CCTV-data for the demonstration catchment. The majority of the 1114 observed construction anomalies consist of improperly installed house connections and visible backwater in the sewer. For these damage types, estimates of the associated size of the defects are difficult and have been omitted at this stage. For most of the other 262 anomalies (cracks, broken shards, joint displacement), information on the extent (e.g. width) of the damage was noted in the TV-protocols and an area of defect could be estimated.

3.2 *Ranges of input parameters*

For the extrapolation to a catchment, the total defect area as listed in Table 1 is used for the calculation of exfiltrating sewage volumes. As all laboratory experiments described a strong variation between high flow and low flow conditions, separate calculations are done for these two cases. Different hydraulic conductivities are assumed for the colmation layer as well as different fill levels in the sewer, leading to different hydraulic gradients.

The total average exfiltration for a single leak is then defined as (5):

$$q_{Leak} = k_{LowFlow} \cdot A_{Leak} \cdot I_{LowFlow} \cdot \alpha + k_{HighFlow} \cdot A_{Leak} \cdot I_{HighFlow} \cdot \frac{1}{\alpha} \tag{5}$$

Table 1. Analysis of CCTV-data for the catchment Danziger Strasse.

Anomaly	Count	Estimated area [m^2]
House connection, improper	237	
House connection, proper	80	
House connection, cracks	10	
House connection, intruding	120	
Backwater	291	
Soil visible	40	
Corrosion	74	
Cracks, transversal	46	0.34
Cracks, longitudinal	52	0.27
Broken shards	28	0.08
Root intrusion	62	0.31
Joint displacement	30	0.86
Missing pieces	44	0.44
Total	1114	2.30

Table 2. Uncertainty ranges in independent input parameters.

Parameter	Unit	Min	Max
Pipe thickness/thickness of colmation layer	[m]	0.01	0.04
Size of an individual defect	[m^2]	0.004	0.018
Ratio high flow/ low flow	[-]	0.01	0.10
K-value colmation layer low flow	[m/s]	1×10^{-8}	1×10^{-5}
Average fill level low flow	[m]	0.01	0.06
K-value colmation layer high flow	[m/s]	1×10^{-6}	1×10^{-4}
Average fill level high flow	[m]	0.06	0.60

With:

α [-]: Ratio between dry weather flow and stormwater flow.
$k_{LowFlow}$ [m/s]: Hydraulic conductivity of the clogging layer during dry weather flow.
$k_{HighFlow}$ [m/s]: Hydraulic conductivity of the clogging layer during stormwater events.
A_{Leak} [m^2]: Total area of open sewer defect/soil interface in the catchment.
$I_{LowFlow}$ [-]: Hydraulic gradient across the colmation layer during dry weather conditions.
$I_{HighFlow}$ [-]: Hydraulic gradient across the colmation layer during dry weather conditions.

Based on the different results from experiments described in the literature as well as the measurements performed directly in Rastatt, minimum and maximum assumptions for seven parameters were determined.

Pipe wall thickness may vary in the catchment area from 0.01 to 0.04 m, corresponding to different pipe materials (concrete, ceramics) and different pipe diameters (from 0.2 to 1 m).

The total defect area in the catchment is the most uncertain parameter besides the hydraulic conductivity. The estimation of the total defect size is influenced by:

- Undetected defects at the pipe joints (detection not possible with CCTV).
- Backwater which remains in the sewer and obscures the sewer bottom during the CCTV inspection. Defects at the sewer bottom are underestimated.
- No CCTV-data on private house connections is available. Private sewer networks may constitute twice the length of the public network. Furthermore, the private sewer networks are likely to be in a worse condition due to the absence of regular inspections.

- CCTV inspection only observes pipe surface but defects may not penetrate the pipe wall.
- Subjectivity of the inspector (crack diameters are frequently only estimates).
- Wrong assumptions in defect area calculations (e.g. 2 m length for a typical longitudinal crack).

As shown in Table 1, the total defect area in the catchment was determined as 2.3 m^2 from the CCTV inspection. This corresponds to an average defect size of 0.008 m^2 which is a rather large value. In most of the literature test sites, the typical defect size is 0.002 m^2. The large average defect size may also be explained by the fact that only critical defects were used for the evaluation. In the absence of reliable information and in order to account for the uncertainties listed above, it is assumed that the total defect area in the demonstration catchment Rastatt-Danziger Strasse may have any dimension between 50% of the calculated value and 200% of the calculated value.

The ratio between high and low water levels in the sewer was assessed based on the evaluation of rainfall data and on direct water level measurements in the sewer. However, no reliable data exist yet on the sewer water level required for a substantial destabilisation of the clogging layer and on the switching between high and low conductivity values of the clogging layer in the leak. In other words: it is unknown, how often and at which water levels the colmation layer is removed or damaged. As a worst case, it is assumed that the colmation layer will be removed during 10% of the year and that high fill levels will be present.

The hydraulic conductivities of the colmation layer are subject to: (a) the range of values from different laboratory experiments, and (b) varying soil types in the area. For this reason, the whole range of values measured in different experiments has been applied for this calculation.

3.3 *Monte Carlo simulations*

A simple kind of Monte Carlo simulation was performed by calculating 2000 randomly selected combinations of the seven independent input parameters within the specified ranges (Table 1) of each parameter. The results are displayed in Figures 5 through 8.

Figures 5 and 6 show the possible exfiltration rates for a single leak during dry weather flow and wet weather flow/storm events. The highest probability for dry weather flow is 11% at 4 l/d. With a probability of 95%, the exfiltration will be below 34 l/d at an individual leak. The highest probability for exfiltration rates during storm events is 10.55% at 100 l/d. With a probability of 95% the exfiltration will be below 2400 l/d at an individual leak.

Figure 7 depicts the probability density function for the yearly average of the exfiltration from a single leak. The highest probability of 23% is found at 10 l/d exfiltration from a single leak. With a probability of 95%, the average annual exfiltration will be below 155 l/d.

For upscaling this information to the catchment, there are two different possibilities:

(a) Applying the probability density function in Figure 7 (derived from the simulation of 2000 leaks) proportionally to the 262 observed sewer leakages in the catchment area of Danziger Strasse. The summed leakage amounts to 5.4 m^3/d (1.6% of the typical dry weather flow of 320 m^3/h), equivalent to 8.8 mm/a groundwater recharge from leaky public sewer systems.

(b) Assuming that all 262 leaks have the same exfiltration characteristics (Figure 8). The result is a broad range of possible groundwater recharge rates in the catchment.

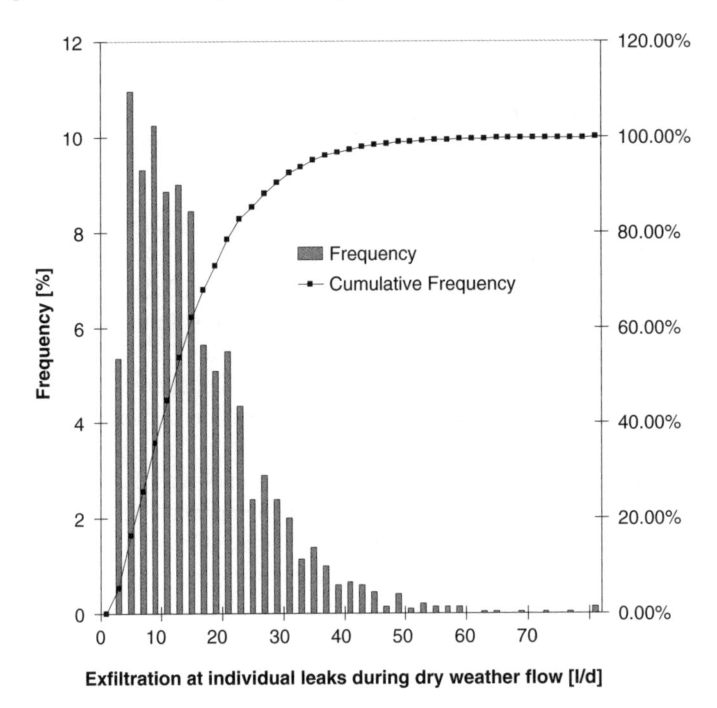

Figure 5. Dry weather flow, individual leak.

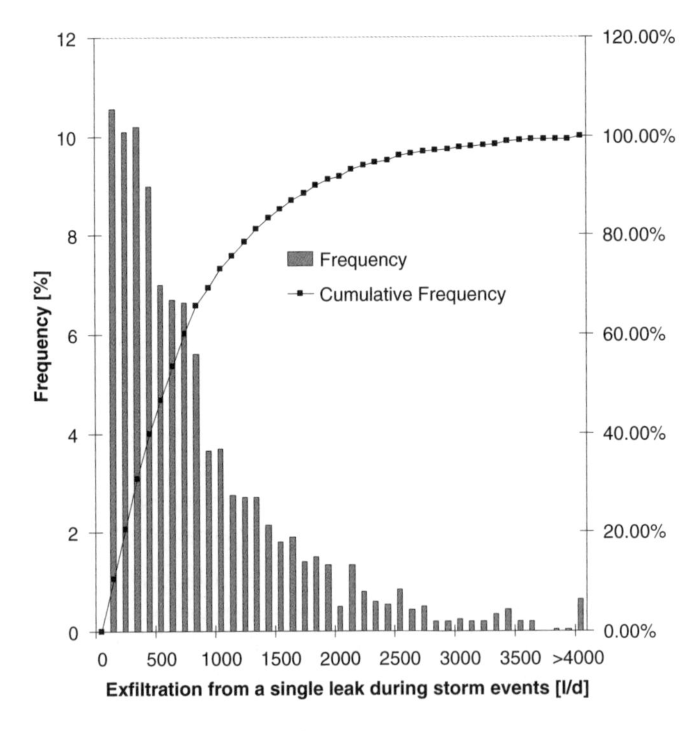

Figure 6. Wet weather flow, individual leak.

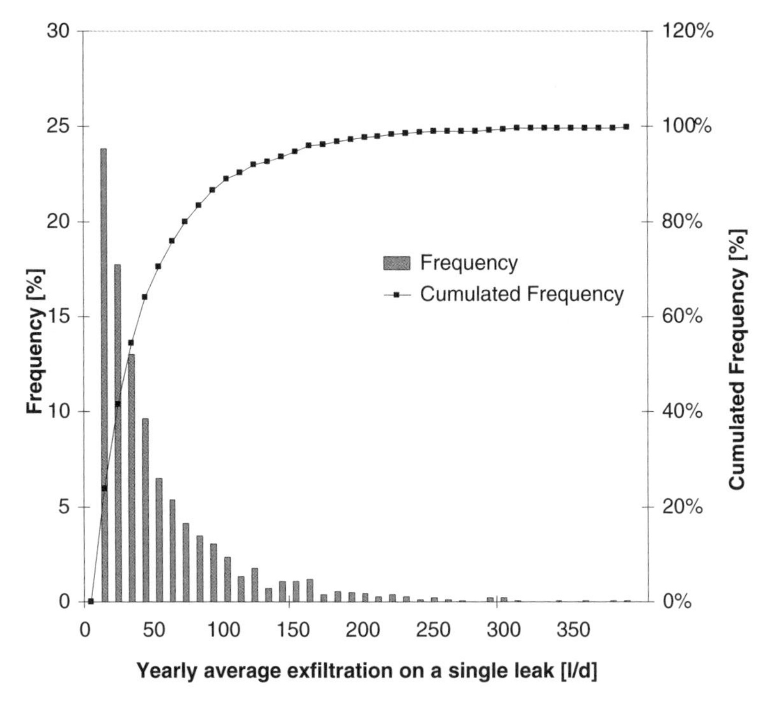

Figure 7. Combined yearly average, individual leak.

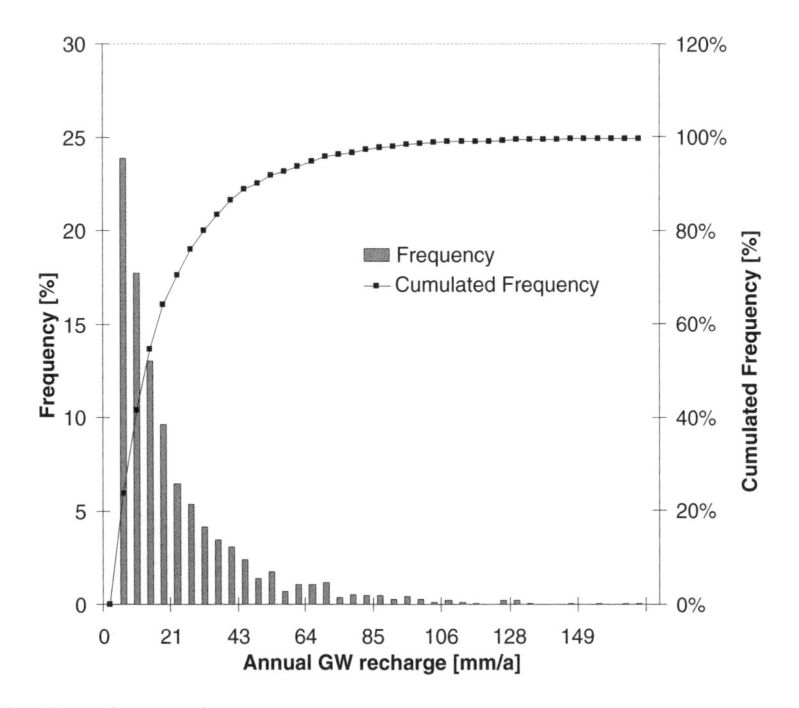

Figure 8. Groundwater recharge.

The highest probability of 23% is found at 4.2 mm/a groundwater recharge from sewers (total leakage of 2.57 m^3/d or 0.8% of the typical dry weather flow of 320 m^3/h). With a probability of 95%, the groundwater recharge will be below 65 mm/a.

4 SUMMARY AND CONCLUSIONS

Exfiltration from leaky sewers can be a significant factor in the urban water balance. However, the direct assessment of the exfiltrating volumes based on TV-camera inspections has large uncertainties due to the imprecise knowledge of the input parameters. Results from several experiments in the literature were compared with new field measurements in Rastatt. Based on the application of Darcy's law to the colmation layer an extrapolation to a 220 ha sized catchment in Rastatt (Germany) was attempted. The calculations took the detailed interpretation of the existing CCTV inspections into account. The CCTV inspections in the study catchment showed a very high frequency of large defects in comparison with other catchments. In order to account for the uncertainties in the input parameters, a Monte Carlo Simulation was applied to the data set of a real sewer network and resulted in most probable exfiltration rates of 10 l/d/leak, equivalent to 4 mm/a artificial groundwater recharge from sewers. However the uncertainty involved is very significant. With a probability of 95% the exfiltration will be below 155 l/d/leak and below 65 mm/a. This large range of possible exfiltration rates underscores the necessity for groundwater quality monitoring and the use of marker species approaches besides the CCTV-based assessments.

5 OUTLOOK

The quantification of sewer leakage is currently being refined as part of the AISUWRS project (Assessing and Improving the Sustainability of Urban Water Resources and Systems; www.urbanwater.de), which is currently running (Burn et al., 2005). A special model has been developed for the quantification of pipe leakage (DeSilva et al., 2004) as well as a contaminant transport model for the unsaturated zone.

The broad range of uncertainty could be significantly reduced in future if the individual input parameters (e.g. water level) were described by appropriate probability density functions.

ACKNOWLEDGEMENTS

This work would not have been possible without the kind and strong support which Dr. Matthias Eiswirth provided until his tragic death on the 30th of December 2003. In him, we have lost a good friend and untiring initiator of urban hydrogeological research far before his time. We will keep him in our memories and remain thankful for everything he has given us.

This work was supported by the European Community in the course of the AISUWRS project (EVK1-CT-2002-00110) and the authors would like to thank all project partners for good cooperation. Special thanks to Ilka Neumann and Brian Morris from the British

Geological Survey for insisting on dealing with uncertainty and thereby enforcing this paper. Special acknowledgements also to Dhammika DeSilva and Ray Correl (CSIRO). Additional funding was provided by the German Science Foundation (DFG) within the research group "Risk Potential from Leaky Sewers for Soil and Groundwater". The authors would like to express their thanks to J. Kramp, P. Polak and V. Tropf from the City of Rastatt for fruitful collaboration.

Further acknowledgements are due to Inka Held and Jochen Klinger for years of fruitful collaboration on the topic of sewer leakage in Rastatt.

REFERENCES

ATV. 1999. *M 143 Advisory Leaflet – Inspection of Sewers and Drains, Part 2: Optical Inspection*. Hennef, Germany: ATV-DVWK.

Barret, M.H., Hiscock, K.M., Pedley, S., Lerner, D.N., Tellam, J.H. and French, M.J. 1999. *Marker species for identifying urban groundwater recharge sources: a review and case study in Nottingham, UK*. Water Research 33, **14**, 3083–3097.

Burn, S., Eiswirth, M., Correll, R., Cronin, A.A., DeSilva, D., Diaper, C., Dillon, P., Mohrlok, U., Morris, B., Rueedi, J., Wolf, L., Vizintin, G. and Vött, U. 2005. *Urban infrastructure and its impact on groundwater contamination*. In: Howard K (Ed.), Matthias Eiswirth Memorial Volume, Rotterdam: Balkema.

DeSilva, D., Burn, S., Tjandraatmadja, G., Moglia, M., Davis, P., Wolf, L., Held, I., Vollertsen, J., Williams, W. and Hafskjold, L. 2004. *Sustainable Management of Leakage from Wastewater Pipelines*. Marrakesh, Morroco.

Dohmann, M., Decker, J. and Menzenbach, B. 1999. *Untersuchungen zur quantitativen und qualitativen Belastung von Boden-, Grund- und Oberflächenwasser durch undichte Kanäle*. In: Dohmann M (Ed.), Wassergefährdung durch undichte Kanäle, Berlin-Heidelberg-New York: Springer, pp 1–82.

Eiswirth, M., Held, I., Hötzl, H., Klinger, J. and Wolf, L. 2004. *Rastatt Interim City Assessment Report*. AISUWRS Deliverable D 9-1. Department of Applied Geology, University of Karlsruhe, Germany.

Eiswirth, M. and Hötzl, H. 1997. *The impact of leaking sewers on groundwater*. In: Chilton J (Ed.), Groundwater in the urban environment, Rotterdam: Balkema, pp 399–404.

Ellis, J. 2001. *Sewer infiltration/exfiltration and interactions with sewer flows and groundwater quality*. Proc. Interurba II, 19.–22. Feb, Lisbon, Portugal.

Härig, F. and Mull, R. 1992. *Undichte Kanalisationssysteme – die Folgen für das Grundwasser*. gwf Wasser – Abwasser 133, **4**, 196–200.

Held, I., Klinger, J., Wolf, L. and Hötzl, H. 2005. *Direct measurements of exfiltration st a sewer test site under operating conditions*. In: Howard K (Ed.), Matthias Eiswirth Memorial Volume, Rotterdam: Balkema.

Forschergruppe Kanalleckagen. 2002. *Gefährdungspotential von Abwasser aus undichten Kanälen für Boden und Grundwasser*. Zwischenbericht an die DFG http://www.rz.uni-karlsruhe.de/~iba/kanal/zwischenbericht.pdf

Keitz, S. von. 2002. *Handbuch der EU-Wasserrahmenrichtlinie*. Berlin: Erich Schmidt, 447 p.

Klinger, J. and Wolf, L. 2004. *Using the UVQ Model for the sustainability assessment of the urban water system*. 19 EJSW Workshop on Process Data and Integrated Urban Water Modelling, Meaux aux Montagne, France: INSA Lyon http://www.citynet.unife.it.

Rauch, W. and Stegner, T. 1994. *The colmation of sewer leaks during dry weather flow*. Wat. Sci. Tech. 30, **1**, 205–210.

Rott, U. and Zacher, K. 1999. *Entwicklung von Verfahren zur Quantifizierung des Wasseraustritts und der Wasser- und Stoffausbreitung in der Umgebung undichter Kanäle*. In: Dohmann M (Ed.), Wassergefährdung durch undichte Kanäle, Berlin-Heidelberg-New York: Springer, pp 213–247.

Sacher, F., Gabriel, S., Metzinger, M., Stretz, A., Wenz, M., Lange, F.T., Brauch, H.-J. and Blankenhorn, I. 2002. *Arzneimittelwirkstoffe im Grundwasser – Ergebnisse eines Monitoring-Programms in Baden-Württemberg*. Vom Wasser, **99**, 183–196.

Vollertsen, J. and Hvitved-Jacobsen, T. 2003. *Exfiltration from gravity sewers – a pilot scale study*. Water Science and Technology 47, **4**, 69–76.

Wolf, L., Eiswirth, M. and Hötzl, H. 2005. *Assessing sewer-groundwater interaction at the city scale based on individual sewer defects and marker species distributions*. Environmental Geology In Press.

Wolf, L., Held, I., Eiswirth, M. and Hötzl, H. 2004. *Environmental impact of leaky sewers on groundwater quality*. Acta Hydrochim. Hydrobiol, 32: 361–373.

Assessment of contaminant impacts

CHAPTER 8

Groundwater flow velocities indicated by anthropogenic contaminants in urban sandstone aquifers

R.G. Taylor[1], A.A. Cronin[2] and J. Rueedi[2]
[1]*Department of Geography, University College London, UK*
[2]*Robens Centre for Public and Environmental Health, University of Surrey, UK*

ABSTRACT: Anthropogenic solutes and microorganisms detected at depth-specific intervals within actively pumped, urban sandstone aquifers in the UK, provide coarse estimates of aquifer penetration rates (sets of vertical groundwater velocities). Stable isotope tracers ($^{15}N/^{14}N$, $^{34}S/^{32}S$) indicate that solutes derive primarily from anthropogenic sources whereas detected bacteria (*Escherichia coli*, faecal streptococci) and viruses (coliphage, enterovirus) reflect a faecal origin. Low concentrations of these microorganisms are detected sporadically though often proximate to aquifer heterogeneities whereas concentrations of anthropogenic solutes gradually decrease with increasing depth from the surface. Observed contaminants trace different sets of vertical groundwater velocities as faecally derived microorganisms indicate aquifer penetration rates ($m \cdot d^{-1}$) that are orders of magnitude more rapid than aquifer penetration rates suggested by anthropogenic solutes ($m \cdot a^{-1}$). Further research is required to test conceptual models of microbial transport and to resolve further the distribution of groundwater velocities using a range of environmental tracers.

1 INTRODUCTION

Groundwater sampled from a particular location or depth represents a range of residence times (i.e. groundwater ages) that occur in the subsurface from source (recharge) to the point of collection. The distribution of groundwater flow velocity sets that gives rise to these groundwater ages, remains, however, difficult to resolve and poorly understood (Kinzelbach and Herzer, 1983; Edmunds and Smedley, 2000) particularly under heterogeneous hydrogeological conditions. As a result, a normal distribution of groundwater velocities around a mean ($\overline{v}$) is regularly assumed (Figure 1) and underlies current strategies, that are based upon natural attenuation of contaminants over distances defined by $\overline{v}$, to protect groundwater sources from contamination (USEPA, 1991; Adams and Foster, 1992; Keating and Packman, 1995; DWAF, 1997). This assumption disregards alternative, multimodal distributions (Figure 1) that may occur in heterogeneous aquifers (Matthai and Belanyeh, 2004) and underestimates the importance of statistically extreme velocities (Taylor et al., 2004).

Efforts to resolve the distribution of velocity sets in fast-flowing environments (e.g., gravel aquifers) typically involve deliberate application of artificial groundwater tracers (e.g., dyes) because the interval between tracer application and detection (i.e. groundwater residence time) is of the order of days to weeks (Flynn, 2003). In less rapid flow systems

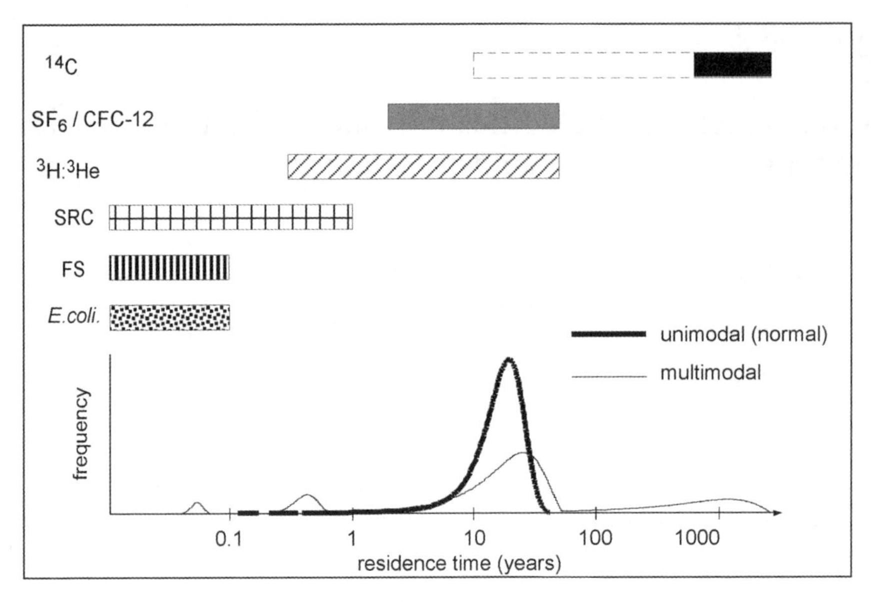

Figure 1. Schematic plot showing the relationship between environmental tracers and residence-time distributions in groundwater (i.e. inferred sets of groundwater velocities).

(e.g. sandstone) where mean groundwater residence times can be in the order of years to decades, artificial tracers are practical only over short distances (<10 m). At larger scales, reliance is placed upon environmental tracers comprising anthropogenic gases (CFCs, SF_6), radioactive isotopes (^3H, ^{14}C) and surface-loaded contaminants such as faecal micro-organisms and anthropogenic solutes (Figure 1). In this paper, we briefly review the results of recent depth-specific monitoring of solutes and faecal bacteria and viruses in Permo-Triassic sandstone aquifers underlying the cities of Nottingham and Birmingham in central England (Powell et al., 2003; Cronin et al., 2003a; Taylor et al., in review). We then consider the range of aquifer penetration rates (sets of vertical groundwater velocities) that are implied by concentrations of surface and near-surface loaded microorganisms and solutes observed with depth in these urban, unconfined sandstone aquifers. Analysed bacteria (*Escherichia coli*, faecal streptococci) and viruses (coliphage, enterovirus) have a faecal origin whereas the anthropogenic source of anionic solutes (NO_3^-, SO_4^{2-}) is confirmed by stable isotope ratios of nitrogen and sulphur in aqueous nitrate and sulphate, respectively.

2 DEPTH-SPECIFIC MONITORING IN SANDSTONE AQUIFERS

Depth-specific monitoring of groundwater chemistry and microbiology was conducted using dedicated multilevel piezometers that were installed to depths of up to 60 mbgl (metres below ground level) in unconfined aquifers of the Sherwood Sandstone Group (SSG) underlying the mature conurbations of Birmingham and Nottingham (Taylor et al., 2003) (Figure 2). Detailed descriptions of methodologies and results from sampling microorganisms, solutes and stable isotopes are provided by Powell et al. (2003) and Taylor et al. (in review), respectively. Field observations are briefly reviewed at one location in Nottingham (Old Basford) in order to highlight contrasting sets of vertical groundwater velocities

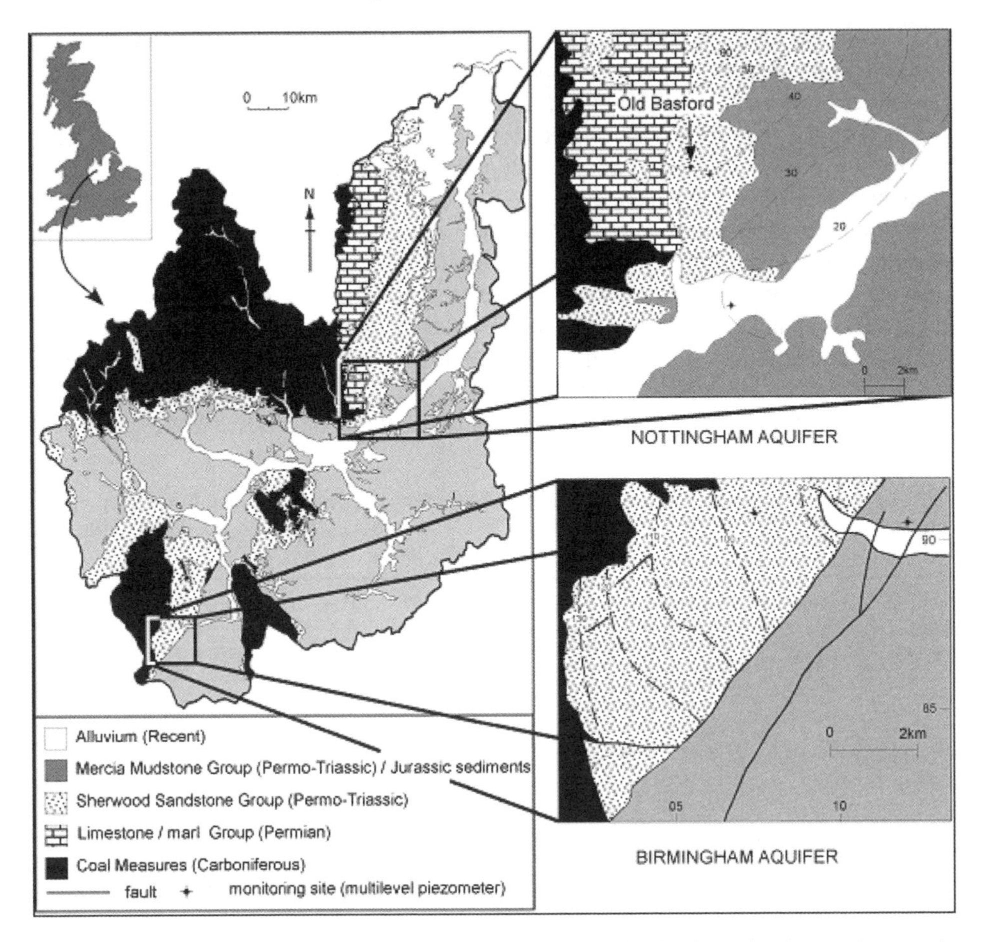

Figure 2. Geology of the Trent River Basin in central United Kingdom showing study areas in Nottingham and Birmingham. Regional gradients are indicated by contours of hydraulic head in 1989 for the Nottingham Aquifer (Charlsley et al., 1990) and Birmingham Aquifer (Ford and Tellam, 1994).

implied by the penetration of anthropogenic microorganisms and solutes in the SSG. Isotopic ratios in anionic solutes are expressed in delta notation relative to international reference standards for S (Canon Diablo Troilite) and N (Vienna Standard Mean Ocean Water) according to:

$$\delta^{14}N \text{ or } \delta^{34}S = [(R_{sample} - R_{standard})/R_{standard}] \times 1000$$

where R = heavy isotope/light isotope (e.g. $^{15}N/^{14}N$, $^{34}S/^{32}S$).

2.1 *Faecal microorganisms*

Faecal bacteria (*Escherichia coli*, faecal streptococci, sulphite-reducing clostridia) and viruses (coliphage, enteroviruses) were regularly detected in low concentrations ($<500\,cfu \cdot L^{-1}$ for bacteria, $<10\,pfu \cdot L^{-1}$ for viruses) at a range of depth intervals up to 60 mbgl in the unconfined sandstone at Old Basford over a 15-month period from May 2000 to August

2001 (e.g. Table 1). The frequency and magnitude of observed contamination show an association with the presence of geological heterogeneities (e.g. fissures) identified by downhole geophysics (Taylor et al., 2003) within sealed sampling intervals (e.g. 13.5 to 16.4 mbgl in Figure 3). Highest bacterial counts in groundwater were recorded in November 2000 (Table 1) following exceptionally heavy rainfall (188 mm) in October 2000, more than twice the monthly average of 87 mm. Similar increases in the concentration of faecal bacteria and viruses in groundwater in response to heavy (seasonal) rainfall, have been observed in Gambia (Wright, 1986), Taiwan (Jean, 1999), and Uganda (Howard et al., 2003).

Rapid vertical penetration and subsequent detection of low numbers of microorganisms in the SSG is highlighted by temporal correlations between enteroviruses, identified at several depth intervals down to 39 mbgl in the unconfined sandstone at Old Basford (Table 2), and those predicted in sewers. Enteroviruses were characterised by Reverse Transcriptase Polymerase Chain Reaction (RT-PCR) during three sampling intervals in March, June, and

Table 1. Number of colony-forming units of thermotolerant coliforms (TTC) and faecal streptococci (FS) detected in 100 mL groundwater samples (TTC/FS) with depth in Old Basford, Nottingham from June 2000 to August 2001 (data from Powell et al., 2003).

Sample depth (mbgl)	Jun. '00		Nov. '00		Jan. '01		Feb. '01		Mar. '01		June '01		Aug. '01	
	TTC	FS	TTC	FS	TTC	FS	TTC	FS	TTC	FS	TTC	FS	TTC	FS
8.0 to 8.1	<1	<1	2	50	<1	<2	<1	NA	<1	<1	<1	<1	<1	<1
11.0 to 11.2	1	<1	<1	50	<1	<2	<1	<1	<1	<1	<1	<1	2	2
15.0 to 15.1	1	1	1	<10	<1	27	1	1	<1	<1	<1	1	<1	1
18.0 to 18.2	<1	<1	2	<10	<1	5	<1	<1	<1	<1	<1	<1	1	1
22.1 to 22.2	<1	<1	<1	<2	<1	1	<1	<1	<1	<1	<1	<1	<1	<1
25.9 to 26.1	<1	<1	4	14	<1	3	<1	1	<1	<1	<1	<1	<1	<1
30.2 to 30.3	<1	<1	2	16	<1	3	<1	<1	<1	<1	<1	<1	<1	<1
35.1 to 35.3	2	<1	<1	2	<1	2	<1	<1	<1	<1	<1	7	<1	<1
39.1 to 39.3	2	<1	1	16	1	<2	<1	<1	<1	<1	<1	<1	1	1

NA: not analysed.

Table 2. Number of plaque-forming units per litre ($pfu \cdot L^{-1}$) of enteroviruses detected in groundwater in Old Basford, Nottingham from June 2000 to August 2001 (data from Powell et al., 2003).

Sample depth (mbgl)	June 2000	Nov. 2000	Jan. 2001	Feb. 2001	Mar. 2001	June 2001	Aug. 2001
8.0 to 8.1	<1	<1	NA	NA	<1	<1	<1
11.0 to 11.2	<1	NA	NA	NA	D[b]	D[c]	<1
15.0 to 15.1	NA	<1	NA	NA	D[b]	<1	<1
18.0 to 18.2	NA	NA	NA	NA	<1	D[c]	<1
22.1 to 22.2	NA	<1	NA	NA	D[b]	<1	<1
25.9 to 26.1	NA	NA	<1	NA	<1	<1	<1
30.2 to 30.3	NA	NA	<1	5[a]	D[b]	<1	<1
35.1 to 35.3	NA	NA	NA	NA	<1	<1	<1
39.1 to 39.3	NA	<1	NA	<1	D[b]	D[c]	<1

NA: not analysed; D: detected by PCR (not quantified); a: unidentified enterovirus (plaque assay); b: Norwalk-Like Virus; c: Coxsackievirus B4.

August 2001. Detection of Norwalk-Like Virus (NLV) in March 2001 but not in June or August 2001 coincides with the expected discharge of higher numbers of NLV to sewers as a result of increased incidence of NLV illnesses in cold months (Mounts et al., 2000; Wyn-Jones et al., 2000; Sellwood, 2001). Coxsackievirus B4 which was detected in groundwater at three depth intervals down to a depth of 39 mbgl in June 2001, was also the most common enterovirus serotype isolated from UK sewage in 2001 (Sellwood, 2001).

Apart from dedicated multilevel piezometer installations, enterovirus and rotavirus have also been recovered from the discharge of industrial boreholes drawing from the SSG in Nottingham and Birmingham over a period from May 1999 to March 2000 (Powell et al., 2000). Sampled boreholes are commonly open over much of the aquifer's thickness but casing depths of approximately 30 m provide a minimum depth of entry into the borehole (i.e. aquifer penetration). Slade (1985) similarly isolated human enteroviruses from an industrial borehole cased to a depth of 31 mbgl in the chalk aquifer underlying London (UK).

2.2 Anthropogenic solutes

Aqueous sulphate concentrations decrease gradually from 116 mg·L^{-1} in the shallow sandstone and then substantially with depth from 22.2 to 35.1 mbgl to 33 mg·L^{-1} (Figure 3). Stable isotope ratios of sulphur in dissolved sulphate range from +4.1 to +2.1‰ over this depth interval and are consistent with anthropogenic sources including both atmospheric deposition from the combustion of fossil fuels and sewage (Bottrell et al., 2000; Moncaster et al., 2000). The absence of isotopic enrichment with depth and existence of oxidising conditions in groundwater (redox potential ranges from +374 to +294 mV) refute the possibility that sulphate reduction is responsible for the observed decrease in aqueous concentrations with depth. Groundwater is also unsaturated with respect to sulphate minerals such as gypsum and

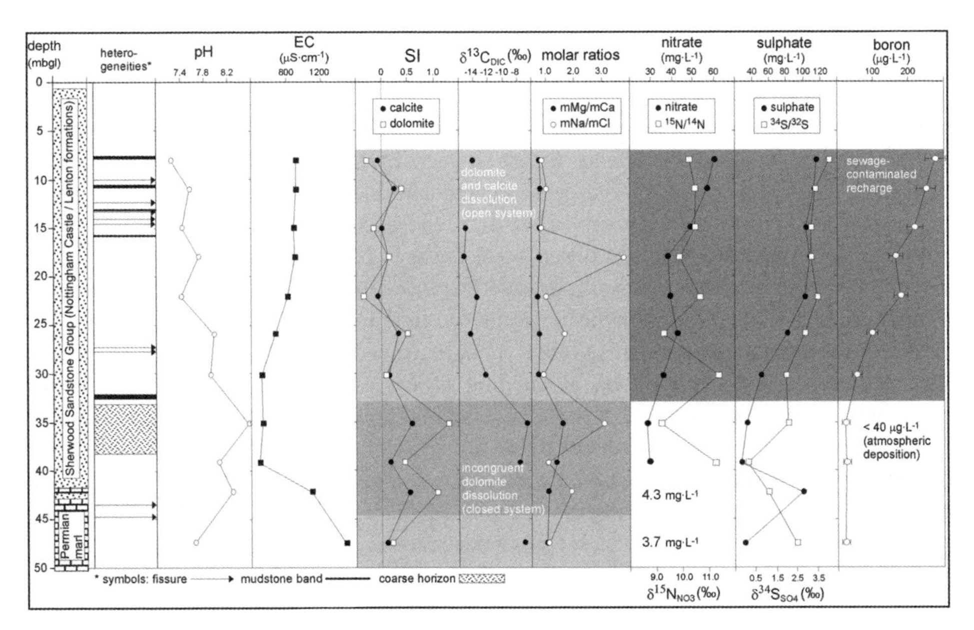

Figure 3. Hydrochemical profile in Permo-Triassic sandstone in Old Basford, Nottingham (from Taylor et al., in review). Vertical exaggeration is ×500. Reference standards for $\delta^{34}S$ and $\delta^{14}N$ are Canon Diablo Troilite and Vienna Standard Mean Ocean Water, respectively.

anhydrite throughout the hydrochemical profile (Taylor et al., in review). At the base of the sandstone (39.1 to 39.3 mbgl), aqueous sulphate is depleted in the heavy isotope ($\delta^{34}S = 0.1‰$) indicating that sulphate derives, in part, from the oxidation of sulphides such as pyrite.

Concentrations of nitrate gradually decrease with depth through the Sherwood Sandstone from 60 to 29 $mg \cdot L^{-1}$ (Figure 3). As the $\delta^{15}N$ of aqueous nitrate derived from sewage is enriched in the heavy isotope through ammonia volatilisation, stable isotope ratios of between +9.2‰ and +11.3‰ clearly implicate sewage as the source of nitrate (Fukada et al., 2004). Sewage contamination of groundwater at this multilevel monitoring site is supported by coincidental detection of sewage-derived microorganisms discussed above. Oxidising conditions in groundwater at this site and observed consistency in the isotopic ratios of nitrogen are incompatible with significant denitrification. Low levels of nitrate around 4 $mg \cdot L^{-1}$ in the underlying Permian marl (Taylor et al., in review) reflect pre-industrial recharge to the SSG in the Nottingham area (Edmunds and Smedley, 2000).

Similar to major-anion chemistry, elemental boron concentrations decrease with depth in the SSG (Figure 3) from 270 $\mu g \cdot L^{-1}$ in shallow groundwaters to a pre-industrial (atmospheric) level of 22 $\mu g \cdot L^{-1}$ (Edmunds et al., 1982) toward the base of the sandstone. Without an obvious geochemical source of boron in the SSG (Edmunds et al., 1982), elevated boron concentrations at shallow depths are assumed to derive from anthropogenic loading at the surface (e.g. sewage, metalworking industries). As such, these data are consistent with the observed depth (30.2 mbgl) to which anthropogenic sources of nitrogen and sulphur have clearly penetrated the sandstone at this site.

3 GROUNDWATER VELOCITY IMPLICATIONS

3.1 *Aquifer penetration rates – sewage-derived microorganisms*

Aquifer penetration rates, implied by the detection of sewage-derived bacteria and viruses, are estimated crudely from the depth below the water table at which microorganisms are detected and the estimated residence time of detected viruses and bacteria in the aquifer. Residence times of selected sewage-derived organisms in groundwater are estimated from the period required to reduce the number of viable bacteria and viruses (i.e. colony or plaque forming units) commonly observed in sewage (10^7 pfu or cfu$\cdot$100 mL^{-1}) to their detection limit in groundwater (1 pfu or cfu$\cdot$100 mL^{-1}) using published inactivation rates (Taylor et al., 2004). These estimates of residence time from experimental data vary considerably for individual organisms (Table 3) but represent a maximum (conservative) residence time since dispersive and attenuating processes, associated with contaminant transport, are ignored. Implied residence times of microorganisms detected at Old Basford range from between two and six weeks for *E. coli* and Coxsackievirus to eight months (240 days) for faecal streptococci (Table 3). Regular detection of these microorganisms at the base of the SSG at this location (39.2 mbgl), which lies 32.7 m below the water table at the time of sampling (Taylor et al., 2003), suggest aquifer penetration rates of between 0.2 and 2 $m \cdot d^{-1}$.

3.2 *Aquifer penetration rates – anthropogenic solutes*

Aquifer penetration rates implied by anthropogenic solute concentrations (Figure 3), are approximated by relating their depth of penetration to patterns of loading. For sulphate, a

Table 3. Observed inactivation rates (λ) for pathogenic and indicator microorganisms in groundwater. Data from Taylor et al. (2004) and references therein. Note: environmental conditions are not specified.

Microorganism	λ^a (day^{-1})	$t_{Nt=1}$^{b} (days)
Bacteria		
Escherichia coli	0.36 to 0.98	16 to 45
Faecal streptococci	0.066 to 0.85	19 to 240
Salmonella typhimurium	0.30 to 0.43	37 to 54
Shigella spp.	0.62 to 0.74	22 to 26
Klebsiella sp.	0.072	220
Viruses		
Poliovirus 1	0.081 to 1.6	10 to 200
Echovirus 1	0.12 to 1.4	12 to 130
Coxsackievirus B3	0.44	37

a: first-order inactivation rate, $N_t/N_0 = e^{-\lambda t}$.
b: $t_{Nt=1}$ = time required for inactivation to reduce the number of colony or plaque forming units in groundwater to 1 in the absence of dilution, adsorption/filtration or dispersion assuming an initial (sewage) concentration (N_0) of 10^7pfu or cfu·100 mL^{-1} (Geldreich, 1991).

well defined peak sulphur emissions occurs in the UK during the late 1960s to early 1970s (Zhao et al., 1998) and coincides with a similar peak in atmospheric emissions in Canada and the United States (Robertson et al., 1989). In a recharge area (i.e. downward vertical hydraulic gradients) of a rural, unconfined sand aquifer in Ontario, Canada, Robertson et al. (1989) found peak groundwater sulphate concentrations occur at depth and coincide with tritium-dated groundwaters that were recharged during the reported peak in anthropogenic S emissions. The multilevel piezometer in Old Basford is similarly located in a recharge area with a downward Darcian flux estimated between 1 and 3 m·a^{-1} at shallow depths (Taylor et al., 2003) that is partly influenced by local abstraction. Upward flow occurs from a coarse-grained layer at the base of the sandstone (Figure 3) to a transmissive, fissured zone at 27 mbgl (Taylor et al., 2003) and complicates interpretations at these depths where contaminated urban recharge may be mixing with regional groundwater flow. Highest groundwater sulphate concentrations are observed at the most shallow sampling interval in the aquifer.

Records of groundwater quality from monitoring wells in the Nottingham aquifer from 1975 to 2000 (Figure 4) reveal a trend toward higher sulphate concentrations (>150 mg·L^{-1}) at the end of the period despite reductions in S emissions since the early 1970s. Post-1981 median concentrations in this dataset derive, however, from fewer observations than median values in the late 1970s (Figure 4). Symmetry among depth profiles of aqueous sulphate, nitrate and elemental boron concentrations (Figure 3) points to a common source. Although sulphuric, nitric and boric acids are frequently used in metalworking (Ford and Tellam, 1994), δ^{15}N signatures of aqueous nitrate trace sewage as its primary source. B is also found in sewage as it is a common constituent in detergents and in a bleach, sodium perborate, which has been used in the UK over much of the 20th century (Barrett et al., 1999). Median concentrations of nitrate in monitoring wells in Nottingham aquifer have remained relatively stable, just above 50 mg·L^{-1}, since the early 1980s (Figure 4) but represent a rise from median concentrations at the start of the record in 1975. Groundwater nitrate concentrations influenced by human

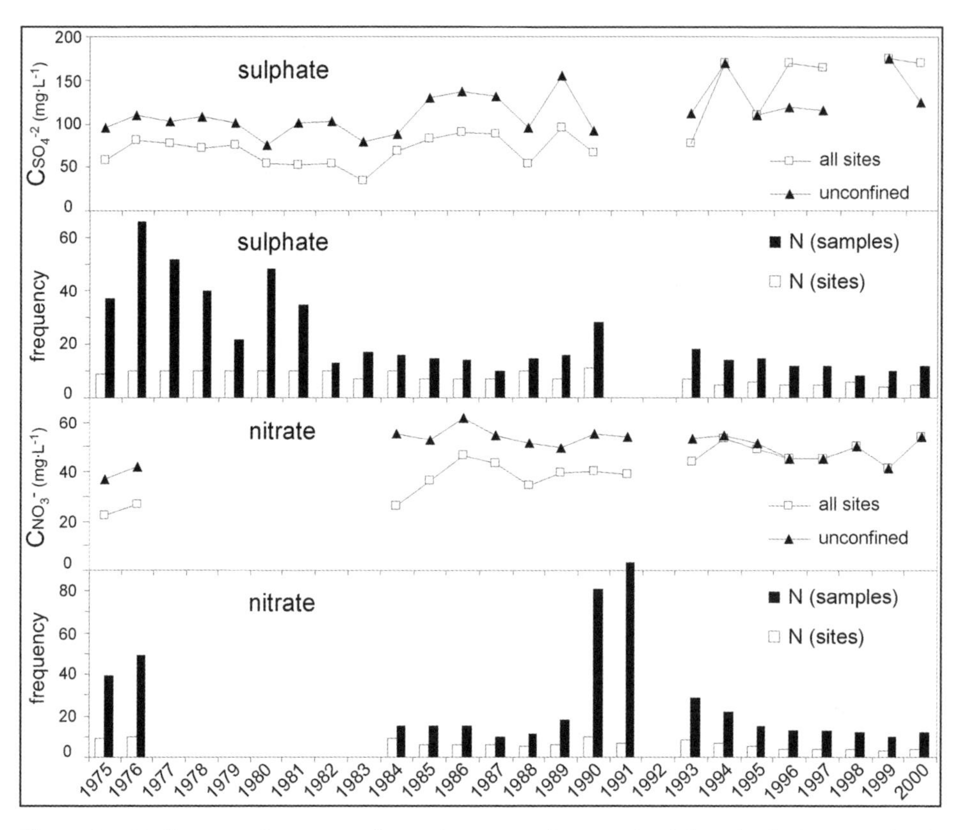

Figure 4. Median annual concentrations of aqueous nitrate and sulphate in the Permo-Triassic sandstone of Nottingham (data derived from monitoring wells of the Environmental Agency of England and Wales). Variations in the number of monitoring wells and collected samples reflect available records. Plots are redrawn from Cronin et al. (2003b).

activity, be it urban sewage or agricultural activities upgradient of the city, extend back to the late 19th century when, in 1874, the median concentration of nitrate in Nottingham wells was $36\,mg\cdot L^{-1}$ (Cronin et al., 2003b).

At the base of the sandstone in Old Basford (Figure 3), boron concentrations reflect pre-industrial conditions (Taylor et al., in review). Aqueous sulphate similarly exists in relatively low concentrations, approximately a quarter of that recorded at the shallowest sampling depth, and stable isotope ratios indicate a primarily mineral origin. In contrast, nitrate concentrations at the base of the sandstone in Old Basford in 2000 are $28.9\,mg\cdot L^{-1}$ and are assumed to reflect a longer history of loading as neither denitrification nor sulphate reduction is indicated by stable isotope tracers and prevailing redox conditions. Assuming that boron concentrations and sulphate represent more recent loading over the 20th century, penetration of these anthropogenic solutes over the saturated thickness of the aquifer to a depth of 30 mbgl suggests very coarse aquifer penetration rates of 0.2 to $0.5\,m\cdot a^{-1}$. This is generally consistent with recent, bulk estimates of (i) groundwater residence times in the unconfined Nottingham aquifer of up to 230 years from point of recharge to abstraction wells (Trowsdale and Lerner, 2003) and (ii) urban recharge ($0.21\,m\cdot a^{-1}$) in Nottingham (Yang et al., 1999).

4 CONCLUDING DISCUSSION

Depth-specific sampling of urban groundwaters in Permo-Triassic sandstone at a multilevel piezometer site in Nottingham (UK) reveals low concentrations of sewage-derived bacteria and viruses that generally coincide with aquifer heterogeneities, and anthropogenic solutes with concentrations that gradually decrease with increasing depth. Aquifer penetration rates (i.e. sets of vertical groundwater velocities) indicated by the presence of sewage-derived microorganisms ($m \cdot d^{-1}$), are orders of magnitude more rapid than aquifer penetration rates suggested by concentrations of anthropogenic solutes ($m \cdot a^{-1}$). This difference is explained by the principle of "size exclusion" (McKay et al., 1993), also known as "differential advection" (Zhang et al., 2001), that recognises the exclusion of particles such as microorganisms from aquifer pores that are too small to allow passage either through adsorption or straining. As a result, microorganisms sample a subset of the total pathways through the medium that favour higher velocities. In contrast, conservative solutes sweep through all pathways of the porous medium and reflect bulk (mean) groundwater velocities. The distribution of groundwater velocities in this environment remains, however, poorly constrained not only by the absence of a reliable estimate of mean groundwater residence time but also due to limitations in the range of groundwater ages indicated by anthropogenic contaminants. The results provide, however, an indication of the range of vertical groundwater velocity sets that are represented within an individual sample of groundwater – even one derived from a well-defined, narrow sampling port of multilevel piezometer.

In terms of source protection, the risk that sewage-derived microorganisms can be transported by groundwater in pathogenic doses more rapidly than predicted by mean groundwater velocities clearly exists and is supported, in part, by widespread detection of pathogens in groundwater sources (Goss et al., 1998; Hancock et al., 1998; Macler and Merkle, 2000; Borchardt et al., 2003). As proposed by Taylor et al. (2004), the movement of sewage-derived microorganisms at velocities exceeding groundwater flows traced by solutes is considered to result from statistically extreme sets of groundwater flow velocities along a selected range of linked microscopic pathways. They argue that rapid aquifer penetration rates can be detected because of the high concentrations of bacteria and viruses that enter near-surface environments in sewage and low detection limits of these microorganisms in water. Further research is required to test this conceptual model and to resolve further the distribution of groundwater velocities that occur in fissured sandstone using a range of environmental tracers including both faecal microorganisms and reliable residence-time indicators such as ^3H:^3He ratios.

ACKNOWLEDGEMENTS

This paper is based upon research funded by the Natural Environment Research Council (UK), Environment Agency (UK), and a Post-Doctoral Fellowship from Natural Sciences and Engineering Research Council (Canada) to R. Taylor.

REFERENCES

Adams, B. and Foster, S.S.D. 1992. Land-surface zoning for groundwater protection. *Journal of the Institution of Water and Environmental Management* 6: 312–319.

Barrett, M.H., Hiscock, K.M., Pedley, S., Lerner, D.N., Tellam, J.H. and French, M.J. 1999. Marker species for identifying urban groundwater recharge sources – A review and case study in Nottingham, UK. *Water Research* 33(14): 3083–3097.

Borchardt, M.A., Bertz, P.D., Spencer, S.K. and Battigelli, D.A. 2003. Incidence of enteric viruses in groundwater from household wells in Wisconsin. *Applied and Environmental Microbiology* 69(2): 1172–1180.

Bottrell, S.H., Webber, N., Gunn, J. and Worthington, R.H. 2000. The geochemistry of sulphur in a mixed allogenic-autogenic karst catchment, Castleton, Derbyshire, UK. *Earth Surface Processes and Landforms* 25: 155–165.

Cronin, A.A., Taylor, R.G., Powell, K.L., Trowsdale, S.A., Barrett, M.H. and Lerner, D.N. 2003a. Temporal variations in the depth-specific hydrochemistry and microbiology of an urban sandstone aquifer – Nottingham, UK. *Hydrogeology Journal* 11: 205–216.

Cronin, A.A., Taylor, R.G. and Fairbairn, J. 2003b. Rising solute trends from regional groundwater quality monitoring in an urban aquifer, Nottingham, UK. *RMZ-Materials and Geoenvironment* 50(1): 97–100. *Proc. XXXIII Cong. Int. Assoc. Hydrogeol.* (Slovenia).

DWAF. 1997. A protocol to manage the potential of groundwater contamination from on site sanitation. Department of Water Affairs and Forestry (Pretoria).

Edmunds, W.M.E., Bath, A.H. and Miles, D.L. 1982. Hydrochemical evolution of the East Midlands Triassic Sandstone aquifer, England. *Geochemica Cosmochimica Acta* 46: 2069–2081.

Edmunds, W.M. and Smedley, P. 2000. Residence-time indicators in groundwater: the East Midlands Triassic Sandstone Aquifer. *Applied Geochemistry* 15: 737–752.

Flynn, R. 2003. Virus transport and attenuation in perialpine gravel aquifers. Doctoral Dissertation, University of Neuchâtel (Switzerland).

Ford, M. and Tellam, J.H. 1994. Source, type and extent of inorganic contamination within the Birmingham urban aquifer system. *Journal of Hydrology* 156: 101–135.

Fukada, T., Hiscock, K.M. and Dennis, P.F. 2004. A dual-isotope approach to the nitrogen hydrochemistry of an urban aquifer. *Applied Geochemistry* 19: 709–719.

Geldreich, E.E. 1991. Microbial water concerns for water supply use. *Environmental Toxicology and Water Quality* 6: 209–233.

Goss, M.J., Barry, D.A.J. and Rudolph, D.L. 1998. Contamination in Ontario farmstead domestic wells and its association with agriculture: 1. Results from drinking water wells. *Journal of Contaminant Hydrology* 32: 267–293.

Hancock, C.M., Rose, J.B. and Callahan, M. 1998. Crypto and giardia in US groundwater. *Journal of the American Water Works Association* 90: 58–61.

Howard, G., Pedley, S., Barrett, M., Nalubega, M. and Johal, K. 2003. Risk factors contributing to microbiological contamination of shallow groundwater in Kampala, Uganda. *Water Research* 37: 3421–3429.

Jean, J.-S. 1999. Outbreak of enteroviruses and groundwater contamination in Taiwan: concept of biomedical hydrogeology. *Hydrogeology Journal* 7: 339–340.

Keating, T. and Packman, M.J. 1995. Guide to groundwater protection zones in England and Wales. National Rivers Authority (London).

Kinzelbach, W. and Herzer, J. 1983. Application of contaminant arrival distributions to the simulation and design of hydraulic decontamination measures in porous aquifers. In: *Groundwater in Water Resources Planning; proc. intern. symp.*, Koblenz, pp. 1147–1158.

Macler, B.A. and Merkle, J.C. 2000. Current knowledge of groundwater microbial pathogens and their control. *Hydrogeology Journal* 8: 29–40.

Matthai, S.K. and Belanyeh, M. 2004. Fluid flow partitioning between fractures and a permeable rock matrix. *Geophysical Research Letters* 31(7): L07602, doi:10.1029/2003GL019027.

McKay, L., Cherry, J.A., Bales, R.C., Yahya, M.T. and Gerba, C.P. 1993. A field example of bacteriophage as tracers of fracture flow. *Environmental Science and Technology* 27: 1075–1079.

Moncaster, S.J., Bottrell, S.H., Tellam, J.H., Lloyd, J.W. and Konhauser, K.O. 2000. Migration and attenuation of agrochemical pollutants: insights from isotope analysis of groundwater sulphate. *Journal of Contaminant Hydrology* 43: 147–163.

Mounts, A.W., Ando, T., Koopmans, M., Bresee, J.S., Noel, J. and Glass, R.I. 2000. Cold weather seasonality of gastroenteritis associated with Norwalk-like viruses. *Journal of Infectious Diseases* 181(suppl. 2): 284–287.

Powell, K.L., Barrett, M.H., Pedley, S., Tellam, J.H., Stagg, K.A., Greswell, R.B. and Rivett, M.O. 2000. Enteric virus detection in groundwater using a glasswool trap. In: O. Sililo et al. (eds.) *Groundwater: Past Achievements and Future Challenges*. Balkema, Rotterdam, pp. 813–816, ISBN 9058091597.

Powell, K.L., Taylor, R.G., Cronin, A.A., Barrett, M.H., Pedley, S., Sellwood, J., Trowsdale, S.A. and Lerner, D.N. 2003. Microbial contamination of two urban sandstone aquifers in the UK. *Water Research* 37: 339–352.

Robertson, W.D., Cherry, J.A. and Schiff, S.L. 1989. Atmospheric sulphur deposition 1950–1985 inferred from sulphate in groundwater. *Water Resources Research* 25(6): 1111–1123.

Sellwood, J. 2001. Communicable Water Disease Report. *CDR Weekly* 11, Feb 2001.

Slade, J.S. 1985. Viruses and bacteria in a chalk well. *Water Science and Technology* 17: 111–125.

Taylor, R.G., Cronin, A.A., Trowsdale, S.A., Baines, O.P., Barrett, M.H. and Lerner, D.N.L. 2003. Vertical groundwater flow in Permo-Triassic sediments underlying two cities in the Trent River Basin (UK). *Journal of Hydrology* 284: 92–113.

Taylor, R.G., Cronin, A.A., Pedley, S., Atkinson, T.C. and Barker, J.A. 2004. The implications of groundwater velocity variations on microbial transport and wellhead protection: review of field evidence. *FEMS Microbiology Ecology* 49: 17–26.

Taylor, R.G., Cronin, A.A., Lerner, D.N., Tellam, J.H., Bottrell, S.H., Rueedi, J. and Barrett, M.H., *in review*. Penetration of anthropogenic solutes in sandstone aquifers underlying two mature cities in the UK. *Journal of Hydrology.*

Trowsdale, S.A. and Lerner, D.N. 2003. Implications of flow patterns in the sandstone aquifer beneath the mature conurbation of Nottingham (UK) for source protection. *Quarterly Journal of Engineering Geology and Hydrology* 36: 197–206.

USEPA. 1991. Handbook – Ground Water and Wellhead Protection. United States Environmental Protection Agency, Rep. No. EPA/625/R-94/001 (Washington).

Wright, R.C. 1986. The seasonality of bacterial quality of water in a tropical developing country. *Journal of Hygiene (Cambridge)* 96: 75–82.

Wyn-Jones, A.P., Pallin, R., Dedoussis, C., Shore, J. and Sellwood, J. 2000. The detection of small round-structured viruses in water and environmental materials. *Journal of Virological Methods* 87: 99–107.

Yang, Y., Lerner, D.N., Barrett, M.H. and Tellam, J.H. 1999. Quantification of groundwater recharge in the city of Nottingham, UK. *Environmental Geology* 38: 183–198.

Zhang, P., Johnson, W.P., Piana, M.J., Fuller, C.C. and Naftz, D.L. 2001. Potential artefacts in interpretation of differential breakthrough of colloids and dissolved tracers in the context of transport in a zero-valent iron permeable barrier. *Ground Water* 39: 831–840.

Zhao, F.J., Spiro, B., Poulton, P.R. and McGrath, S.P. 1998. Use of sulphur isotope ratios to determine anthropogenic sulphur signals in a grassland ecosystem. *Environmental Science and Technology* 32: 2288–2291.

CHAPTER 9

Development of a GIS model for assessing groundwater pollution from small scale petrol spills

Abraham Thomas[1] and John Tellam[2]
[1]*Department of Earth Sciences, University of the Western Cape, P Bag X17, Bellville 7535, Cape Town, South Africa*
[2]*Hydrogeology Research Group, Earth Sciences, School of Geography, Earth and Environmental Sciences, University of Birmingham, Edgbaston, Birmingham B15 2TT, United Kingdom*

ABSTRACT: An ArcView GIS based "Petrol Station BTEX Pollution Model" has been developed to assess the reactive dissolved phase migration from ground level to the water table for cases where spills of (multi-compound) non-aqueous phase liquids (NAPLs) have occurred. It uses inputs from an urban recharge GIS model. The model calculation is undertaken in seven steps, viz. 1) input of data; 2) estimation of volumetric water content in the unsaturated zone; 3) calculation of soil/water, air/water, and NAPL/water partitioning coefficients; 4) multiphase partitioning of spilled NAPLs into the four phases; 5) calculation of initial leachate concentrations for each NAPL component; 6) calculation of retardation factors; 7) calculation of final concentrations and fluxes reaching the water table taking into account degradation. The main input data required are: locations, volumes, and areas of spills; soil texture, hydraulic properties, and organic carbon content; recharge rates; and water table depths. The model has been trialled on the Birmingham (UK) aquifer for the case of petrol stations. In this aquifer, the risks to groundwater appear to be minimal for small-scale spills, except where the unsaturated zone thickness is limited close to the main river. The program could be used for vulnerability mapping, determining threats from existing NAPL storage sites, and in planning locations for future sites.

1 INTRODUCTION

Petrol spill incidences are quite common in fuelling stations and can happen either from leaking underground petrol storage tanks or from surface spills while transferring petrol. Spill incidences occur mainly due to overfilling during vehicle fuelling or from fuel hose breaks. Such spills or releases are uncommon but can involve the release of many litres of fuel. When the spilled fuel moves downward through the soil a portion of it is left behind in the soil as residual non-aqueous phase liquid (NAPL) that is immobile and discontinuous (Figure 1). The residual phase petrol thus formed serves as a long term, continuous source of aqueous contaminants to urban groundwater. In this environment, the main pollutants will include benzene, toluene, ethyl benzene and xylene (BTEX).

Many groundwater resource evaluation studies in urban (or agricultural) catchments are targeted towards assessment of the impacts of diffuse pollutant sources on regional water

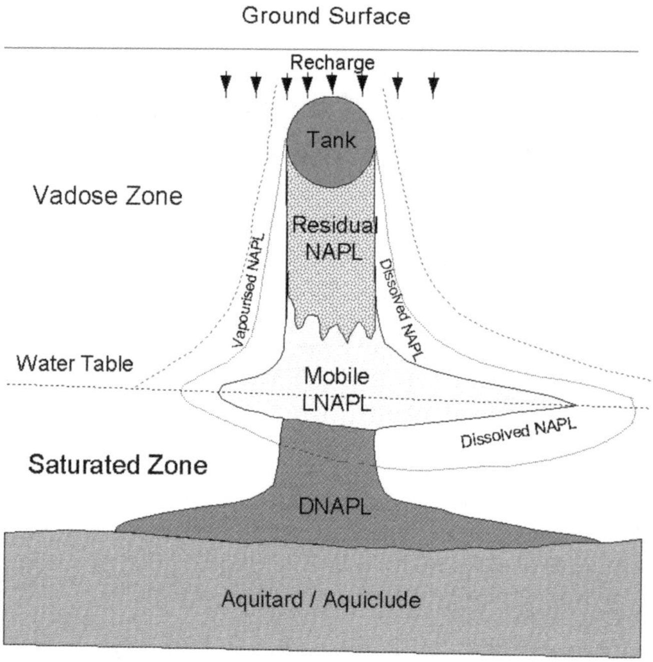

Figure 1. Conceptual model of fate and transport of NAPLs in the subsurface environments. NB: The NAPLs can be of two types: 1) light non-aqueous phase liquid (LNAPLs) which can float on water and 2) dense non-aqueous phase liquid (DNAPLs), which are heavier/denser than water.

quality only. However, there is often a need to assess the impact of frequently occurring point source pollution on regional-scale groundwater quality. It is unlikely that accurate predictions for pollution at a specific site will ever be possible without very detailed local site investigation. However, leakage or spills from many point sources can have significant impacts on the regional water quality. Therefore it is essential to have some tool that can help in studying the impact of frequently occurring point source pollution on regional water quality. Accordingly an effort was made to develop a GIS-based point source pollution model, taking the example of small-scale BTEX spills at petrol stations. This paper presents the principles of the model, which was developed within a geographic information system (GIS), and demonstrates its application in assessing the impact of small-scale spills from petrol stations to the Birmingham (United Kingdom) unconfined sandstone aquifer.

2 OVERVIEW OF MODEL

The model is developed within ArcView GIS (version 3.2). The interface of the model is shown in Figure 2. The model first simulates the multiphase partitioning in the residual BTEX zone below the spill. Steady-state leaching of this zone is assumed, and the subsequent migration of the contaminated aqueous phase is then simulated using analytical expressions which take into account sorption and degradation. The main outputs from the model are pollutant concentrations in recharge water and contaminant flux at the water table.

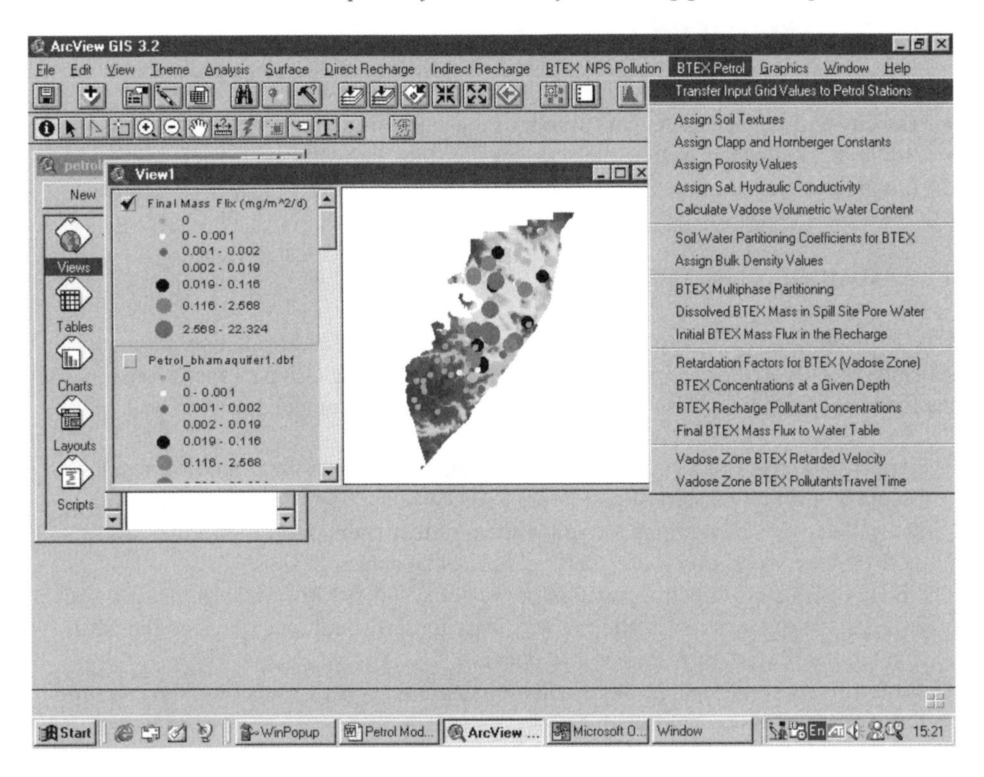

Figure 2. Interface of the GIS model.

3 MODEL INPUT PARAMETERS

The major input data required for this model are the following:

Spatial input data (user supplied data).

- Locations of the petrol stations or storage tanks (spill site).
- Subsurface soil/geological information (lithology, soil texture, fraction of organic carbon).
- Recharge rates and water table depths at the locations of the petrol stations.
- Depth of the underground storage tanks.
- Horizontal cross-sectional area of spills.

Flow and transport parameters/soil hydraulic properties (user supplied data)

- Porosity of the underlying sequence.
- Bulk density of the underlying sequence.
- Saturated hydraulic conductivity of the underlying sequence.
- Clapp and Hornberger constant of the underlying sequence.
- Residual saturation of petrol or diesel in vadose zone and saturated zone.

Chemical parameters of the BTEX compounds (default values available in the model)

- Weight percentage of the BTEX compounds in petrol or diesel.
- Density and average molecular weight of the petrol or diesel.
- Solubility, molecular weight, Henry's law constant, half-life and organic carbon partitioning coefficient of the BTEX compounds.

4 ASSUMPTIONS OF THE MODEL

Data available in urban areas on pollution events are often limited. In addition, detailed representation of all processes becomes difficult to implement using GIS packages. Consequently, the following simplifying assumptions have been made in the model described here.

The upper part of the vadose zone soil underneath a leaking underground storage tank is contaminated by a spill of petroleum hydrocarbons. Capillary forces immobilize the spilled or leaked petrol in the soil with no further lateral migration and it remains as a residual NAPL over a user-defined horizontal area.

The BTEX pollutants in petrol partition between the soil, NAPL, water in the soil, and soil gas. Linear sorption isotherms with equilibrium conditions describe the partitioning of pollutant between the four phases.

The soil/drift properties can be described by a single set of properties from the soil surface to the bottom of vadose zone.

Flow processes in the vadose zone are downward under steady state conditions without any lateral flow or interflow. The recharge flux and moisture content are uniform throughout the vadose zone, i.e. the recharge flux is in steady-state.

Hydrodynamic dispersion or diffusion is negligible, except as it contributes to volatilization.

First order degradation of the pollutant is assumed. The degradation rate does not change with soil depth and time.

5 MODELLING STEPS/PROCEDURE

5.1 *Overview*

In this model, the BTEX recharge pollutant fluxes from petrol station spills are estimated through seven stages:

1. transfer of spatial input data including geology, recharge, water table depths, and hydraulic properties;
2. estimation of volumetric water content in the unsaturated zone using the Clapp and Hornberger method;
3. calculation of soil-water partitioning coefficients, K_d;
4. multiphase partitioning into the air, water, soil and NAPL phases using the respective partitioning coefficients for BTEX compounds and estimation of their respective concentrations;
5. calculation of initial BTEX leachate concentrations (the concentrations immediately underneath the zone of residual NAPL);

6. calculation of unsaturated zone retardation factor for BTEX compounds; and
7. calculation of final BTEX concentration and BTEX pollutant mass flux reaching the water table following degradation.

These modelling steps are described in turn below.

5.2 *Display of petrol station theme and transfer of input data*

The first operation completed by the model is the generation of an event theme (or display of a feature theme) showing the locations of the petrol stations of interest. If a feature theme showing locations of petrol stations is not available, then a dbase format input data table having locations of petrol stations, represented by X and Y coordinate values, can be imported to the model. After adding this table, an event theme can be generated from it using the "add event theme" menu in the ArcView "view" interface. The next step is to transfer the following input data values: recharge rate, depth to water table, lithology, soil texture, porosity, saturated hydraulic conductivity, and Clapp and Hornberger constant. Recharge rate, vadose zone depth, and lithology can be transferred from respective floating-point grid themes to the petrol station locations by the first sub-menu option in the model interface (Figure 2).

5.3 *Estimation of volumetric water content in the vadose zone*

The volumetric water content in the vadose zone is calculated using the Clapp and Hornberger method (Clapp and Hornberger, 1978). The required input data for this sub-model are soil texture, porosity, saturated hydraulic conductivity, and Clapp and Hornberger constant value. The various sub programmes/codes provided in this model (Figure 2) help in assigning input data and later calculate the volumetric water content.

Clapp and Hornberger method

According to Clapp and Hornberger (1978), the volumetric water content of a soil, θ_w is given by:

$$\theta_w = \theta_s \left[\frac{V_d}{K_s} \right]^{1/(2b+3)} \tag{1}$$

where θ_s is the saturated water content of the soil (total porosity) [-];
V_d is the recharge rate [L/T];
K_s is the saturated hydraulic conductivity at the saturated water content θ_s [L/T]; and
b is the Clapp and Hornberger constant for the soil [-].

The Clapp and Hornberger constant (b) is the constant in the Clapp and Hornberger equation relating the relative saturation of the soil to the relative conductivity of the soil (Clapp and Hornberger, 1978).

$$\frac{\theta_w}{\theta_s} = \left[\frac{K}{K_s} \right]^{2b+3} \tag{2}$$

where K is the hydraulic conductivity of the soil at a volumetric water content θ_w. If b is not known, it can be estimated using the values presented by Clapp and Hornberger for different soil textures (Clapp and Hornberger, 1978).

5.4 *Calculation of soil-water partitioning coefficient, K_d*

The soil-water partitioning coefficient, K_d, is estimated in the usual way using the following relationship

$$K_d = K_{oc} f_{oc} \quad (3)$$

where K_d = soil-water partitioning coefficient
K_{oc} = organic carbon partitioning coefficient
f_{oc} = fraction of organic carbon within the soil matrix.

5.5 *Multiphase system partitioning of BTEX compounds in the unsaturated zone*

5.5.1 *BTEX properties*

In a BTEX-contaminated unsaturated zone, the contaminant will exist in all of the four phases – soil, water, air and NAPL – as shown in Figure 3. The partitioning of a particular BTEX compound in the petrol among the four phases, and hence its subsequent migration and fate in the subsurface environment, depends on its chemical properties, in particular volatility, solubility, air-gas diffusion coefficient, water-liquid diffusion coefficient, organic carbon partition coefficient, Henry's law constant, and the rate of chemical decay. The chemical properties and other characteristics of BTEX compounds used in the model runs described below are shown in Tables 1, 2 and 3.

5.5.2 *Initial BTEX contaminant bulk concentration before partitioning*

The *initial BTEX contaminant concentration* ($C_{benzene}$) in the petrol is calculated using Fuel density × Mass fraction of BTEX present in the petrol. Once the petrol has spread through the soil profile, the benzene mass concentration in the NAPL in the vadose zone = $m_{benzene} = \theta_o C_{benzene}$ where θ_o is the volumetric content of the NAPL in the soil. This amount of benzene will be shared between the different phases present in the subsurface environment.

5.5.3 *Multiphase solute partitioning*

The porosity, n, of the unsaturated soil can be divided as follows;

$$n = \theta_w + \theta_a + \theta_{NAPL} \quad (4)$$

where θ_w is volume of water/total volume;
θ_a is volume of air/total volume; and
θ_{NAPL} is volume of the NAPL/total volume.

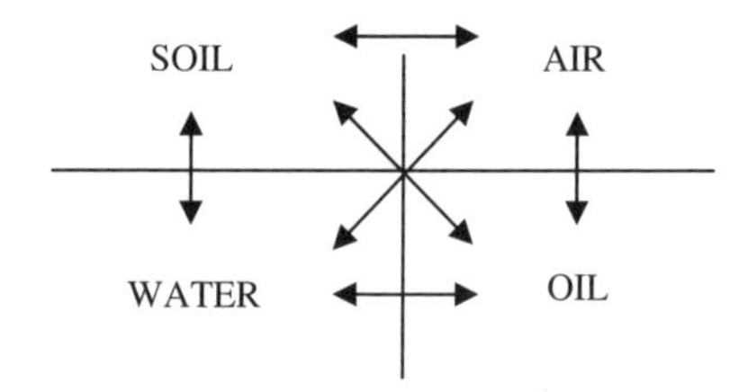

Figure 3. Solute partitioning in a multiphase system. Arrows indicate mass transfers in response to movement towards equilibrium.

Table 1. Weight percentage of BTEX mixture in petrol.

Alkyl benzenes	No. of carbons	Equivalent carbon no.	Weight percent in petrol	Weight percent in diesel
Benzene	6	6.50	0.12–3.50	0.003–0.10
Toluene	7	7.58	2.73–21.80	0.007–0.70
Ethyl benzene	8	8.50	0.36–2.86	0.007–0.20
o-Xylene	8	8.81	0.68–2.86	0.001–0.085
m-Xylene	8	8.60	1.77–3.87	0.018–0.512
p-Xylene	8	8.61	0.77–1.58	0.018–0.512

Source: Gustafson et al. (1997).

Table 2. Chemical properties of BTEX compounds at 20°C used in the model runs.

BTEX constituent	Weight percentage in petrol (%)	Molecular weight (g/mol)	Aqueous solubility (mg/L)	Vadose zone half life (days)
Benzene	1.9	78.11	1780	720
Toluene	8.1	92.14	515	28
Ethyl benzene	1.7	106.17	152	228
Xylenes	9.0	106.17	175	360

Table 3. Partitioning characteristics for BTEX compounds.

Constituent	Water solubility (mg/L)	Vapour pressure at 25°C (atm)	Henry's law constant K'_H (atm-m^3/mol)	Organic carbon partition coefficient K_{oc} (L/kg)
Benzene	1750	0.125	5.59 E−03	8.30 E+01
Toluene	535	0.037	6.37 E−03	3.00 E+01
Ethyl benzene	152	0.013	6.43 E−03	1.10 E+03
o-Xylene	175	0.012	5.10 E−03	8.30 E+02

If the constituent concentrations in the water, air, and NAPL phases are designated C_w, C_a, and C_{NAPL} respectively (all on a mass per unit volume basis), and C_s is the soil phase concentration (mass of chemical sorbed per unit mass of soil), then the *bulk concentration*, m (which is the mass of constituent per bulk volume of soil) can be specified by the following equation:

$$m = \theta_w c_w + \theta_a c_a + \theta_{NAPL} C_{NAPL} + \rho_b C_s \tag{5}$$

where ρ_b is the soil bulk density. Estimation of the values of volumetric NAPL content, θ_o and volumetric air content, θ_a, are considered in the following subsections.

5.5.4 *Estimation of volumetric NAPL content*

Following a small petrol spill in the subsurface, the volumetric content of petrol in the vadose or saturated zone can be estimated using the following relationship:

$$\text{Volumetric NAPL content} = \text{porosity} \times \text{residual saturation (\%)}/100. \tag{6}$$

Residual saturation in the subsurface is highly variable and its magnitude is affected by many factors including properties of the medium and fluid, hydraulic gradients, ratios of fluid viscosities and densities, gravity, and buoyancy forces (e.g. Mercer and Cohen, 1990). Although field-scale values for residual saturation are difficult to measure and accurately estimate, data compiled by Mercer and Cohen (1990) indicate that residual saturation values of most NAPLs in the unsaturated zone range from 10% to 20%. A study of LNAPL in fine-grained sandstone of San Diego (USA) showed typical LNAPL saturations between 5% and 20%, with free phase LNAPL in several wells, even in zones with LNAPL saturations less than 25% (Huntley et al., 1994). Saturation values of free-phase NAPLs have been reported as low as 15% to 25% (Huntley et al., 1994). For the purpose of demonstrating the model for the Birmingham aquifer, a value of 10% of total pore volume or porosity will be assumed since currently no site specific reporting of residual saturations are available.

5.5.5 *Estimation of volumetric air content after a spill*

After a spill, some of the void spaces that were previously occupied by air will now be occupied by the petrol. Assuming the Clapp and Hornberger method to be valid when NAPL is present, the volumetric water content in the unsaturated soil, θ_w can be estimated, and hence the volumetric air content:

$$\theta_a = n - (\theta_{NAPL} + \theta_w) \tag{7}$$

Condition:

$n > (\theta_{NAPL} + \theta_w)$ and vadose depth > 0.

When vadose depth = zero

$$\theta_a = \text{zero} \tag{8}$$
$$\theta_a = n - \theta_{NAPL} \tag{9}$$

After assigning residual LNAPL, if the sum of the volumetric NAPL content and the volumetric water content is greater than porosity, then the volumetric air content is treated as zero assuming the available air space is occupied by NAPL in such cases.

5.5.6 *Contaminant concentrations in various phases*

When analysing solute transport in a multiphase system, it is usual to express concentrations in each phase relative to concentrations in a reference phase. In the analysis of solute leaching it has been found convenient to consider the water phase as the reference phase (Charbeneau and Daniel, 1992). When the water phase is the reference phase, one can calculate concentrations in air, NAPL and soil respectively as:

$$C_a = K_H C_w \tag{10}$$

$$C_{NAPL} = K_o C_w \tag{11}$$

$$C_s = K_d C_w \tag{12}$$

where K_H is the air water partitioning constant [-];
K_o is the NAPL-water partitioning coefficient [-];
K_d is the soil-water partitioning coefficient [L^3/M]; and
ρ_b is the soil bulk density [M/L^3].

Substituting the above three equations into the bulk concentration equation (Equation 5 in Section 5.5.3) results in

$$m = (\theta_w + \theta_a K_H + \theta_{NAPL} K_o + \rho_b K_d)C_w = B_w C_w \text{ or } C_w = m/B_w \quad (13)$$

where B_w = bulk water partitioning coefficient ($= (\theta_w + \theta_a K_H + \theta_{NAPL} K_o + \rho_b K_d)$).

Using the initial bulk concentrations of BTEX in the soil and the partition coefficients, the concentrations of each contaminant in each phase can be calculated. The corresponding percentage of the constituents in the water, air, soil and NAPL phase can also be calculated using the following equations:

$$\text{Percentage of the constituent in the water phase, w} = 100(\theta_w/B_w) \quad (14)$$

$$\text{Percentage of the constituent in the air phase, a} = 100(\theta_a K_H/B_w) \quad (15)$$

$$\text{Percentage of the constituent in the soil phase, s} = 100(\rho_b K_d/B_w) \quad (16)$$

$$\text{Percentage of the constituent in the NAPL phase, NAPL} = 100(\theta_{NAPL} K_o/B_w) \quad (17)$$

5.5.7 *Estimation of the air-water partitioning coefficient, K_H*

In the model, Henry's Law is applied for estimating the concentration of BTEX in the air phase. In order to calculate K_H ([-]) values in the model, the user has to input the K'_H (atm-m^3/mol) values of BTEX compounds (Table 3) and the corresponding vadose zone temperature (in Celsius) i.e.,

$$K_H = K'_H/RT, \quad (18)$$

where R is the universal gas constant (atm-m^3/mol-K), and T is temperature [K].

5.5.8 *Estimation of NAPL-water partitioning coefficient, K_o*

The major factors controlling the rate of dissolution are the effective solubility of the hydrocarbons, NAPL saturation in the subsurface, and the groundwater seepage velocity (e.g. Bedient et al., 1999).

The effective solubility of a particular organic chemical in a NAPL mixture is estimated by multiplying its mole fraction in the NAPL mixture by its pure phase aqueous solubility, i.e., using Raoult's law (Feenstra et al., 1991):

$$C \cong XS \quad (19)$$

where C is the concentration of contaminant in soil leachate (g/L);
X is the mole fraction of contaminant in NAPL [-]; and
S is the aqueous solubility of contaminant (g/L).

This effective solubility relationship will over-estimate leachate concentrations as it represents the maximum theoretical concentration that can be achieved when a NAPL mixture exists in the subsurface: in practice, in a NAPL zone some groundwater will not actually come into contact with the NAPL.

Raoult's law can be used to define K_o at any time given the concentration of different constituents in the NAPL phase. For a multi component NAPL mixture, Baehr and Corapcioglu (1987) write the Raoult's law expression in the form:

$$K_o = \frac{m_k \sum_{j=1}^{N} (c_{NAPLj}/m_j)}{S_k \gamma_k} \tag{20}$$

where k is one species out of N species which make up the NAPL;

m_j is the molecular weight of the jth constituent (g/mol);

c_{NAPLj} is concentration of the jth constituent in the NAPL phase (g/L);

S_k is solubility of species k in water (g/L);

γ_k is activity coefficient of the kth species, assumed here to be unity [-].

As the NAPL composition changes through dissolution, volatilization, degradation and sorption, the K_o also will change with time (Charbeneau and Daniel, 1992). However, in the present model, to save involving an iterative calculation, K_o is assumed to be a constant in time. Hence, for a given time, the above equation can be written simply as

$$K_o = \frac{\text{Molecular weight of species} \times \text{Molar concentration of NAPL}}{\text{Solubility of species in g/L}} \tag{21}$$

where Molar concentration of NAPL (mol/L) is $= \sum_{j=1}^{N} (c_{NAPLj}/m_j)$.

Since petrol contains over 200 hydrocarbon species in it, calculation of the molar concentration is laborious to do precisely and requires knowledge of the exact composition of the petrol. However, if the average molecular weight of petrol is known, the molar concentration of NAPL can be estimated as its fuel density divided by its average molecular weight. In this model, a typical literature value of 104.8 g/mol is taken as the average molecular weight of petrol (Fetter, 1999).

The effective solubility of BTEX in the NAPL can be calculated using:

$$\text{Effective Solubility}_i = \frac{\begin{array}{c}\text{Percent Weight}_i \text{ in NAPL} \times \\ \text{Av. Molecular Weight of NAPL} \times \text{Solubility}_i\end{array}}{\text{Molecular Weight}_i \times 100} \tag{22}$$

where i represents a BTEX compound.

5.6 *Estimating NAPL leachate concentrations*

5.6.1 *Estimation of the volume of pore water in the NAPL source zone*

In the model, it is assumed that the recharge water passing through the residual petrol is in chemical equilibrium with the NAPL. Because of this assumption, the initial concentration in the leachate is independent of recharge rate (though the initial pollutant mass flux will vary with recharge rate). If petrol is released in the unsaturated zone, it will contaminate a volume of rock dependent on the amount and area of release. The depth of penetration of the NAPL spill L_{NAPL} can be estimated using:

$$L_{NAPL} = \frac{V_{spill}}{\theta_{NAPL} A_{spill}} \tag{23}$$

where V_{spill} is volume of spilled petrol;
θ_{NAPL} is volumetric content of petrol in the vadose or saturated zone;
A_{spill} is area of spill.
Knowing the residual saturation in each type of geological unit and the volume of the spill, the volume or depth of the zone affected by residual NAPL can be calculated.

In the model, it is assumed that the water in the NAPL source zone is in chemical equilibrium with the NAPL. Accordingly, at any given time, the volume of pore water containing aqueous phase BTEX in the NAPL source zone is estimated using:

$$\text{Pore water volume} = L_{NAPL} \cdot \theta_w \cdot A_{spill} \tag{24}$$

Calculation of the petrol spill area requires information on the dimensions and number of leaking tanks. Normally petrol filling stations in UK urban areas have four to six underground storage tanks, each having a storage capacity ranging from 12000 litres to 34000 litres of petrol. Their total horizontal cross sectional area is a good estimate of the maximum spill area. Newer petrol stations in UK cities have total storage capacities of around 10000 litres made up from four tanks. New big petrol filling stations on motorways or main roads can store up to 500000 litres of fuel. An underground storage tank having a storage capacity of 25000 litres will have a horizontal cross-sectional area of around 16 m^2 (~8 m long and 2 m in diameter). Based on this, in the absence of site-specific data, the effective spill area in urban petrol filling stations can be assumed to be around 48 m^2.

5.6.2 *Estimating the concentration of dissolved BTEX in the pore water*

The total dissolved BTEX contaminant mass in the petrol-affected pore water can be estimated using:

$$\text{Dissolved BTEX Mass in Pore Water} = \text{Pore Water Volume} \times \text{Leachate Concentration} \tag{25}$$

Substituting for pore water volume ($= L_{NAPL}\theta_w A_{spill}$) and L_{NAPL} gives:

$$\text{Dissolved BTEX Mass in Pore Water} = \frac{V_{spill} \times \theta_w \times C_w}{\theta_{NAPL}} \tag{26}$$

Knowing the net recharge rate passing through the soil column, the volume of the leachate produced per unit time can be estimated as recharge rate multiplied by the effective area of petrol spill.

The initial BTEX mass flux in the recharge is estimated as recharge rate multiplied by the leachate concentration. It can also be estimated using the following equation:

$$\text{Initial BTEX Mass Flux in Recharge} = \text{Initial Leachate Mass in given time period/Area of spill.}$$

5.7 *Transport of aqueous phase contaminants in vadose zone*

This sub-model simulates the downward movement of the dissolved pollutant and estimates the concentration of pollutants and pollutant flux from the vadose zone.

In a steady-state flow system the average linear groundwater velocity, V_a, is given by:

$$V_a = \frac{V_d}{\theta_w} \tag{27}$$

where V_d is the recharge.

The velocity of a retarded pollutant in the vadose zone is given by:

$$V_p = \frac{V_a}{R_f} \tag{28}$$

where R_f is the retardation factor for the pollutant in the vadose zone.

$$\text{i.e., } V_p = \frac{V_d}{\theta_w R_f}. \tag{29}$$

The retardation factor, R_f, for the vadose zone is calculated using the following equation:

$$R_f = 1 + (\rho_b K_d + (\theta_s - \theta)K_H)/\theta \tag{30}$$

where ρ_b is the bulk density of the soil;
θ is the water content of the soil on a volume basis;
θ_s is the saturated water content of the soil on a volume basis;
K_d is the partition coefficient for pollutant in the soil; and
K_H is the dimensionless value of Henry's law constant (C_a/C_w).

The incorporation of Henry's law in the above equation accounts for the loss of dissolved volatile organic compounds like BTEX through volatilisation. However, it has been noted that if volumetric water content is low the use of the above equation might lead to the prediction of retardation factors which are too high.

Ignoring dispersion, the leading edge of the contaminated pulse will reach the water table at a time T that can be calculated as:

$$T = \frac{z\theta_w R_f}{V_d} \tag{31}$$

where z is the depth of the vadose zone.

5.8 *Estimation of pollutant concentrations and fluxes at the water table*

In an aerobic environment, biodegradation of fuels generally follows a first order relationship (Salinitro, 1993):

$$C_2 = C_1 \exp(-T\lambda) \tag{32}$$

where

C_1 is the initial concentration of chemical applied at the ground surface or in the soil through leaks;

C_2 is the concentration of chemical exiting the vadose zone; and

λ is the first order degradation rate coefficient for the chemical with time $[T^{-1}]$.

Substituting the expression for the travel time and writing in terms of half life, this becomes:

$$C_2 = C_1 \exp\left[\frac{-0.693 R_f z \theta_w}{V_d T_{1/2}}\right] \tag{33}$$

This is the equation used in the model to calculate the final concentration reaching the water table.

The one dimensional pollutant mass flux due to advection is equal to the quantity of the water infiltrating times the concentration of solutes. Therefore, the pollutant flux to the water table from the vadose zone due to a petrol spill is given by:

$$\text{Pollutant Flux} = V_d \times C_2$$

$$\text{i.e., Recharge Pollutant Flux} = V_d C_1 \exp\left[\frac{-0.693 R_f z \theta_w}{V_d T_{1/2}}\right] \tag{34}$$

6 DEMONSTRATION OF THE MODEL FOR THE BIRMINGHAM AQUIFER

6.1 *Introduction*

Model trials have been conducted using data for the Birmingham urban aquifer, UK. The part of the aquifer considered here is unconfined and is composed of Triassic Sandstone (Jackson and Lloyd, 1983; Rivett et al., 1990; Ford and Tellam, 1994). Quaternary deposits overlie the aquifer, and modify the recharge amounts.

6.2 *Acquisition and processing of petrol station data*

A petrol station database for the Birmingham area was acquired from "Catalyst Limited" a company which provides retail market information on diesel and petrol sales. The Catalyst petrol station database includes information on the names and addresses of the petrol stations with their locations marked in six digit national grid co-ordinates. The database also contains information on the brand, combined total annual sales volume of petrol and diesel fuel, and petrol station plot area. The combined total annual sales volume of petrol and diesel fuel from each filling station ranges from 200 kilolitres per year to 9000 kilolitres per year. There is no direct relationship between the fuel sales volume and the plot area.

The original Catalyst petrol station data table was supplied in Microsoft Access format, and for the ease of using it in ArcView GIS, it was first converted into dbase format within Access. By using the x and y co-ordinates of each petrol station, this data table was viewed as an event theme in ArcView using the "add event theme" menu option. The total number of petrol stations in Birmingham and available in this data table is 90. Those petrol stations falling in the unconfined area of the aquifer were selected using the "Select by theme" option, and a shape file of these petrol station locations was created. Only 62 of the 90 petrol stations are sited on the unconfined part of the aquifer (Figure 4).

6.3 *Example model run*

A model was run assuming 25 litres of petrol leaks at all petrol filling stations over a spill area of one square metre. After incorporating geology, recharge, and land use information, the first step performed in this model run was the estimation of volumetric water content underneath the petrol stations using a daily average recharge rate input. The latter was for the autumn and winter period of 1980, and was calculated using the recharge model of Thomas (2001). Using the lithological descriptions of the geological units underneath the petrol stations, typical input values of hydrological parameters were chosen from the literature. The assigned soil textures, porosity, Clapp and Hornberger constants and saturated hydraulic conductivity values selected for the model run are given in Table 4. Using the daily recharge rate, the above soil property parameters, and an assumed residual saturation of 10%, the model calculated volumetric NAPL content at each petrol station (Table 5 and Figure 4).

Assuming a BTEX concentration in fresh petrol as shown in Table 2 with a fuel density of 729 kg/m^3 and an average molecular weight of 104.8 g/mol, a simulation of multiphase partitioning was performed using the model. For an assumed residual saturation of 10% for petrol in the unsaturated zone, the model calculated the site-specific bulk concentration

Table 4. Physical properties of various geological units underlying petrol stations on the Birmingham aquifer.

Geological unit	Texture	Porosity	Bulk density (kg/L)	Permeability (m/day)	f_{OC} (%)	Clapp & Hornb. const
Alluvium	Silty clay	0.40	1.26	6.00E−02	1.15	10.40
Bromsgrove Sandstone Formation	Sand	0.27	2.12	2.60E+00	0.05	4.05
Glaciofluvial Deposits	Sand	0.35	1.90	8.64E+00	0.05	4.05
Glaciolacustrine Deposits	Silty clay loam	0.40	1.40	9.00E−03	0.34	7.75
Head	Clay	0.42	1.35	3.00E−04	0.55	11.40
Hopwas Breccia	Sand	0.24	2.50	2.00E+00	0.05	4.05
Kidderminster Formation	Sand	0.25	2.38	4.10E+00	0.05	4.05
River Terrace Deposits, First Terrace	Sand	0.35	1.80	8.64E+00	1.18	4.05
River Terrace Deposits, Second Terrace	Sand	0.35	1.80	8.64E+00	1.18	4.05
Till	Clay	0.25	1.40	1.00E−02	0.67	11.40
Wildmoor Sandstone Formation	Sand	0.26	2.21	2.75E+00	0.05	4.05

Sources: Allen et al. (1997); Bridge et al. (1997); Kruseman and de Ridder (1994); Freeze and Cherry (1979); Shepherd (2000) and Clapp and Hornberger (1978).

and bulk water partitioning coefficients of BTEX compounds (Table 6) using the bulk densities and organic carbon contents estimated using the information on underlying rock types. The estimated air-water partitioning coefficients and NAPL-water partitioning coefficients are given in Table 7.

Table 5. Calculated volumetric water, volumetric NAPL and volumetric air content at each [example] petrol station.

Petrol station number	Porosity (n)	Vol. water (θ_w)	Vol. NAPL (θ_{NAPL})	Vol. air (θ_a)
1	0.25	0.1081	0.0250	0.1169
2	0.35	0.1422	0.0350	0.1728
3	0.35	0.1424	0.0350	0.1726
4	0.40	0.3135	0.0400	0.0465
5	0.25	0.2157	0.0250	0.0093
6	0.25	0.2134	0.0250	0.0116
7	0.25	0.2133	0.0250	0.0117
8	0.35	0.1424	0.0350	0.1726
9	0.26	0.1167	0.0260	0.1173
10	0.25	0.2133	0.0250	0.0117
11	0.35	0.1381	0.0350	0.1769
12	0.25	0.2156	0.0250	0.0094
13	0.35	0.1424	0.0350	0.1726
14	0.35	0.1465	0.0350	0.1685
15	0.42	0.3780	0.0420	0.0000

Table 6. Calculated bulk concentrations and bulk water partitioning coefficients of BTEX compounds in soils at example petrol stations on the Birmingham aquifer for a vadose zone residual LNAPL (petrol) saturation of 10%.

Petrol station number	Initial bulk BTEX concentration in soil (mg/L)				Bulk water partitioning coefficient, B_w			
	Benzene	Toluene	E. benzene	Xylene	Benzene	Toluene	E. benzene	Xylene
1	346.28	1476.23	309.83	1640.25	7.86	31.60	122.91	106.62
2	484.79	2066.72	433.76	2296.35	12.62	50.11	193.60	165.51
3	484.79	2066.72	433.76	2296.35	10.94	44.02	171.29	148.67
4	554.04	2361.96	495.72	2624.40	13.73	54.45	210.61	181.16
5	346.28	1476.23	309.83	1640.25	8.63	34.15	132.00	113.51
6	346.28	1476.23	309.83	1640.25	8.63	34.14	132.00	113.51
7	346.28	1476.23	309.83	1640.25	8.63	34.14	132.00	113.51
8	484.79	2066.72	433.76	2296.35	10.94	44.02	171.29	148.67
9	360.13	1535.27	322.22	1705.86	8.17	32.83	127.69	110.78
10	346.28	1476.23	309.83	1640.25	8.63	34.14	132.00	113.51
11	484.79	2066.72	433.76	2296.35	10.93	44.02	171.28	148.67
12	346.28	1476.23	309.83	1640.25	8.63	34.15	132.00	113.51
13	484.79	2066.72	433.76	2296.35	10.94	44.02	171.29	148.67
14	484.79	2066.72	433.76	2296.35	10.94	44.03	171.29	148.67
15	581.74	2480.06	520.51	2755.62	13.81	54.88	212.61	183.79

Table 7. Calculated molar concentration, partitioning coefficients and effective solubility of BTEX compounds in fresh petrol.

BTEX constituent	Molar conc. (mol/L)	Oil-water partitioning coefficient, K_o	Air-water partitioning coefficient, K_H	Effective solubility (mg/L)
Benzene	0.18	305.25	0.18	45.38
Toluene	0.64	1244.54	0.22	47.45
Ethyl benzene	0.12	4858.75	0.24	2.55
Xylenes	0.62	4220.17	0.19	15.55

Table 8. Calculated NAPL penetration depths when 25 litres of petrol are released over an area of 1 m^2, and leachate concentrations in the ground below example petrol stations on the Birmingham aquifer.

Petrol station no.	Oil penetration depth (m)	Recharge rate (m/day)	Leachate concentration (mg/L)			
			Benzene	Toluene	E. benzene	Xylene
1	1.00	3.72E−04	44.06	46.71	2.52	15.38
2	0.71	3.94E−04	38.41	41.24	2.24	13.87
3	0.71	3.99E−04	44.33	46.94	2.53	15.45
4	0.63	1.82E−04	40.34	43.38	2.35	14.49
5	1.00	2.21E−04	40.14	43.23	2.35	14.45
6	1.00	1.68E−04	40.15	43.24	2.35	14.45
7	1.00	1.67E−04	40.15	43.24	2.35	14.45
8	0.71	4.00E−04	44.33	46.94	2.53	15.45
9	0.96	3.79E−04	44.10	46.76	2.52	15.40
10	1.00	1.67E−04	40.15	43.24	2.35	14.45
11	0.71	2.84E−04	44.34	46.95	2.53	15.45
12	1.00	2.20E−04	40.14	43.23	2.35	14.45
13	0.71	4.00E−04	44.33	46.94	2.53	15.45
14	0.71	5.49E−04	44.31	46.94	2.53	15.45
15	0.60	2.43E−04	42.11	45.19	2.45	14.99

6.3.1 *Estimation of pollutant fluxes in the recharge*

Assuming that the petrol spilled at a depth of 2 m from ground surface, the calculated NAPL penetration depths at each of the petrol stations is shown in Table 8. Using the average daily recharge rate during autumn and winter 1980, the initial BTEX leachate concentrations in the recharge (Table 8) and the initial BTEX mass flux in the leachate were calculated. The calculated initial leachate concentration and the initial mass flux in the leachate for benzene are shown in Figure 5. Assuming a first order decay biodegradation rate under an aerobic environment (see Table 2), the final mass flux and concentration reaching the water table were then calculated (Table 9). From the model results obtained, the final concentration and mass flux of a BTEX pollutant, for example benzene, reaching the water table is shown in Figures 6 and 7.

The results from the model show that only very few petrol stations in Birmingham pose a threat of point source BTEX contamination to the aquifer provided they leak only a small

Table 9. Calculated vadose zone retardation factors and final BTEX pollutant concentration in recharge waters at example petrol stations on the Birmingham aquifer when 25 litres of petrol are released over an area of 1 m^2.

Petrol station no.	Vadose zone retardation factor					Final pollutant concentration in recharge (mg/L)			
	Benzene	Toluene	Ethyl benzene	Xylene	Vadose depth (m)	Benzene	Toluene	Ethyl benzene	Xylene
1	2.15	4.59	13.43	10.39	7	2.17E+00	0.00E+00	0.00E+00	0.00E+00
2	13.66	46.14	165.66	125.25	1	3.84E+01	4.12E+01	2.24E+00	1.39E+01
3	1.82	3.33	8.69	6.81	15	1.32E−02	0.00E+00	0.00E+00	0.00E+00
4	4.89	14.93	51.91	39.42	2	4.03E+01	4.34E+01	2.35E+00	1.45E+01
5	4.64	14.08	48.87	37.12	57	0.00E+00	0.00E+00	0.00E+00	0.00E+00
6	4.68	14.22	49.39	37.52	51	0.00E+00	0.00E+00	0.00E+00	0.00E+00
7	4.68	14.23	49.41	37.53	43	0.00E+00	0.00E+00	0.00E+00	0.00E+00
8	1.82	3.33	8.69	6.81	17	3.88E−03	0.00E+00	0.00E+00	0.00E+00
9	2.01	4.11	11.71	9.09	15	1.92E−02	0.00E+00	0.00E+00	0.00E+00
10	4.68	14.23	49.41	37.53	27	0.00E+00	0.00E+00	0.00E+00	0.00E+00
11	1.85	3.41	8.94	7.00	27	0.00E+00	0.00E+00	0.00E+00	0.00E+00
12	4.64	14.09	48.90	37.14	33	0.00E+00	0.00E+00	0.00E+00	0.00E+00
13	1.82	3.33	8.69	6.81	5	6.84E+00	0.00E+00	0.00E+00	1.30E−05
14	1.79	3.25	8.47	6.65	12	4.46E−01	0.00E+00	0.00E+00	0.00E+00
15	2.48	6.35	20.61	15.79	4	1.17E−02	0.00E+00	0.00E+00	0.00E+00
16	2.12	4.50	13.09	10.14	69	0.00E+00	0.00E+00	0.00E+00	0.00E+00

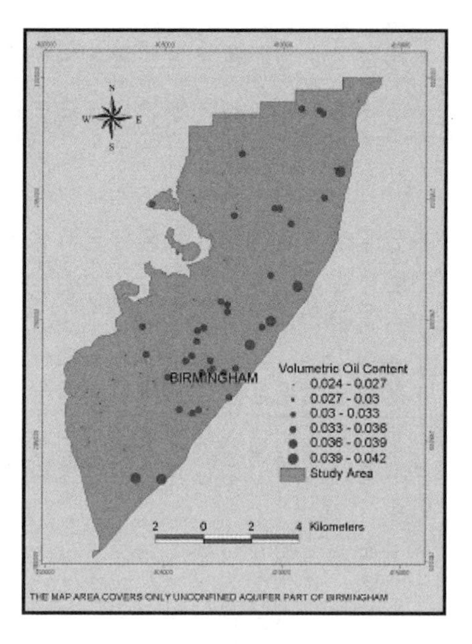

Figure 4. Volumetric NAPL content of petrol under residual NAPL saturation at petrol stations in the Birmingham area.

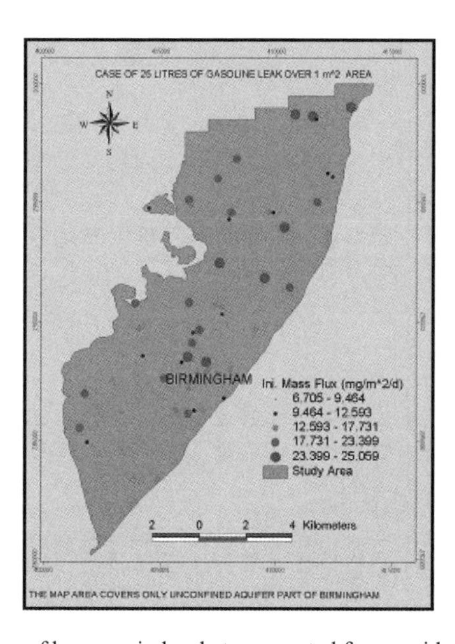

Figure 5. Initial mass flux of benzene in leachate generated from residual NAPL (petrol) at petrol stations in the Birmingham area.

amount of petrol. Most of the BTEX is degraded within the unsaturated zone before it reaches the water table.

The model results show that the most "vulnerable" petrol stations over the unconfined aquifer are those located on the alluvium where the depths to the water table are as little as

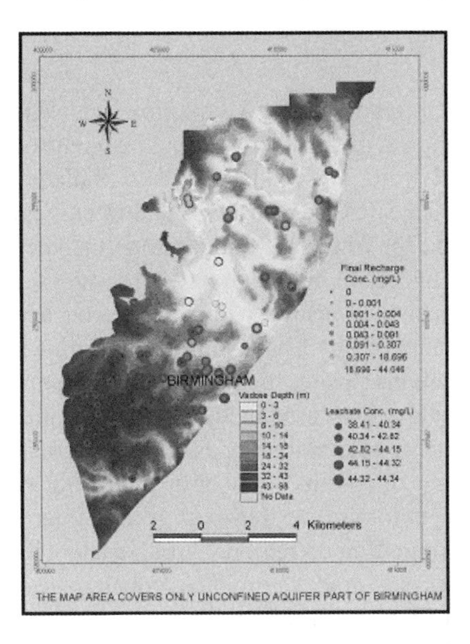

Figure 6. Initial leachate concentration and final recharge concentration of benzene at petrol stations in the Birmingham area.

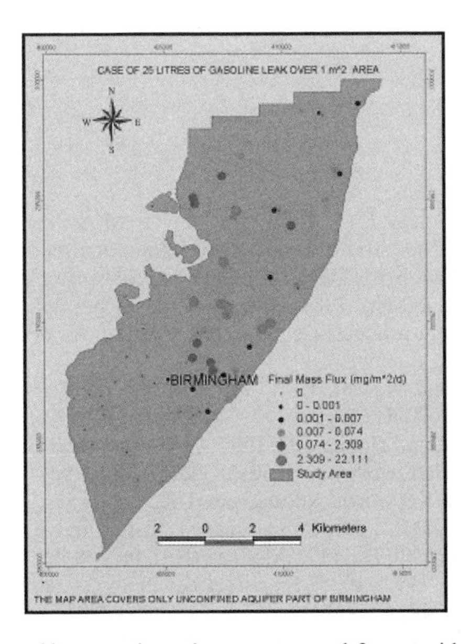

Figure 7. Final mass flux of benzene in recharge generated from residual petrol at petrol stations in the Birmingham area.

2 m. Under such shallow depths, the BTEX pollutants are not totally degraded and the non-degraded part ultimately reaches the water table. This model has significance in the sense that precautionary measures may be required in the case where spills occur from such "vulnerable" petrol stations.

7 CONCLUSIONS

Using ArcView GIS, a "Petrol Station BTEX Pollution Model" has been developed which can help in assessing the impact of frequently occurring point source pollution on groundwater quality at a regional-scale. The model has been trialled on the Birmingham (UK) aquifer, and the results indicate that the risks to groundwater in this system are minimal for most small-scale spills, except where the vadose zone thickness is limited close to the main river and at stations underlain by alluvium where water table depths are less than 2 meters. Data on the effects on the sensitivity of recharge concentrations to spill properties at existing sites can be obtained easily. The results from the model calculations can be used to identify those petrol stations posing most threat to groundwater contamination from BTEX chemicals. They could also be used in identifying potentially vulnerable areas when selecting locations for new petrol stations or other fuel storage facilities: this could be done by investigating a range of possible sites, or by producing a map. The output from the model could be used as input to a solute transport groundwater flow model to investigate the effects of spill plume migration on regional groundwater chemistry. The model could in principal be used in stochastic mode to produce probabilistic output. It can also be used to simulate the fate of BTEX in a variety of environmental conditions. For instance, the user can change the hydraulic property parameters and run the model to simulate the effects in various situations. The basic model could be used for any point source organic contaminant spill or leak provided its chemical parameters (density, solubility, half life period, molecular weight, Henry's law constant, and organic carbon partitioning coefficient) are known.

REFERENCES

Allen, D.J., Brewerton, L.J., Coleby, L.M., Gibbs, B.R., Lewis, M.A., MacDonald, A.M., Wagstaff, S.J. and Williams, A.T. 1997. The physical properties of major aquifers in England and Wales. BGS Technical Report WD/97/34. British Geological Survey, Keyworth, Nottingham, UK.

Baehr, A.L. and Corapcioglu, M.Y. 1987. A compositional multiphase model for groundwater contamination by petroleum products–2, numerical solution: *Water Resources Research*, 23(1), 201–213.

Bedient, P.B., Rifai, H.S. and Newell, C.J. 1999. *Groundwater Contamination, Transport and Remediation.* Prentice Hall PTR. 604p.

Bridge, D. McC, Brown, M.J. and Hooker, P.J. 1997. Wolverhampton Urban Environmental Survey: An Integrated Geoscientific Case Study. British Geological Survey Technical Report WE/95/49. British Geological Survey, Keyworth, Nottingham, UK.

Charbeneau, R.J. and Daniel, D.E. 1992. Contaminant Transport In Unsaturated Flow. *Handbook of Hydrology* edited by Maidment, David R. McGraw-Hill, Inc. pp. 15.1–15.53.

Clapp, R.B. and Hornberger, G.M. 1978. Empirical equations for some soil hydraulic properties. *Water Resources Research*, 14, 601–604.

Feenstra, S., Mackay, D.M. and Cherry, J.A. 1991. Presence of Residual NAPL Based on Organic Chemical Concentrations in Soil Samples. *Ground Water Monitoring Review*, II(2), 128–136.

Fetter, C.W. 1999. *Contaminant hydrogeology*, 2nd edition. Prentice Hall Inc. 500p.

Ford, M. and Tellam, J.H. 1994. Source, type and extent of inorganic contamination within the Birmingham urban aquifer system, UK. *Journal of Hydrology*, 156, 101–135.

Freeze, R.A. and Cherry, J.A. 1979. *Groundwater.* Prentice-Hall Inc., Englewood Cliffs, N.J. 604p.

Gustafson, J.B., Tell, J.G. and Orem, D. 1997. Selection of Representative TPH Fractions Based on Fate and Transport Considerations. Volume 3, Total Petroleum Hydrocarbon Criteria Working

Group Series. July 1997. Amherst Scientific Publishers, 150 Fearing Street, Amherst, Massachusetts 01002. ISBN 1-884-940-12-9.

Huntley, D., Hawk, R.N. and Corley, H.P. 1994. Non-aqueous Phase Hydrocarbon in a fine-grained sandstone: 1. Comparison between measured and predicted saturations and mobility. *Ground Water*, 32(4), 626–634.

Jackson, D. and Lloyd, J.W. 1983. Groundwater chemistry of the Birmingham Triassic Sandstone aquifer and its relation to structure. *Q. J. Engineering Geology. London*, 16, 135–142.

Kruseman, G.P. and de Ridder, N.A. 1994. Analysis and Evaluation of Pumping Test Data. Second Edition. Publication 47, International Institute for Land Reclamation and Improvement, P.O. Box 45, 6700 AA Wageningen, The Netherlands.

Mercer, J.W. and Cohen, R.M. 1990. A review of immiscible fluids in the subsurface: Properties, models, characterization, and remediation. *Journal of Contaminant Hydrology*, 6, 107–163.

Rivett, M.O., Lerner, D.N., Lloyd, J.W. and Clark, L. 1990. Organic contamination of the Birmingham aquifer. *Journal of Hydrology*, 113, 307–323.

Salinitro, J.P. 1993. The role of bio-attenuation in the management of aromatic hydrocarbon plumes in aquifers. Ground Water Monitoring Review – Dublin, OH: Ground Water Publishing Co. Volume 13, 4/FALL, pp. 150–161.

Shepherd, K. 2000. The Geochemistry of Urban Groundwater Pollutants. Second Year PhD Report. School of Earth Sciences, University of Birmingham.

Thomas, A. 2001. A Geographic Information System Methodology For Modelling Urban Groundwater Recharge And Pollution. PhD Thesis. The School of Earth Sciences, The University of Birmingham, Birmingham, United Kingdom.

CHAPTER 10

Assessment of groundwater contaminant vulnerability in an urban watershed in southeast Michigan, USA

D.T. Rogers[1], K.S. Murray[2] and M.M. Kaufman[2]
[1]Amsted Industries Incorporated, USA
[2]University of Michigan, USA

ABSTRACT: Groundwater contaminant vulnerability was evaluated in the Rouge River watershed in southeast Michigan, USA which includes metropolitan Detroit. The evaluation was performed using an analytical model of solute transport in soil combined with calculated surface and subsurface risk factors to quantify the potential impacts to groundwater and surface water quality. Four categories of contaminants (DNAPLs, LNAPLs, PAHs, and lead) were evaluated at 83 sites across varied soil types. The results indicate that risk to groundwater differs significantly depending on the soil types and contaminants, with DNAPL compounds posing greatest risk in soils within the sand and moraine surface geology of the watershed. Significant differences in the cost of remediation also exist between the different contaminant categories. These findings are significant because DNAPLs are persistent in the environment, and this contamination, coupled with the watershed's hydraulic connection with the Great Lakes creates the potential for ecological and public health impacts at broad geographic scales. The results demonstrate the need to conduct contaminant vulnerability analyses in urban regions and enact contaminant control measures for those contaminants that are mobile and persistent in the urban environment.

1 INTRODUCTION

To maintain a sustainable urban environment, a thorough understanding of groundwater contaminant vulnerability is crucial. Groundwater is an important mechanism that transports contaminants from release points to locations of human and ecological exposure, thereby greatly reducing the potential sustainability of a given urban region even if the surface water and groundwater are not used as a human resource.

Groundwater contamination results from a sequence of anthropogenic and natural activities and processes involving a contaminant source, its mobilization through the soil (transport action), and its ultimate fate (sinks). Sources of contamination include human activities performed primarily on the surface that release toxic substances into the environment. Such substances may be transported over time, or remain relatively close to their source. Because contaminants released into the environment are often mixtures of chemical compounds (e.g. gasoline), an understanding of the physical chemistry of each of the chemical constituents and the geologic and microbiologic environment into which they are released is necessary to evaluate their transport and fate. During transport, these compounds remain at various locations or sinks for different time periods. Examples of temporary sinks include the near-surface soil, and groundwater. The mobility of contaminants in groundwater depends upon (1) the specific physical chemistry of the contaminant and (2) local aquifer characteristics. Subsequent or eventual sinks include inland lakes and streams, or the ocean.

In a recent study of groundwater vulnerability within the Rouge River watershed in southeast Michigan, Murray and Rogers (1999) evaluated the geologic and contaminant characteristics at over 400 sites of environmental concern. Over 90% of the sites showing groundwater contamination derived from anthropogenic sources were located on either glacial outwash, moraines, or beach deposits, while fewer than 1% of the contaminated sites were located on clay-rich glacial lacustrine or lodgment till units. Murray and Rogers (1999) also determined that the costs of redevelopment were highly dependant on the extent of the contamination. The average cost to remediate sites located on the outwash, moraine, or beach deposits was $57,273 per kilogram of contaminant, while the cost to remediate sites located on the less sensitive clay-rich units averaged only $127 per kilogram of contaminant. Fortuitously, the majority of metropolitan Detroit's estimated 10,000 brownfield sites are located on these clay-rich deposits (Michigan Department of Environmental Quality [MDEQ], 2002).

In a quantitative study of surface and subsurface risk factors to groundwater within the Rouge River watershed, Kaufman et al. (2003) discovered clear differences in risk to groundwater depending on whether or not clay was present. This study also identified a significant increase in risk of groundwater contamination associated with geologic units composed predominantly of coarse-grained soil, such as sand, and a lower overall risk associated with geologic units composed of fine-grained soils such as clay. Furthermore, a strong association was observed between remediation costs and surface and subsurface risk factors, with remediation costs increasing with increased risk.

The research study presented here further considers groundwater vulnerability in the Rouge River watershed by examining the transport behaviour of specific anthropogenic contaminants under various geological conditions. Contaminant types evaluated include: (1) volatile organic compounds (including both dense nonaqueous phase liquids (DNAPLs) and light nonaqueous phase liquids (LNAPLs)); (2) polynuclear aromatic hydrocarbons (PAHs); and (3) lead.

The concept of groundwater vulnerability originated in France during the 1960s and was introduced into the scientific literature by Albinet and Margat (1970). Groundwater vulnerability has evolved considerably since this time and often incorporates "risk assessment" considerations in addition to "intrinsic" vulnerability characteristics defined as a function of the natural properties of the overlying soil, unsaturated zone, and rock column (Foster, 1987; Foster and Hirata, 1988; Robins et al., 1994).

In the present study, emphasis is placed on the specific type of contamination identified, this is important yet is often overlooked or under represented. Considering specific types of contamination into groundwater vulnerability models is important because each contaminant has unique physical chemical characteristics that can significantly influence its behaviour when released into the sub-surface environment. It is well known that contaminant fate and transport evaluations require interdisciplinary analysis that involves chemical, geologic, hydrologic and biologic factors (USEPA, 1989, 1992). It therefore follows that the development of groundwater vulnerability models for specific contaminants should also require an interdisciplinary process.

The critical physical/chemical attributes that influence contaminant risk are associated with migration potential, and include (USEPA, 1989, 1996a; Wiedemeier et al., 1999):

1. solubility,
2. vapor pressure,
3. density,
4. chemical stability,
5. persistence, and
6. adsorption potential.

While toxicity is also an important factor for evaluating risk, mobility and persistence are especially critical since these factors dictate the chemical's ability to migrate from its point of release in the environment to a distant point of human exposure, such as a drinking water supply or surface water body.

2 STUDY AREA

The study area is the Rouge River watershed in southeastern Michigan (Figure 1), USA. This watershed is the most populated in Michigan, with 1.5 million residents and high levels

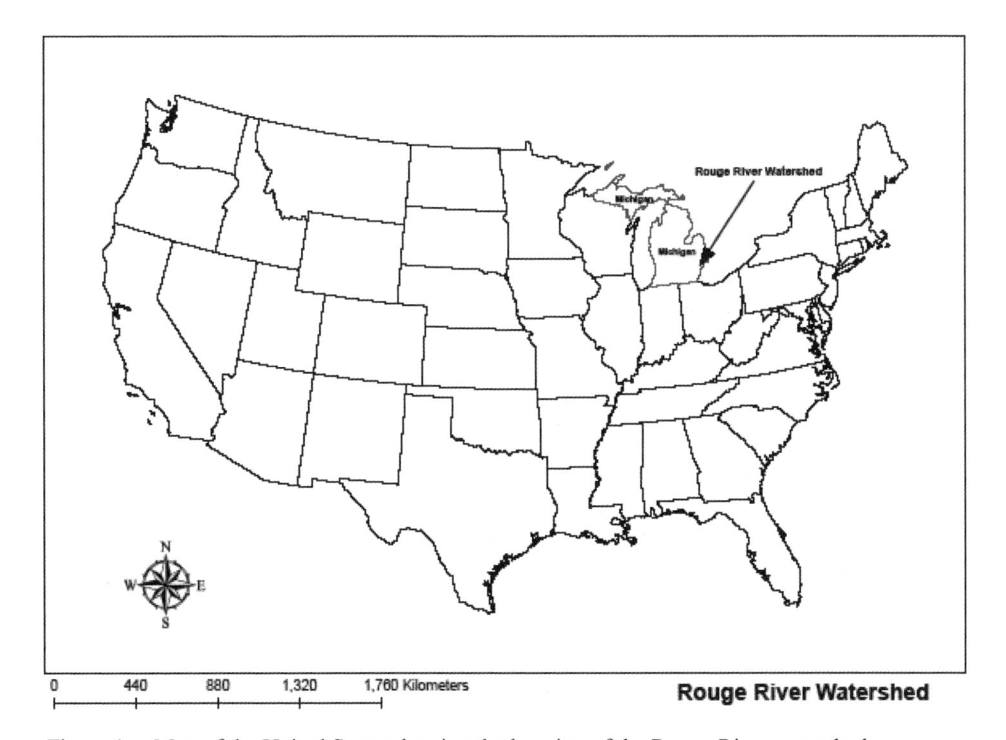

Figure 1. Map of the United States showing the location of the Rouge River watershed.

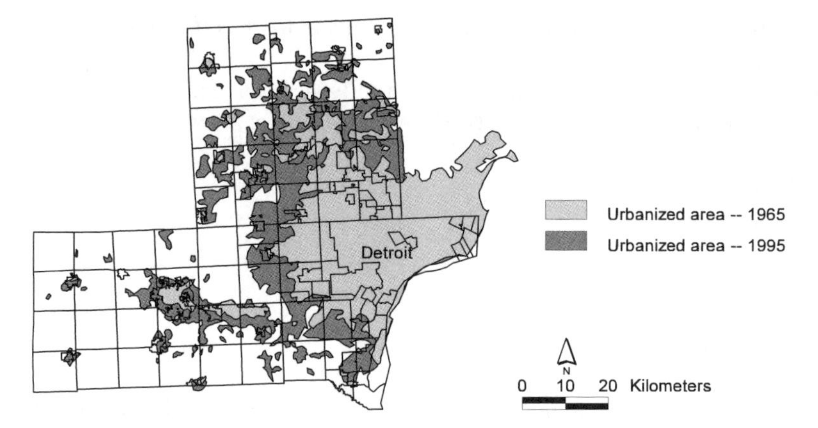

Figure 2. Progress of urbanisation in southeast Michigan (SEMCOG, 1997).

of urbanisation. In 2000, 99% of the watershed's population lived within the U.S. Census-defined urbanised area, and the population density was higher than that of any other watershed in the eastern United States. Water quality of the Rouge River is severely degraded. In recent surveys, the Rouge River was ranked as one of the worst toxic sites in Michigan (MDEQ, 2002).

Situated within the humid microthermal climate zone of the Midwestern United States, the Rouge River is effluent, meaning it is fed by groundwater entering as baseflow. Thus any significant contamination of groundwater that does not readily degrade or attenuate will ultimately be delivered to the surface water drainage network (Murray et al., 2000).

The Rouge River watershed consists of over 200 km of streams, tributaries, lakes and ponds within a 700 km^2 area. The fan-shaped watershed includes all or part of 47 different municipalities and is currently the focus of intense scientific study and restoration. It has been identified as an "Area of Concern" by the International Joint Commission (Hartig and Zarull, 1991) and cited as a significant source of pollution to the lower Great Lakes (Murray and Bona, 1993).

Over 66 percent of the Rouge River watershed has been developed with land uses ranging from residential and commercial to heavy industrial. The remaining land, largely present in the western portion of the watershed, is primarily rural and undeveloped, interspersed with small farms. By the year 2010, the population of southeast Michigan is projected to increase by 500,000 and much of the rural part of the watershed is expected to undergo rapid urban growth (SEMCOG, 1997). Figure 2 shows the progression of urbanisation within the watershed.

The projected growth will result in significant environmental pressure on southeast Michigan and creative, scientifically informed environmental and land use planning techniques will be required to prevent degradation of soil and water resources. Although these techniques will rely largely on infrastructural, demographic, historic and economic information, they must also include an understanding of the following:

1. physical and chemical aspects of soil and water quality,
2. relationship between surface water and groundwater quality,
3. movements of pollutants in the soil, and
4. preexisting pollutant impacts.

3 GEOLOGIC AND HYDROGEOLOGIC SETTING

The study area is located on the southeastern edge of the Michigan Basin, covering an area of 316,000 km^2 and comprises sedimentary rocks including primarily limestones, shales, and sandstones. These sedimentary rocks are Paleozoic in age, and rarely outcrops because of the presence of a thick deposit of glacial drift in the region. Beneath the study area, the Paleozoic rocks range from 425 to 730 metres thick and gently dip toward the center of the basin to the northwest. The depth to bedrock ranges from more than 110 metres in the northwest of the study area to less than 15 metres in the southeast (Rieck, 1981a).

Sediments of Pleistocene age unconformably overlie the Paleozoic rocks of the Michigan Basin. The missing 280 million years of sedimentary record can be attributed to uplift and widespread erosion of the Michigan Basin following the late Paleozoic (Dorr and Eschman, 1988).

The surficial geology in Michigan is dominated by glacial sediments that are typically over 60 metres thick and, at some locations more than 300 metres thick (Rieck, 1981a). These sediments were deposited during the Pleistocene Epoch by the Wisconsinan stage of glaciation, and consist of outwash, moraine, and beach, bar and lake deposits (Farrand, 1988). Varied and complex lithologies are exhibited within these deposits and include coarse gravels, fine-grained sands, and clays (Bergquist and MacLachlan, 1951; Mozola, 1969; Rieck, 1981a, 1981b).

Five distinct near-surface geologic units have been identified within the study area (Leverett, 1911; Sherzer, 1916; Farrand, 1982; Rogers, 1996). The units are classified by their composition and include moraine, sandy clay, sand, sandy and silty clay, and clay (Figure 3). Summaries of the key hydrogeologic features within each unit are based on the results of investigations at over 500 contaminated sites, with the following classification adopted from Rogers (1996):

- The *moraine* unit contains interbedded sands, gravels, and clays with occasional glacial erratics. The surface elevation of this unit ranges from 290 masl (metres above sea level) in the extreme northwestern corner of the study area to 250 masl along its contact with the sandy clay unit. The thickness of the unit is generally less than 45 metres. Groundwater encountered within the moraine unit exists under both confined and unconfined conditions. Multiple zones of saturation are present, ranging in thickness from a metre to over 10 metres. Public and private entities use the groundwater within this unit as a potable water source.
- Southeast of the moraine unit is a unit of *sandy clay* comprising light-brown to gray sandy clay with occasional pebbles. The surface elevation of this unit ranges from 215 to 250 masl. In the northern part of the study area the thickness of the sandy clay unit tends to be less than 2 metres, but this increases to 6 metres in the south. Groundwater in this unit is unconfined, locally discontinuous and is unavailable in sufficient volumes for pumping at a sustained rate.
- The *sand* unit is characterized as a moderate yellowish-brown to light olive-gray, fine to coarse grained quartz sand that becomes finer with depth. This textural change with depth is attributed to its deposition during lake regression. The sand is 90% quartz. At many locations, the sand unit is stratified and has well developed cross-bedding, ripple marks, and scour and fill features. Localized evidence of reworking and eolian deposition is also present within the upper portion of the sand unit. The surface elevation of this unit

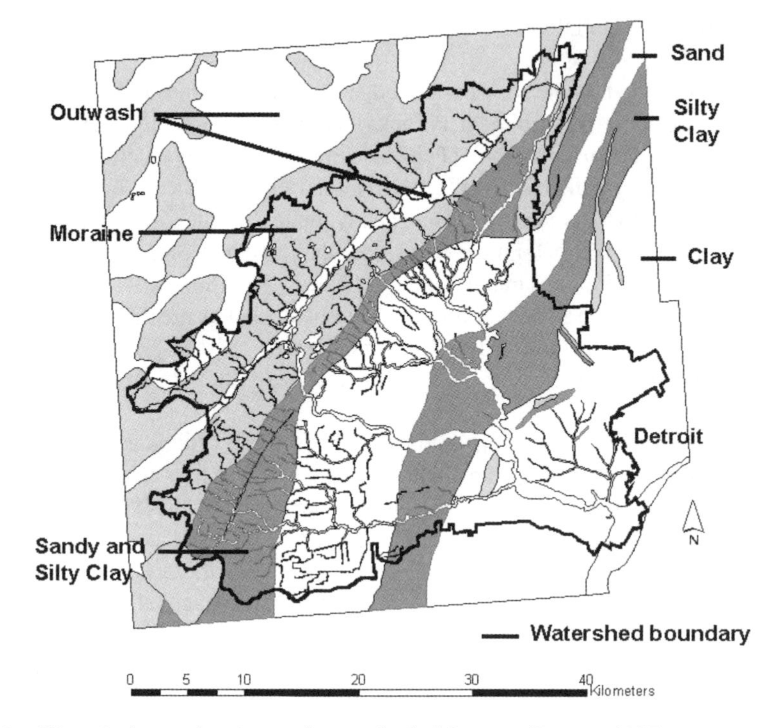

Figure 3. Watershed map showing surface geological features (Rogers, 1997).

ranges from 195 to 220 masl, and its thickness varies between 1.5 and 10 metres, generally increasing to the east. Groundwater in the sand unit is unconfined with saturated thicknesses varying from one metre to more than 6 metres. Based on evidence from recovery wells installed at contaminated sites, the average yield from the sand unit is 0.02 to 0.04 cubic metres per minute. Consequently, the unit has been used historically as a source of potable water.

- To the east of the sand unit is the *sandy and silty clay* unit. This unit is characterized as a medium to light olive-gray, mottled, sandy to silty clay bearing well-rounded pebbles. Discontinuous fine-grained sand lenses are present within this unit, especially along its western margin with the Rouge River. Well developed varves have also been observed along the western margin of this unit where it is either interbedded or lies beneath the sand unit. Individual varves range in thickness from less than 0.25 to nearly 0.65 cm and typically extend vertically for 15 cm. The total thickness of the sandy and silty clay unit ranges from 3 to 6 metres, with a higher sand content present along its contact with the sand unit. The surface elevation of this unit ranges from 190 to 195 masl. Groundwater in this unit is unconfined or semi-confined and occurs, within the discontinuous sand lenses, where present, at the contact with the lower clay unit, or in the reworked river deposits along the Rouge River.
- The *clay* unit forms the eastern boundary of the study area. This unit consists of mottled bluish to medium light olive-gray clays, with trace amounts of well-rounded pebbles. Vertical to sub-vertical hairline fractures are present that reduce in frequency with depth. They are associated with stress changes due to wetting and drying cycles and/or

freezing and thawing (Freeze and Cherry, 1979; Murray et al., 2000). Along its western margin, the clay unit is 3 metres thick; it thickens to more than 10 metres along its eastern boundary with the Detroit River. Surface elevation of this unit ranges from 175 to 190 masl. The characteristics of groundwater within this unit are similar to those found within the sandy and silty clay unit.

4 GROUNDWATER VULNERABILITY METHODS

The research presented here combines the risks associated with surface conditions, subsurface conditions, and specific contaminants in the development of a single integrated model for groundwater vulnerability assessment. This is achieved by combining the properties of toxicity, mobility, and persistence for specific contaminants within four contaminant categories (DNAPLs, LNAPLs, PAHs, and lead) to create a Contaminant Risk Factor (CRF) score. This score is then added to existing risk scores associated with the surface and subsurface conditions.

To evaluate the model, a Welch analysis of variance (ANOVA) was used to test the means of the total risk score and total remediation cost, for a range of soil groups and contaminant categories. Bartlett's test of homogeneity of variance, which produces an F ratio, was used to confirm the unequal variances of the total cost and total risk across the different soil groups and contaminant categories. All statistical tests were performed at the 0.05 level of significance.

4.1 *Surface risk*

Evaluations of surface risk have traditionally used spatially generalized categories of land use, such as "industrial, commercial, or residential" to represent various levels of risk (Barringer et al., 1990; Eckhardt and Stackelberg, 1995; Secunda et al., 1998). The use of general land use categories is problematic for groundwater vulnerability assessment because of their inadequate spatial resolution: spatial resolution is defined as the smallest identifiable element in a sequence (Tobler, 1988). In urban and urbanising areas, mixed land uses within small areas such as city blocks are common, such that the variable risks may be obscured by generalization when the capture zones for wellhead protection are delineated. For example, the "10-year capture zone" is the sub-surface and surface areas from which water (and any contamination it carries) will reach the well over a time period of 10 years. Broad generalisations of land use within these areas will lead to unreliable assessments of risk.

The use of the Standard Industrial Classification (SIC) code for each establishment within a general land use category permits an improved parcel-level spatial resolution of the relative risks to groundwater. The SIC code is a four digit code: the first two digits identify a major group, such as agriculture, retail trade, and manufacturing; the third digit denotes industry groups within each major group, such as agricultural crop production; and the fourth digit identifies a specific industry code, such as grain storage (United States Office of Management and Budget, 1987).

A comparable measure of risk between different establishment types is achieved through the use of contamination incidence rates (Kaufman, 1997). Incidence rates are obtained by:

1. assigning a SIC code to each source of contamination appearing on the current Michigan Environmental Response Act list (MDEQ, 2002);

Table 1. Vulnerability matrix of subsurface risks.

Parameter identification number	Parameter description	Rating strength
1	*Depth to groundwater*	
	Less than 3 metres	10
	3 to 10 metres	5
	Greater than 10 metres	1
2	*Occurrence of groundwater*	
	Greater than 75%	10
	25% to 74%	5
	Less than 25%	1
3	*Area of groundwater recharge*	
	Significant area of recharge	10
	Insignificant area of recharge	1
4	*Potential discharge of groundwater to surface water*	
	Significant area of discharge	10
	Insignificant area of discharge	1
5	*Composition of geologic units*	
	Gravel	10
	Sand	9
	Silt	5
	Clay	1
6	*Potential travel time to point of exposure*	
	Less than 10 years	10
	10 to 25 years	5
	Greater than 25 years	1
7	*Source of potable water*	
	Source of potable water	10
	Not a source of potable water	1

2. obtaining the total number of establishments in Michigan for each SIC code from the County Business Patterns (US Bureau of the Census, 1997); and
3. dividing the number of contaminated sites with a specific SIC code by the total number of establishments in Michigan with that same SIC code.

To provide numerical equivalency with scores calculated for subsurface conditions (see below) the incidence rates are multiplied by 10, converted to scores of between 0 and 10 and summed for a circular area around each water well. This total score represents the relative risk of the surface activities.

4.2 *Subsurface risk*

To measure groundwater vulnerability, Murray and Rogers (1999) developed a numerical rating system using a modified form of DRASTIC (Aller et al., 1987). The model uses weighting coefficients similar to the DRASTIC model for the geologic and hydrologic parametres (Table 1). The first parameter shown in Table 1 is the depth to the shallow water

Table 2. Sensitivity scoring of subsurface risk.

Geologic unit	Parameters (from Table 1)							Total score†	Relative ranking‡
	1	2	3	4	5	6	7		
Sand	10	10	10	10	9	5	10	64	High
Moraine	5	5	10	10	5	5	10	50	Medium
Sandy clay	5	1	1	1	5	1	1	15	Low
Sandy and silty clay	5	1	1	1	5	1	1	15	Low
Clay	5	1	1	1	1	1	1	11	Low

†Total score is the sum of the values of the seven parameters.
‡Rankings are based on ranges for the total score: Low = 0–20; Medium = 21–50; High >50.

table aquifer within the glacial deposits. The second parameter, occurrence of groundwater, is based on data collected from sites of environmental concern within the study area (Rogers and Murray, 1997). For example, a geologic unit with a less than 25% occurrence of groundwater means there is less than a one in four chance of encountering groundwater in a well drilled into this unit to a depth of 6 metres. The third and fourth parameters, areas of groundwater recharge and discharge, provide a means of representing the horizontal and vertical components of groundwater movement within the study area. Both of these parameters are related to parametres 5 and 6, the textural characteristics and travel time, which can be used to assess soil and solute transport properties of the near-surface geologic materials. The seventh parameter, source of potable drinking water, refers to whether the geologic unit is, or can be, a source of drinking water.

The weighted rating strengths for each geologic unit are summed to determine a final relative index of unit sensitivity (Table 2). Where units (e.g. the moraine) comprise a range of geologic materials, the rating strengths represent composite scores. For example, for parameter 5 (composition), the moraine is classified as silt with a midpoint value of 5 to reflect its heterogeneous mixture of sand, gravel, and clay. The sand and moraine units are substantially more vulnerable to groundwater contamination than the other units in the watershed because they are composed of highly permeable sediments, contain groundwater, and occur in areas of the watershed subject to groundwater recharge.

4.3 *Contaminant risk*

Contaminant risk associated with the presence of a specific chemical compound can be developed as a function of three criteria (USEPA, 1989):

1. toxicity or potency,
2. mobility, and
3. persistence.

These criteria recognise that contaminants released into the environment only present a risk to humans if there is a completed exposure pathway. While toxicity is clearly an important consideration, the risk to humans becomes high only if the chemical is both mobile and persistent. In many cases, a chemical that is extremely toxic to humans may not present as much risk as a chemical that is only moderately toxic but is mobile and has a high propensity to migrate and potentially contaminate a public water supply.

The contaminants types evaluated in this study include:

1. DNAPLs,
2. LNAPLs,
3. PAHs, and
4. lead.

These contaminants are common within the Rouge River watershed and are used widely in the urban setting in products such as gasoline, diesel fuel, fuel oil, lubricants, solvents, cleaners, paints, pigments, and metal products (Rogers, 1996).

Toxicity values were obtained from the USEPA (2004) Integrated Risk Information System (IRIS), which was updated in October 2004. Toxicity values selected were the more conservative value listed for either carcinogenic or chronic effects for the oral pathway. The oral pathway was selected as opposed to the dermal and inhalation pathway because for this study, oral ingestion of groundwater is expected to be the dominant exposure pathway.

To obtain a value to represent the mobility of each compound, two variables were used: Henry's Law constant (H) and the retardation factor (R). Henry's Law constants were obtained from USEPA (1996a) and Wiedemeier et al. (1999). The retardation factor was calculated by first calculating the distribution coefficient using Equation 1.

$$K_d = (Foc)(Koc) \tag{1}$$

where: K_d = distribution coefficient ($mL \cdot g^{-1}$); Koc = organic carbon partition coefficient (l/mg); Foc = fraction of organic carbon in soil (mg/mg).

Values for the organic carbon partition coefficient were obtained from Wiedemeier et al., (1999) and the fraction of organic carbon in soil was obtained from Wiedemeier et al. (1999) and USEPA (1996b).

Once the distribution coefficient was calculated, the Retardation Factor was calculated using Equation 2 (Wiedemeier et al., 1999).

$$R = 1 + \frac{(\rho_b)(K_d)}{\eta} \tag{2}$$

where: R is the retardation factor; ρ_b = bulk density of aquifer matrix (mg/l); K_d = distribution coefficient ($mL \cdot g^{-1}$); η = effective porosity.

Henry's Law constant (H) ($atm. \cdot mol^{-1} \cdot m^{-3}$) is a measure of the tendency for organic solutes to volatilize. It is related to vapor pressure (VP) (atm.), molecular weight (MW) (g/mol); solubility in water (W_s) (g/L) according to Equation (3):

$$H = (VP)/(MW)(W_s) \tag{3}$$

With Henry's Law constant and the retardation factor in place, the mobility of a specific compound can be expressed as Equation (4):

$$M = (H)(R) \tag{4}$$

where: M = mobility; H = Henry's Law constant; R = retardation factor.

Persistence values were obtained from Howard et al. (1991) and USEPA (1996a, 1996b, 2000) and are expressed as first order decay rates in years. In general, the first order decay rates selected for each compound were chosen as the most conservative of the spectrum of data available.

Finally, the Contaminant Risk Factor (CRF) was calculated by multiplying the chemical compound's Persistence (P) in the environment by the inverse of its Toxicity (T) and the inverse of its Mobility (M). Using the inverse of the chemical compound's toxicity and mobility ensures that the toxicity and mobility values remain a positive number so that appropriate weighting of the values can be achieved. The CRF equation is expressed as Equation (5):

$$\mathrm{CRF} = \frac{1}{(\mathrm{T})}\frac{1}{(\mathrm{M})}(\mathrm{P}) \tag{5}$$

where: CRF = Contaminant Risk Factor; T = toxicity; M = mobility; P = persistence (years).

Contaminant risk factors were calculated for each of the significant compounds in each contaminant group listed in Table 3. The compounds listed include those most commonly detected within the watershed and those required to be evaluated by the Michigan Dept. of Environmental Quality (MDEQ) (1994) as targeted compounds. Once the CRF for each compound was calculated, results were averaged to obtain an overall CRF for each group of compounds for each geologic unit in the watershed. They are included in Table 4.

4.4 *Total environmental risk*

The total environmental risk value for specific compounds in each of the geologic units within the watershed is represented as the sum of the surface risk value, subsurface risk value, and the contaminant risk factor (Equation 6).

$$\text{Total risk} = \text{Surface risk} + \text{Subsurface risk} + \text{Contaminant risk} \tag{6}$$

Table 4 lists the total risk factor for each compound group evaluated. Values for surface risk were obtained from Kaufman et al. (2003) and are average values for surface risk in each geologic unit. Subsurface values were obtained from Murray and Rogers (1999).

Table 3. Chemical list showing four groups of compounds.

DNAPL	LNAPL	PAHs	Lead
Tetrachloroethene	Benzene	Naphthalene	Elemental lead
Trichloroethene	Toluene	Phenanthrene	
Cis 1,2-dichloroethene	Ethylbenzene	Chrysene	
Trans 1,2-dichloroethene	Xylenes (total)	Benzo[a]pyrene	
1,1,1-trichloroethane		Benzo[a]anthrocene	
Vinyl chloride		Benzo[b]fluoranthene	
		Benzo[g,h,i]perylene	
		Benzo[k]fluoranthene	

Table 4. Contaminant risk factors.

Contaminant type	Geologic unit	Mean surface risk value	Subsurface risk value	Contaminant risk factor (CRF)	Total risk factor
DNAPL	Clay	0.04	11	256	267
	SC	33.7	15	647	695.7
	SSC	0.42	15	672	687.4
	Sand	343.64	64	913	1,320.6
	Moraine	240.54	50	856	1,146.5
LNAPL	Clay	0.04	11	0.08	11.12
	SC	33.7	15	0.42	49.12
	SSC	0.42	15	0.47	15.89
	Sand	343.64	64	0.72	408.36
	Moraine	240.54	50	0.65	290.54
PAHs	Clay	0.04	11	0.0001	11.04
	SC	33.7	15	0.0008	48.7
	SSC	0.42	15	0.001	15.42
	Sand	343.64	64	0.002	407.64
	Moraine	240.54	50	0.002	290.54
Lead	Clay	0.04	11	0.03	11.07
	SC	33.7	15	0.3	49.0
	SSC	0.42	15	0.34	15.76
	Sand	343.64	64	1.1	408.74
	Moraine	240.54	50	1.0	291.54

5 DISCUSSION

Calculated contaminant risk factors for DNAPL compounds were significantly higher than the other contaminant groups. DNAPL compounds had higher contaminant risk factors because, in general, DNAPL compounds are more toxic, more mobile, and much more persistent in the environment than the other contaminant groups. LNAPL compounds were much less persistent than DNAPL compounds, in most cases, by more than an order of magnitude. PAH and elemental lead compounds were far less mobile than DNAPL compounds, again, in most cases, by at least an order of magnitude.

The resulting total risk factors listed in Table 4 clearly indicate that

1. total risk increases as the mean grain size of the geologic material increases, and
2. DNAPL compounds represent the highest risk of any contaminant group examined.

To further test the significance of this observation, data on the cost to remediate a kilogram of contaminant was obtained from 83 sites of environmental contamination within the watershed. Table 5 presents a summary of the remedial cost for each contaminant group within the watershed.

The risk variations across the geologic units at the 83 sites of environmental contamination were not statistically significant ($B = 7.7$, $p < 0.0001$; Welch ANOVA $F = 2.6$, $p < 0.06$). However, risk variation across the different contaminant categories produced a strong statistical significance ($B = 300.3$, $p < 0.0001$; Welch ANOVA $F = 88.7$, $p < 0.0001$). The difference between the mean risk scores for DNAPLs and the other three contaminant categories was over 712 points.

Table 5. Remedial cost and total environmental risk.

Contaminant type	Geologic unit	Number of sites (n = 83)	Remedial cost ($/kg)	Total risk factor
DNAPL	Clay	6	721	267
	SC	2	1,366	695.7
	SSC	3	3,263	687.4
	Sand	8	99,127	1,320.6
	Moraine	4	148,648	1,146.5
LNAPL	Clay	7	319	11.12
	SC	3	518	49.12
	SSC	4	778	15.89
	Sand	8	527	408.36
	Moraine	5	6,411	290.54
PAHs	Clay	8	922	11.04
	SC	7	843	48.7
	SSC	2	203	15.42
	Sand	5	444	407.64
	Moraine	0	No Data	290.54
Lead	Clay	4	68	11.07
	SC	3	230	49.0
	SSC	2	538	15.76
	Sand	2	442	408.74
	Moraine	0	No Data	291.54

With respect to the total cost of remediation, there was a significant difference across the 5 soil types ($B = 38.1$, $p < 0.001$; Welch ANOVA $F = 5.2$, $p < 0.003$). The total cost of remediation also varied significantly by contaminant type ($B = 55.8$, $p < 0.001$; Welch ANOVA $F = 10.01$, $p < 0.001$). The difference between the mean total remediation costs in the Sand Unit and the Clay Unit was slightly over $940,000; the difference between the mean cost of remediation with DNAPLs and PAHs was just over $1 million. These figures illustrate the significant differences apparent in remediation costs across the different geologic units and contaminant types. The statistical results also suggest a weighting mechanism for contaminant type may be required to more fully explain the risk score variations.

6 CONCLUSION

The results demonstrate a significant risk differential between specific contaminants released within the Rouge River watershed located in Michigan, USA. The risk differential was confirmed through an examination and comparison of risk and cost of remediation at 83 sites of environmental contamination. DNAPL compounds consistently ranked highest in total risk and cost, followed by LNAPLs, lead, and PAHs, respectively.

The results of this study are significant because within the Rouge River watershed, industrial development of greenfield sites located on materials ranked as highly vulnerable to groundwater contamination is outpacing brownfield redevelopment, despite the potential for the degradation of potable water resources. This trend is occurring because of a perception that the development of older brownfield sites within the City of Detroit, which is located within the clay, would be a long and costly process resulting in continued liability. This study and others conducted by Murray and Rogers (1999), and Kaufman et al. (2003), have demonstrated that older brownfield sites in the City of Detroit are typically located on soils with a much lower vulnerability to groundwater contamination and are therefore far less costly to develop. A comprehensive approach to urban redevelopment should include contaminant groundwater vulnerability studies. These studies can become a basic component of the land use planning process, much as environmental site assessments are to the real estate industry.

This study has also identified DNAPLs as the contaminant group with the highest environmental risk. The high relative risk of DNAPL compounds is due to a combination of toxicity, mobility, and persistence of DNAPL compounds once they have been released into the environment.

The results reveal significant potential for degradation of shallow groundwater in the watershed, especially with respect to chemically persistent DNAPL compounds. This finding is significant because of the watershed's hydraulic connection with the Great Lakes and raises ecological and public health concerns. In addition, the results further underscore the importance of conducting contaminant vulnerability analysis in urban regions and enacting contaminant control measures for those contaminants that demonstrate mobility and persistence in the urban environment, such as DNAPLs. The methodology outlined in this study for contaminant vulnerability analysis should be applied to other urban watersheds. In addition, future studies should include a more comprehensive list of chemical compounds such as polychlorinated compounds and other metals, such as arsenic, cadmium, chromium and mercury.

REFERENCES

Albinet, M. and Margat, J. 1970. Cartographie de la vulnérabilité a la pollution des nappes d'eau souterrains. Bulletin BRGM 2:13–22.

Aller, L., Bennett, T., Lehr, J.H., Petty, R.J. and Hackett, G. 1987. DRASTIC: A standardized system for evaluating ground water pollution potential using hydrogeologic settings. Ada, Oklahoma: U.S. EPA-600/2-87-035.

Barringer, T.H., Dunn, D., Battaglin, W.A. and Vowinkel, E.F. 1990. Problems and methods involved in relating land use to ground-water quality. *Water Resources Bulletin* 26(1):1–9.

Bergquist, S.G. and MacLachlan, D.C. 1951. *Guidebook to the Study of Pleistocene Features of the Huron-Saginaw Ice Lobes in Michigan*. The Glacial Field Trip of the Geological Society of America, Detroit Meeting. 54p.

Dorr, J.A. and Eschman, D. 1988. *Geology of Michigan*. The University of Michigan Press. Ann Arbor, Michigan. 476p.

Eckhardt, D.A.V. and Stackelberg, P.E. 1995. Relation of ground-water quality to land use on Long Island, New York. *Ground Water* 33(6):1019–1033.

Farrand, W.R. 1982. *Quaternary Geology of Southern (& Northern) Michigan*. Michigan Department of Natural Resources, Geological Survey Division. Lansing, Michigan. 1:500,000.

Farrand, W.R. 1988. *The Glacial Lakes Around Michigan*. Michigan Department of Natural Resources Bulletin No. 4. Lansing, MI.

Foster, S.S.D. 1987. Fundamental concepts in aquifer vulnerability, Pollution Risk and Protection Strategy. No 38: 69–87, eds. W. van Duijvenbooden and H.G. van Waegeningh, In: *Vulnerability of Soil and Groundwater to Pollutants, TNO Committee on Hydrological Research*, The Hague, Netherlands.

Foster, S.S.D. and Hirata, R. 1988. *Groundwater Risk Assessment: A Methodology Using Available Data*. CEPIS Technical Report, Lima, Peru.
Freeze, R.A. and Cherry, J.A. 1979. *Groundwater*. Prentice-Hall, Englewood Cliffs, NJ. 604p.
Hartig, J.H. and Zarull, M.A. 1991. Methods of restoring degraded areas in the Great Lakes. *Reviews of Environmental Contamination and Toxicology*, pp. 127–154.
Howard, H.H., Boethling, R.S., Jarvis, W.F., Meylan, W.M. and Michalenko, E.M. 1991. *Environmental Degradation Rates*. Lewis Publishers, New York, 725p.
Kaufman, M.M. 1997. Spatial assessment of risk to groundwater from surface activities. *Papers and Proceedings of the Applied Geography Conferences* 20:135–142.
Kaufman, M.M., Murray, K.S. and Rogers, D.T. 2003. Surface and Subsurface Risk Factors to Ground Water Affecting Brownfield Redevelopment Potential. *Journal of Environmental Quality*. Vol. 32. pp. 490–499.
Leverett, F.B. 1911. *Map of the Surface Formations of the Southern Peninsula of Michigan*. Geological survey of Michigan, Publication 25, Lansing, MI. 1:1,000,000.
Michigan Department of Environmental Quality (MDEQ). 2002. *Michigan Sites of Environmental Contamination*. http://www.deq.state.mi.us/erd/sites/misites.html
Mozola, A.J. 1969. Geology for land and groundwater development in Wayne County, Michigan. Michigan Geologic Survey. 21p.
Murray, J.E. and Bona, J.M. 1993. *Rouge River National Wet Weather Demonstration Project*. Wayne County, Michigan.
Murray, K.S. and Rogers, D.T. 1999. Groundwater Vulnerability, Brownfield Redevelopment and Land Use Planning. *Journal of Environmental Planning and Management*. Vol. 42. No. 6. pp. 801–810.
Murray, K.S., Zhou, X., McNulty, M. and Mazur, D. 2000. Relationship between land use, near surface geology and water quality in an urban watershed, Southeast Michigan, USA. *Proceedings of the International Symposium on Hydrology and the Environment*. Wuhan, China, pp. 67–71.
Rieck, R.L. 1981a. Glacial drift thickness, Southern Peninsula. Plate 15. *Hydrogeologic Atlas of Michigan*. Western Michigan University, Kalamazoo, MI.
Rieck, R.L. 1981b. Community public water supplies. Plate 22. *Hydrogeologic Atlas of Michigan*. Western Michigan University, Kalamazoo, MI.
Robins, N., Adams, B. and Foster, S.S.D. 1994. Groundwater vulnerability mapping: the British perspective. *Hydrogéologie* 3:35–42.
Rogers, D.T. 1996. *Environmental geology of metropolitan Detroit*. Clayton Environmental Consultants, Inc., Novi, Michigan. 152p.
Rogers, D.T. and Murray, K.S. 1997. Occurrence of groundwater in metropolitan Detroit, Michigan, USA., eds. J. Chilton et al., In: *Groundwater in the Urban Environment*. Balkema, Rotterdam, The Netherlands, pp. 155–160.
Rogers, D.T. 1997. *Surface Geological Map of the Rouge River Watershed*. Rouge River National Wet Weather Demonstrations Project. Wayne County, Michigan. 1:62,500. 2 sheets.
Secunda, S., Collin, M.L., Melloul, M.L. and Abraham, J. 1998. Groundwater vulnerability assessment using a composite model combining DRASTIC with extensive agricultural land use in Israel's Sharon region. *Journal of Environmental Management* 54(1):39–57.
SEMCOG – Southeast Michigan Council of Governments. 1997. *Annual Report on population growth*. 37p.
Sherzer, W.H. 1916. *Geologic Atlas of the United States, Detroit Folio No. 205*. United States Geological Survey, Reston, VA. 1:62,500.
Tobler, W. 1988. Resolution, resampling and all that, 129–137, eds. H. Mounsey and R. Tomlinson, In: *Building Database for Global Science*, London, Taylor and Francis.
U.S. Bureau of the Census. 1997. County Business Patterns. http://tier2.census.gov/cbp/index.html
United States Environmental Protection Agency (USEPA). 1989. *Transport and Fate of Contaminants in the Subsurface*. EPA/625/4-89/019. Washington, DC. 148p.
United States Environmental Protection Agency (USEPA). 1992. *Dense Nonaqueous Phase Liquids*. Office of Research and Development. Washington, DC. 81p.
United States Environmental Protection Agency (USEPA). 1996a. *Bioscreen. Natural Attenuation Decision Support System, Version 1.4*. Office of Research and Development. Washington, DC. 65p.

United States Environmental Protection Agency (USEPA). 1996b. *Soil Screening Guidance: Technical Background Document.* Office of Solid Waste and Emergency Response. Washington, DC. 230p.
United States Environmental Protection Agency (USEPA). 2000. *Biochlor, Natural Attenuation Decision Support System, Version 1.0.* Office of Research and Development. Washington, DC. 45p.
United States Environmental Protection Agency (USEPA). 2004. *Integrated Risk Information System.* Washington, DC. http:/USEPA.gov/IRIS.html http://USEPA.gov/IRIS/index.html
United States Office of Management and Budget. 1987. *Standard Industrial Classification Manual.* U.S. Government Printing Office, Washington, DC.
Wiedemeier, T.H., Rifai, H.S., Newell, C.J. and Wilson, J.T. 1999. *Natural Attenuation of Fuels and Chlorinated Solvents in the Subsurface.* John Wiley & Sons, Inc., New York, New York. 617p.

CHAPTER 11

Xenobiotics in urban water systems – investigation and estimation of chemical fluxes

F. Reinstorf, G. Strauch, M. Schirmer, H.-R. Gläser, M. Möder, R. Wennrich, K. Osenbrueck and K. Schirmer
UFZ – Centre for Environmental Research Leipzig-Halle, Germany

ABSTRACT: The behaviour and effects of xenobiotics in urban water systems are not well known. Xenobiotics include pharmaceuticals, fragrances and endocrine disruptors, and there are numerous ways they can enter the environment and the human food chain. Field investigations and modelling of the xenobiotics in urban areas are presented that demonstrate their mobility as well as their potential impacts on human health and ecosystems. Using indicators for anthropogenic impacts on water resources such as Bisphenol A and t-Nonylphenol (endocrine disruptors), Carbamacepine (antiepileptic drug), Galaxolide and Tonalide (fragrances), ^{34}S-sulphate and ^{15}N-nitrate (stable isotopes), and Gadolinium (rare earth element), investigations of the pathways and the behaviour of the substances in the environment have been carried out. In the city of Halle an der Saale in Germany, remarkably high concentrations of the indicators were found in rivers and in groundwater. Using ^{34}S-sulphate and ^{15}N-nitrate analyses, the movement of groundwater to surface water and various sources of nitrate have been investigated. Evidence for the degradation of xenobiotics was found. Based on the measurements, a water balance and mass flux rates were determined for rivers in the city. The calculation of mass loads shows increasing values of Galaxolide, Tonalide and Bisphenol A, Carbamacepine and t-Nonylphenol, is stable or decreases during passage through the city.

1 INTRODUCTION

Urban water systems are frequently polluted with xenobiotics as a direct and/or indirect result of anthropogenic activities (Figure 1). A xenobiotic (Greek, xenos "foreign"; bios "life") is a compound that is foreign to a living organism. Principal xenobiotics include: drugs, carcinogens, and industrial chemicals all of which have been introduced into the environment. Xenobiotics related to human behaviour and activity such as pharmaceuticals, fragrances and endocrine disruptors are increasingly found in urban water systems. Pharmaceutical components such as lipid regulators, analgesics, β-blockers, and anti-epileptics are widespread contaminants in surface waters (Stachel et al., 2003; Ricking et al., 2003; Heemcken et al., 2001). Some of these contaminants have been found to disrupt endocrine function but in most cases, little information exists as to the potential long-term effect of such contaminants on the urban ecosystem and on human health. Currently, a comprehensive risk assessment about possible effects of pharmaceutical residues in the environment is limited because of a lack of valid data, particularly on the potential long-term toxicological effects of pharmaceuticals and their metabolites in environmentally relevant concentrations.

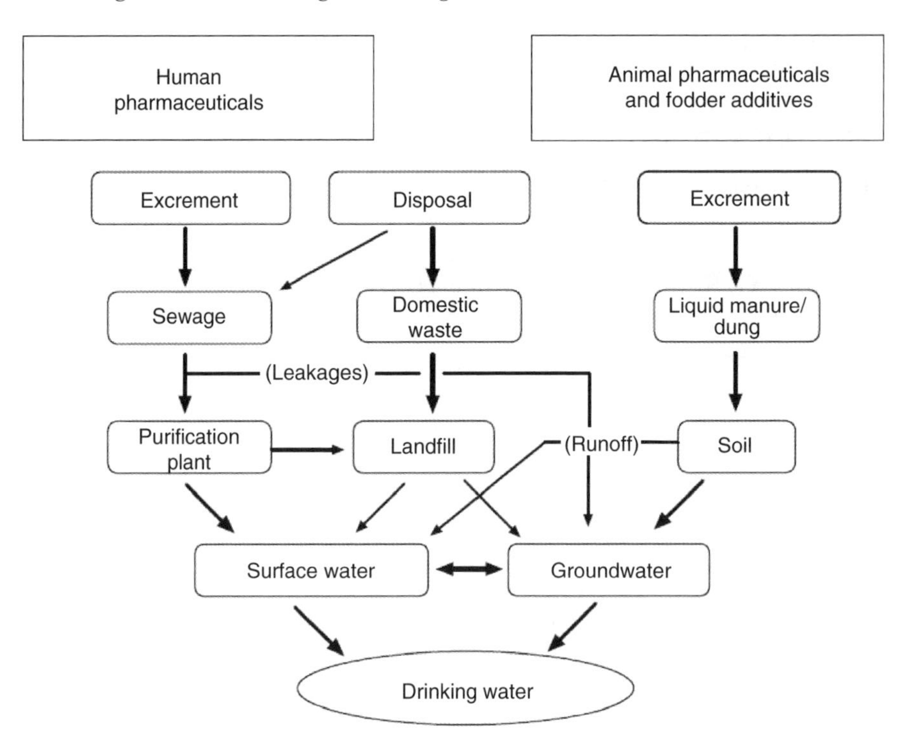

Figure 1. Potential pathways of human and animal pharmaceuticals and fodder additives into the aquatic environment (after Ternes, 1998).

Human pharmaceuticals and their metabolites enter the environment through sewage. After drug ingestion, large parts of the bioactive compounds are generally excreted and reach surface water via municipal sewage plants. Industrial accidents during pharmaceutical production and inappropriate disposal could also be an uncontrolled input of these bioactive components into the environment.

Due to their polarity, some pharmaceutical contaminants are poorly retained in sludge and sediments and may contaminate groundwater. On the other hand, most endocrine active contaminants tend to be lipophilic and accumulate in organic (bio films) and sediment material. In this case, local sewage represents the most important pathway for these components and therefore, remarkably high concentrations of xenobiotics can be expected in surface waters with high proportions of sewage.

Recognising that pharmaceuticals can be encountered at concentrations comparable to plant-protective agents in the aquatic environment and that these substances are biologically very active, the question arises as to how to predict and assess their risk to human health and to ecosystems. An assessment of the risk of adverse effects of drug residues in the environment is only possible if either their concentrations in relevant environmental compartments are known and the environmental concentrations are established at which harmful effects will occur.

To resolve this issue an interdisciplinary project on the assessment of risk of urban water pollution, focussing on xenobiotics, was initiated. The aim of the project was to develop new integrated methodologies for determining the impact of human activities on the urban

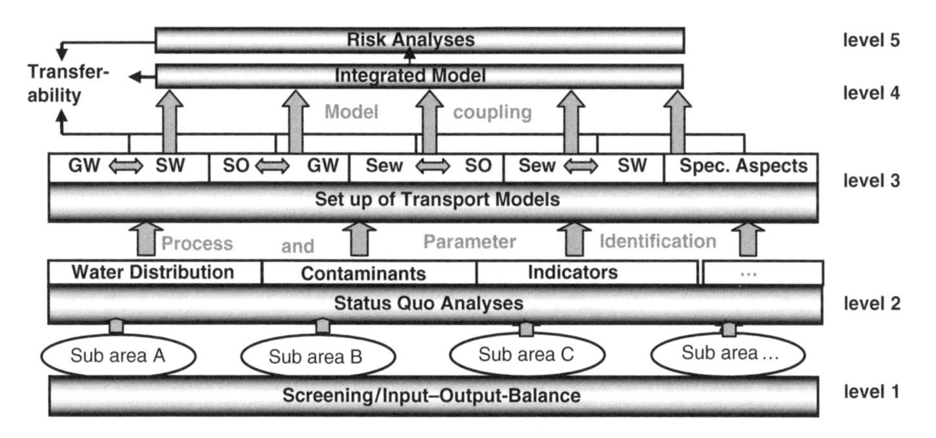

Figure 2. Structure of an integrated model for the management of urban water bounded xenobiotic fluxes (GW: Groundwater; SW: Surface water; SO: Seepage water; Sew: Sewer leakage).

water system and on processes within the urban watershed. The overall goal was to create an integrated model which could be used to assess the risks to humans and ecosystems and would enable the development of suitable management strategies. In this article, sampling results a water balance and a chemical mass balance for the city of Halle an der Saale in Germany are presented.

2 MATERIAL AND METHODS

2.1 *Integrated model conceptualisation*

The conceptual approach to this research is shown in Figure 2 and will be carried out within the next 4 years. Beginning with an initial screening, a water balance and a chemical mass balance (level 1), plans are to perform a status quo analysis of the distribution of known industrial chemicals and xenobiotic indicators within the city's water (level 2).

With the help of process and parameter identification, transport models are then set up (level 3) for various compartments (sewers, soil, groundwater and surface water pathways). Each modelling tool will be examined to determine its transferability to other locations (e.g. arid regions, developing countries, etc.) and will be coupled to an integrated model (level 4) which will also be checked for its transferability and applicability.

The incorporation of various urban land use patterns, different scenarios will lead to a risk analysis (level 5) of the effects these patterns have on water distribution, humans and ecosystems and the corresponding socio-economic consequences.

2.2 *Study area*

The study area is the city of Halle an der Saale in Germany (Figure 3). The city (population 270,000) is located at the edge of the Mitteldeutsches Trockengebiet, a relatively dry region in Germany with precipitation averaging 453 mm/yr (1961–1990) and a mean annual temperature of +9°C. The city lies 96 metres above sea level (masl) and has an area

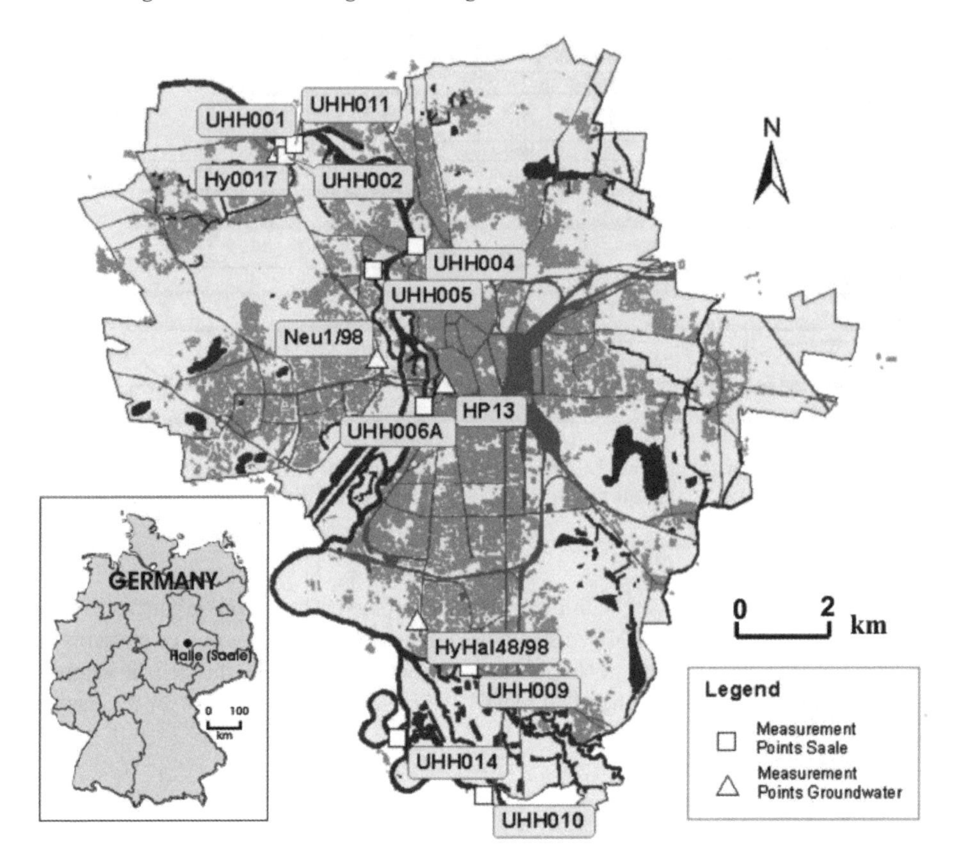

Figure 3. City of Halle an der Saale with sampling sites and location in Germany.

of approximately 135 km^2. The city is located on the River Saale with a mean discharge of $8.5 \times 10^6 m^3/d$ (Gauge Halle-Trotha). The flow direction is from south to north. In the south of the city, Weisse Elster River flows into Saale River. Weisse Elster River is downstream of the city of Leipzig (about 450,000 inhabitants) and has a mean discharge of $2.1 \times 10^5 m^3/d$ (Gauge Oberthau).

As is illustrated on Figure 4, the subsurface of Halle an der Saale is highly complex. There are more than 16 aquifers in the city area. Most important are the Muschelkalk, Zechstein (neither situated in the cross section shown in Figure 4), Buntsandstein and Rotliegendes, as well as the Quaternary and Tertiary sediments. Within the flood plains lies a valley aquifer comprising gravels and sands of the Niederterasse. An interaction between groundwater and surface water of the Saale River can be expected at these areas.

The city of Halle an der Saale has a number of artificial discharges including the Halle-Nord sewage plant and some combined sewer overflows. Trettin (2004) estimated from a simple water balance, that the inflow to the Saale River from these sources is less than 2% of its discharge.

Sewage treatment in the city of Halle an der Saale occurs centrally at one of the most advanced sewage plants (SP) in Europe (year of completion 1998); max. capacity of

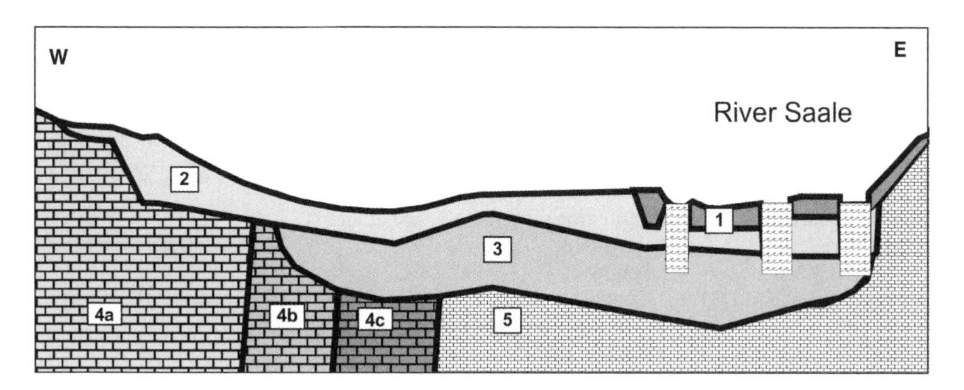

Figure 4. Geological cross section through the centre of the city of Halle (Saale), 1 = Anthropogenic fill, 2 = Alluvial loam, 3 = Quaternary aquifer, 4a, 4b, 4c = Upper, middle, lower Buntsandstein,

75,200 m^3/d), which is located in the north of Halle an der Saale (see Figure 3, UHH011). The water outflow from the SP averages $2.9 \times 10^4 m^3/d$, which is approximately 0.34% of the mean discharge of the Saale River.

Water sampling of the Saale River (see Figure 3) was carried out at a total of 9 sites (UHH001, UHH002, UHH004, UHH005, UHH006A, UHH009, UHH010, UHH011, UHH014) along the river. At these sites, 4 sampling rounds have been carried out. The sampling results presented here are for the years 2002/03 (14/2/02, 16/7/02, 25/3/03, 8/9/03). The groundwater sampling was carried out at 8 monitoring points. Here the results are shown for 4 sites (see Figure 3: HyHal48/98, HP13, Neu1/98, Hy0017) sampled 24/3/04.

2.3 *Indicators*

To evaluate anthropogenic effects on the urban water system, chemical indicators are used. The criteria for selecting good indicators are as follows:

- They should be representative of the type of anthropogenic emission, e.g. hospital waste or residential run-off.
- They should be persistent in the environment.
- They should be readily transported with the water flow.
- They should either have an important impact on human health or the ecosystem, or act as a suitable surrogate for chemicals which are important for human or ecosystem health.
- Their analysis should not be too labour-intensive if possible and the demands of sampling should be no more than moderate.

Based on these criteria the following indicators were selected (Table 1). They are described according to their environmental properties.

Bisphenol A (BPA; 2,2′-Bis-(4-hydroxophenyl)propane) can be degraded relatively quickly in water, sludge or soils by aerobic biological processes (Dorn et al., 1987; Lobos et al., 1992). Anaerobic degradation is presently unknown. Solubility in water is 120–360 mg/l. The acute

Table 1. Selected indicators for tracing of pharmaceuticals, endocrine disruptors and fragrances.

Agent/indicator	Effect/usage	K_{ow}	Detection limits in ng/l
Bisphenol-A (BPA)	Endocrine disruptor	2.2–3.8	0.05
Carbamacepin (CA)	Antiepileptic drug	1.58	2
t-Nonylphenol (NP)	Endocrine disruptor (Xenoestrogen)	4.48	25
Galaxolide (HHCB)	Fragrance	6–7	0.2
Tonalide (AHTN)	Fragrance	6–7	0.2
Gadolinium (Gd); Gd-DTPA	Contrast medium	Unknown	190
Nitrate (NO_3^-)	Fertilizer, sewage constituent	–	1×10^4
Sulphate (SO_4^{2-})	Atmospheric deposition, geogenic	–	5×10^4

toxicity of BPA was determined in fish to range from about 3–5 mg/l and in endotherms (rats, mice) from a dose of about 1.5 mg/kg body mass. BPA can cause reversible skin or mucous membrane irritations and allergic reactions in humans, and both photo- as well as cross-sensitization, for example with diethylstilboestrol (DES) have been found (BUA, 1997; Chahoud et al., 2001).

Carbamacepin (CA; Carbamacepin(10,11) epoxide as the relevant substance) is persistent. A higher elimination of CA can be found, if at all, under aerobic settings. Degradation depends also on the content of total organic carbon (TOC) (Preuß et al., 2001). Natural retention in anaerobic parts of the aquifer can occur only by sorption (Mersmann, 2003). With the help of sorption isotherms, determined in batch experiments, it has been shown that the sorption of CA strongly depends on the organic carbon content of the sediments. According to present results, CA shows degradation and a relatively high mobility (Mersmann, 2003). The logarithm of the octanol/water partition coefficient log K_{ow} is small in comparison with the other indicator substances.

Technical (t-)Nonylphenols (NP; a mixture of various isomers with 4-nonylphenol as the main constituent) have a moderate bio-accumulation compared to the other indicators used. Bio-concentration factors (BCF) are 10–1,300 for fish, 2,000 for mussels, and 10,000 for algae. Persistence is relatively low. NP is inherently biodegradable in sewage treatment plants. The known estimated half lives are 300 days in soil, 150 days in surface water, and 0.3 days in the atmosphere. 96-hour LC50 toxicity in fish was estimated to be 0.128 mg/l. The NOEC (no observable effect concentration) for a 33-day fish embryo study was 0.0074 mg/l.

Fragrances such as Galaxolide (HHCB; 1,3,4,6,7,8-hexahydro-4,6,6,7,8,8-hexamethyl-cyclopenta-gamma-2-benzopyran) and Tonalide (AHTN; 7-Acetyl-1,1,3,4,4,6-hexamethyl tetrahydronaphthalene) are currently in widespread use in personal care and household cleaning products. They are regarded as persistent and readily bio-accumulate. Some fragrances, especially those containing nitrosating agents are suspected of being carcinogenic (Balk et al., 1999).

Gadolinium (Gd; Rare Earth Element (REE)), especially as Gd-DTPA, has been used as a contrast medium in magnetic resonance tomography (MRT) for years. Once used, Gd is excreted un-metabolised in urine. Due to its chemical persistence and high water solubility it readily passes through sewage treatment plants and contaminates surface waters. Its presence is recognised by an increase in the total Gd-concentration that creates a so-called positive Gd-anomaly, e.g. a significantly increased concentration of Gd compared to the North American shale composite (NASC) values. The fact that Gd chelates are very stable and mobile can be useful for the purpose of tracing transport paths in aquatic systems (Möller et al., 2000).

Nitrate (NO_3^-) is the highest oxidized ion within the nitrogen cycle and is an important constituent of natural and artificial fertilizers. Most nitrate found in water is derived from dissolved fertilizer, or enters more naturally via nitrification of organic matter in the soil. The isotopic composition of nitrate in terms of ^{15}N and ^{18}O allows its origin and transformation processes such as denitrification and mixing to be identified. The concentration limit of nitrate in drinking water is set at 50 mg/l in Germany because of the toxic and carcinogenic effects to humans.

Sulphate (SO_4^{2-}) is one of the main dissolved constituents in the study area waters due to the presence of sediments rich in evaporitic sulphate in the catchment area and due to former open cast mining activities upgradient of the urban area. The stable isotopes ^{34}S and ^{18}O allow the source of the sulphate to be determined and can be used as a tracer of groundwater–river water interaction.

2.4 *Analytical methods*

Pharmaceutical analyses were carried out as follows:

1. Solid phase extraction (SPE) and clean-up for sample preparation

One litre water samples (glass fiber filtered, 52 μm) were acidified to pH = 3 and spiked with an internal standard mixture. Commercial SPE cartridges (EASY Chromabond, 6 ml) were conditioned with 2 × 6 ml methanol and 6 ml pure water (double distilled, pH = 3). After sample addition at a flow rate of approximately 10 ml/min, the cartridges were dried for 2 hours with inert gas. The analytes were eluted with 2 × 6 ml methanol. The eluates were unified, evaporated to dryness and redissolved in 200 μl n-hexane for GC-MS measurement. In case of a high matrix load, an additional clean up procedure with silica gel is needed. The eluates (10 ml n-hexane/acetone mixture (65/35; v/v)) were evaporated to a final volume of 200 μl for GC-MS analysis.

2. Derivatization

After fullscan GC-MS analysis (determination of musk compounds) the eluate was evaporated to dryness. One ml acetone, 10% aqueous potassium carbonate (100 μl) and acetone solution of 5% PFBBr reagent (100 μl) were added and allowed to react at 60°C for 1 hour. After cooling 1 ml n-hexane was added, shaken for 30 s and washed with 0.5 ml pure water. The organic phase was separated, dried over sodium sulphate and reduced to 200 μl for GC-negative chemical ionization (NCI)-MS measurements (Nakamura et al., 2000).

3. GC-MS analysis

Gas chromatographic separation was performed with a "HP 6890 series" gas chromatograph equipped with a fused silica capillary column "HP-5MS", 30 m × 0.25 mm id, film thickness: 0.25 μm (Agilent Technologies). The mass spectrometric analysis was undertaken with a mass selective detector 5973 (Agilent Technologies, San José, USA) operated with electron impact (EI, 70 eV) as well as negative chemical ionization (NCI) using methane as a reactant gas.

One μl was injected at 280°C injection temperature. The GC temperature program started at 50°C and the oven temperature was raised to 280°C at a rate of 10°C/minute. The

MS transfer line and the ion source temperatures were set at 200°C and 280°C, respectively (EI mode, 70 eV). Methane was used as NCI reagent gas with a flow rate of 2.5 ml/min.

The Rare Earth Element (REE) analysis was carried out as follows:

The surface water samples were filtered (0.45 μm) and acidified (HNO_3, pH = 2). The chemical analysis of Gd and the other REE were carried out, after appropriate enrichment using ethylhexylphosphates, with an ICP–MS ELAN 5000 (Perkin-Elmer, Sciex) as described by Hennebrüder et al. (2004). The detection limit in 1 M HNO_3^- solution was 90 ng/l and 190 ng/l in natural water.

Stable isotope analysis of nitrate and sulphate were performed as following:

Nitrate in the water samples was separated by ion exchange onto anion resin XAD 8 (BioRad), eluted, neutralized with silver oxide and freeze dried (Chang et al., 1999). The silver nitrate was converted to CO and N_2 by high temperature pyrolysis and measured with a ConFlow-Isotope Ratio Mass spectrometer XL-Plus (Thermo Finnigan). Sulphate was precipitated from the original sample by adding saturated $Ba(OH)_2$ solution, filtered, washed neutral, and dried. The solid $BaSO_4$ was converted to SO_2 and its $^{34}S/^{32}S$ ratio was measured with an Isotope Ratio Mass spectrometer delta S (Thermo Finnigan). The $^{15}N/^{14}N$ and $^{34}S/^{32}S$ ratios of the samples were expressed in the δ-notation (Gehre et al., 2002) in relation to the internationally accepted standards – "air" (N) and "Canyon Diabolo Troilite" CDT (S).

3 RESULTS AND DISCUSSION

3.1 *Examination of the concentration and isotopic measurements*

Nitrate and sulphate concentrations, along with their stable isotopic composition, were studied in the Saale River and its tributary, the Weisse Elster River as they passed through the city (Figure 5). In addition, groundwater samples were taken from sites close to the Saale River as well as at locations in the city center and the city outskirts.

Nitrate dissolved in the river water is associated with at least two distinct sources. At the outflow from the sewage plant, treated waste water enters the River Saale with enriched $\delta^{15}N$ values of >12‰. Such values are typical for septic waste waters (Wassenaar, 1999) where ammonia degassing causes an enrichment of ^{15}N and the residual nitrogen component is subsequently oxidized to nitrate. The samples at the outflow from the sewage plant are characterised by generally higher concentrations of nitrate in relation to the river water.

In contrast to the waste water signature, the nitrate of both rivers varies seasonally in $\delta^{15}N$ between +6‰ in winter and +12‰ in summer and from 40 to 15 mg/l in nitrate concentration, respectively. These values are typical for nitrate input from rural catchments (Mayer et al., 2002). Denitrification processes in the hyporheic zone of rivers during the summer season (Sebilo et al., 2003) most likely contribute to an enrichment in $\delta^{15}N$ of nitrate at low concentration, as observed during the hot summer of 2003. The seasonal variations are small compared to the difference between the waste water and the river water signature in Figure 5a. However, these temporal variations are not correlated to urban activities and therefore constrain the use of nitrate and the ^{15}N isotope ratio for investigating anthropogenic effects on urban river water.

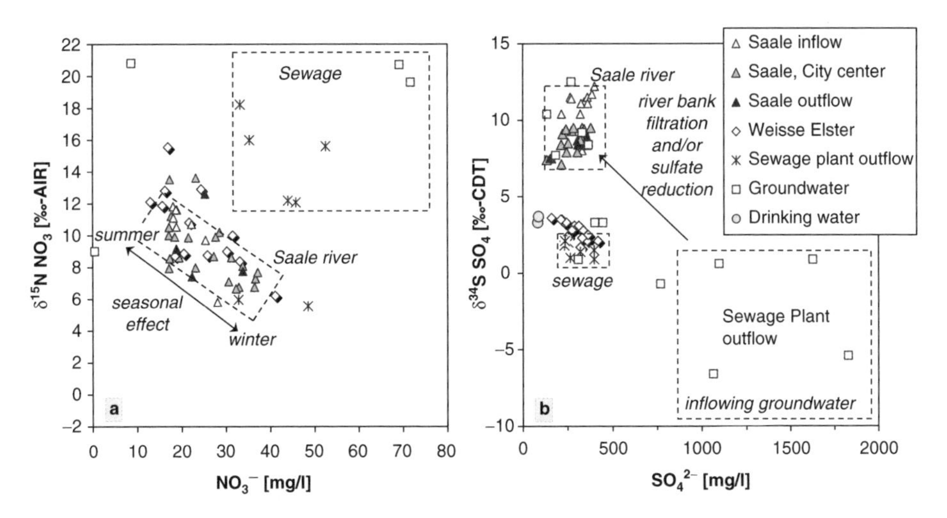

Figure 5. Different sources of nitrate and sulphate in groundwater and water samples from the Rivers Saale and Weisse Elster as indicated by $\delta^{15}N$ (a) and $\delta^{34}S$ (b). In (a) the seasonal variation of $\delta^{15}N\ NO_3^-$ in the Saale River is clearly distinguishable from sewage plant water as well as from groundwater influenced by the latter. In (b) an even greater range of values is covered by the distinct sources of sulphate allowing one to trace and presumably quantify groundwater–river water interactions.

These limitations do not apply to the urban groundwater investigations. In Figure 5a the $^{15}N\ NO_3^-$ signatures of most of the groundwater samples are apparently influenced by nitrate from the urban sewage water system. If the nitrate concentrations and isotopic signatures of the potential sources (e.g. leaking water from the sewer system) are known from measurements, the nitrate stable isotopes can provide an efficient and sensitive indicator for inflow of urban contaminants into the groundwater.

The Saale and Weisse Elster Rivers show a comparable range of sulphate concentrations. In contrast, sharply different stable isotopic compositions in Figure 5b clearly confirm distinctive sources of sulphate within each of the river catchments. Sulphate in the Saale River is dominated by leaching of Triassic evaporites, whereas sulphate in the Tertiary Weisse Elster basin stems predominantly from anthropogenic sources (pyrite oxidation, atmospheric deposition). Temporal variations of sulphate and $\delta^{34}S$ in the river waters are small. The difference between Saale inflow and Saale city center is a result of the confluence of Saale and Weisse Elster and is in agreement with the respective discharge values.

Most urban groundwater samples show higher sulphate concentrations with simultaneously depleted $\delta^{34}S$ values, which most likely is related to the remnants of lignite mining and other anthropogenic activities in the presumed recharge area (Asmussen and Strauch, 1998). Although the outflowing water from the sewage plant reveals depleted isotopic values similar to the groundwater samples, it is characterised by much lower sulphate concentrations and thus can easily be distinguished from groundwater as well as from water in the River Saale. In contrast, sulphate and $\delta^{34}S$ of sulphate in several groundwater samples from the Quaternary aquifer in the vicinity of the Saale River are very similar to those in Saale River water. While this can be attributed to bank filtration of river water into the groundwater, sulphate reduction, causing an enrichment of $\delta^{34}S$ and a lower sulphate concentration, cannot be ruled out completely.

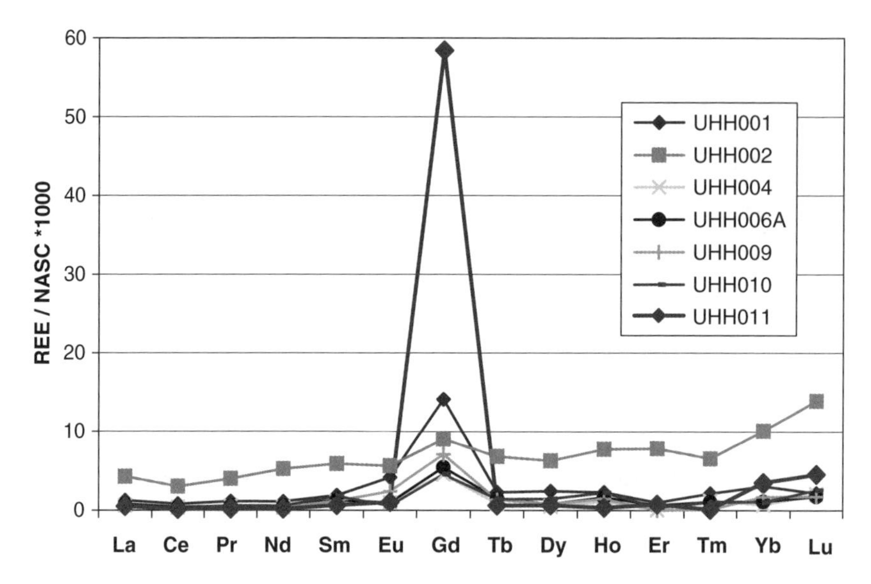

Figure 6. Rare Earth Elements (REE) distribution with the Gadolinium anomaly in the River Saale as it passes through the city. Results were obtained in one sampling campaign in July 2004.

The contrasts in sulphate concentration and sulphate isotopic composition demonstrate their potential for identifying and quantifying the transfer of urban-sourced contaminants between groundwater and surface water.

As indicated above, the inflow of Gd-DTPA into the rivers as they pass through the city was also investigated. For this purpose concentrations of REE, which behave similarly in the aquatic environment, were analysed. Figure 6 shows the concentrations of REE at various sampling locations. A significant increase in the Gd concentration is apparent immediately downstream of the SP (location UHH011). Most of the Gd is bounded in a dissolved complex but is mobile and can be used as an indicator for the emergence of Gd-DTPA from the sewage plant.

Figure 7 shows the concentration of several pharmaceuticals and fragrances in the river water and in the groundwater for the sampling round in March 2004. BPA was found in all Saale River water samples. Concentrations tend to increase along the river within the city (see UHH009, UHH006A, UHH005, and UHH001). A similar trend was observed for NP. The concentration of CA decreased downstream while the fragrances showed no observable trend. Only HHCB showed a slight rise downstream.

BPA and NP were found in groundwater at higher concentrations than observed in the River Saale. However, current observations are limited to the centre and the north of the city (see Neu1/98 and Hy0017) which are characterised by a high percentage by old infrastructure. In the south, where more modern urban infrastructures are situated, only low concentrations of BPA and NP were found (see HyHal48/98 and HP13).

3.2 *The water balance and chemical load calculations*

A mass balance of the water-transported xenobiotics was calculated for the whole city area. Long-term mean values were used for this first evaluation. The calculation began with a

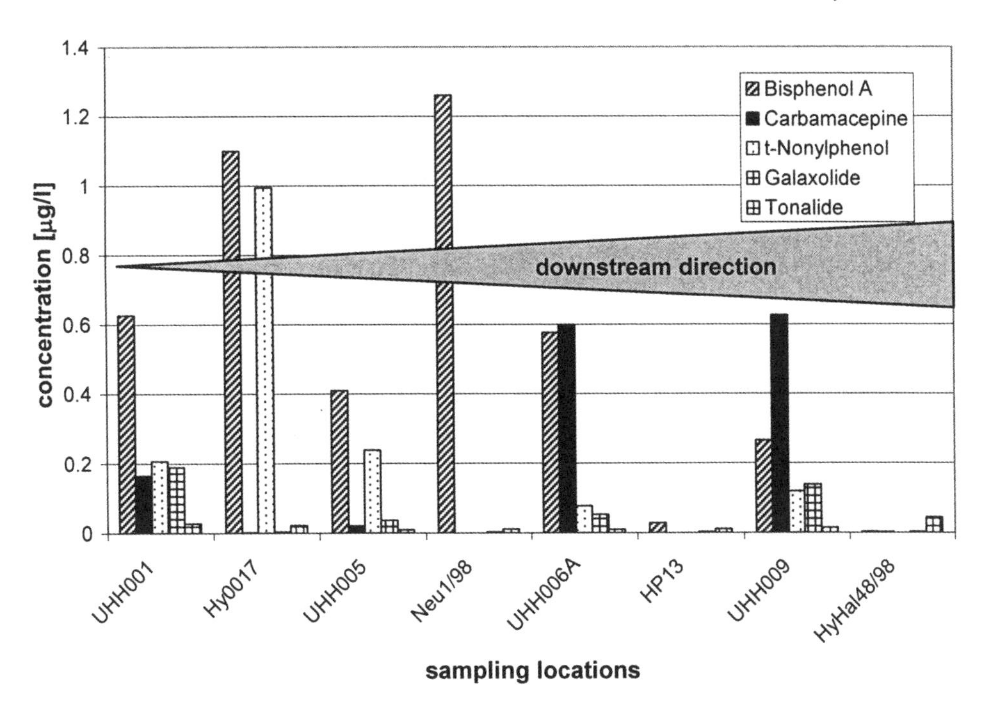

Figure 7. Concentration of xenobiotics [µg/l] at selected sampling sites in groundwater (HyHal48/98, HP13, Neu1/98, and Hy0017) and surface water (UHH001, UHH005, UHH006A, and UHH009) in

water balance and subsequently incorporated chemical concentrations that enabled chemical fluxes to be calculated. A river structure scheme is shown in Figure 8.

The water balance equation is given by:

$$MQ_{out}(UHH001) = MQ_{in}(UHH010/014) + MQ_{in}(UHH009) + MQ_{in,diffuse} + MQ_{in}(UHH011) + MQ_{in}(UHH002) \quad (1)$$

where MQ [m^3/s] = long-term mean discharge of the river/brook; in/out = inflow/outflow of the balance area; and UHHXYZ = the sampling location XYZ.

The only gauge in the study area is located in Halle-Trotha. It is reasonably close to the sampling point UHH001 and can be regarded as representative for this part of the river. To calculate the diffuse groundwater inflow within the city area (also known as the baseflow contribution), a hydrological generation model was applied. Using a grid size of $1{,}000 \times 1{,}000$ m average annual values for key inputs and outputs such as precipitation and evaporation were used to determine a mean diffuse recharge of 2.8 l/s/km^2. This translates to a diffuse inflow to the River Saale of approximately 0.86 m^3/s, which in turn corresponds to a mean groundwater recharge of approximately 66 mm/yr. Due to uncertainties in the model parameters, particularly the urban recharge, it will be necessary to verify this diffuse inflow by independent methods. This will be carried out using mixing calculations

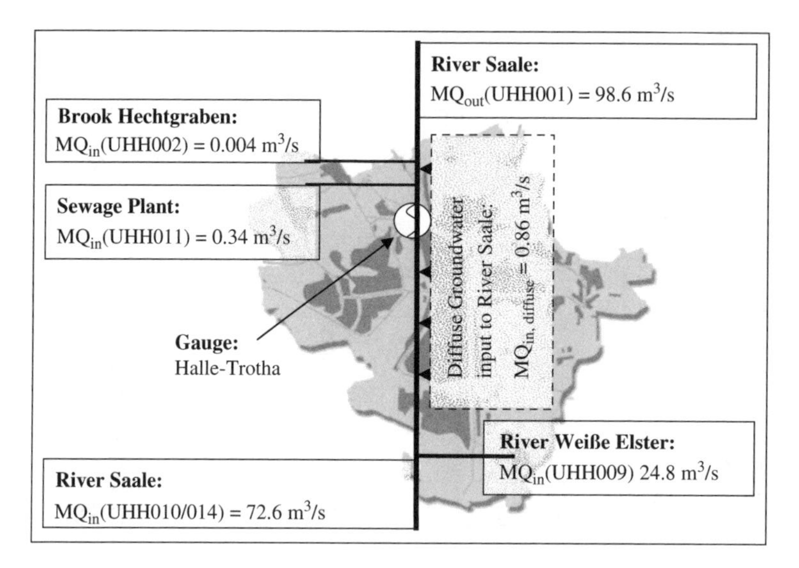

Figure 8. Scheme of the rivers and main inflows to the River Saale within the city of Halle at different river locations.

based on ^{34}S-sulphate analyses. The calculated balance values of the discharges and diffuse inflow are summarised in Figure 8.

In future, the fluxes will be more precisely estimated and incorporate seasonal variations. With the calculated mean annual water flows, the load determination was performed for the indicator substances. The calculation was carried out using mean chemical concentrations derived from the four sampling campaigns. The chemical mass loadings were calculated using the equation:

$$MF_i = \bar{c}_i * MQ * 31.536 \tag{2}$$

where MF_i [kg/yr] = mean load of the substance i over the observation period, $\bar{c}_i$ [μg/l] = mean concentration of the substance i in the observation period; and MQ [m³/s] = long-term mean discharge at the corresponding river section. The factor 31.536 is due to the use of mixed units: 1 μg/l = 1×10^{-6} kg/m³, 1 m³/s = 31,536,000 m³/yr, i.e. 1 μg/l × 1 m³/s = 31,536 kg/yr.

Table 2 shows the mean concentrations of the indicator substances, the mean discharges at the sampling locations and the chemical mean loads. The Weisse Elster River which passes through Leipzig, the neighbouring city, contributes only 30% of the flow in the Saale River but appears to be a major source of its contaminants.

Downstream of its confluence with the Weisse Elster River, the mass transport of xenobiotics in the River Saale rises significantly. Mass loadings of endocrine disruptors (BPA and NP) and the antiepileptic (CA) are of greatest concern.

By aggregating the loads at UHH010/014, UHH009 and UHH011 and comparing this total to the load at UHH001, the contribution of the city can be calculated (Table 3).

Table 2. Indicator substance loads in the Saale River system in the city of Halle an der Saale.

Substance	Mean concentration in the River Saale [ng/l]	Mean discharge of the River Saale [m^3/s]	Mean load of the River Saale [kg/yr]
River Saale (UHH010/014) (upstream of the city)			
Bisphenol-A	209.7		480.1
Carbamacepine	102.2		234.0
t-Nonylphenol	124.0	72.6	283.8
Galaxolide	34.9		79.9
Tonalide	5.3		12.1
River Weisse Elster (UHH009) (mouth)			
Bisphenol-A	255.0		199.4
Carbamacepine	494.0		386.4
t-Nonylphenol	275.9	24.8	215.8
Galaxolide	62.1		48.5
Tonalide	32.8		25.7
Sewage plant (UHH011)			
Bisphenol-A	192.4		2.1
Carbamacepine	1686.4		18.1
t-Nonylphenol	219.7	0.34	2.4
Galaxolide	1814.0		19.5
Tonalide	165.9		1.8
River Saale (UHH001) (downstream of the city)			
Bisphenol-A	214.8		668.0
Carbamacepin	183.2		569.6
t-Nonylphenol	170.4	98.6	529.7
Galaxolide	135.3		420.8
Tonalide	50.2		156.2

Table 3. Contaminant contributions by the city of Halle an der Saale.

	UHH 010/014	UHH 009	UHH 011	Sum of UHH010/014, UHH009, UHH011	UHH 001	Difference [%]	Trends of the loads
BPA	480.1	199.4	2.1	681.6	668.0	−2.0	Stable
CA	234.0	386.4	18.1	638.4	569.6	−10.8	Stable
NP	283.8	215.8	2.4	501.9	529.7	5.5	Stable
HHCB	79.9	48.5	19.5	147.9	420.8	184.5	Increasing
AHTN	12.1	25.7	1.8	39.6	156.2	294.5	Increasing

The chemical loading of HHCB increases by 272.9 kg/yr and AHTN increases by 117.8 kg/yr. This trend is not observed for the other xenobiotics investigated which remain nearly constant: BPA −2%, CA −10.8% and NP +5.5%. It can be concluded either the city has minimal influence on the level of these xenobiotics in the river or that dilution (e.g. by diffuse groundwater inflow) or degradation processes within the river mask the influence of the city.

In terms of the analysis, it should also be noted that the high temporal variability of the hydrological regime and chemical transport brings uncertainty into the calculations. Realistically, only 4 measurements are insufficient for a final evaluation of the chemical transport and a definitive interpretation of the chemical trends within the city. Therefore, further measurements will be undertaken in future and the number of sampling locations will be increased. Also, the processes responsible for biological degradation, chemical precipitation, sorption, and dilution have to be studied intensively in detail at selected river sections.

4 CONCLUSIONS

The purpose of this research was to investigate the occurrence and behaviour of xenobiotics in urban water systems, with a particular focus on transport fluxes. In the city of Halle an der Saale, a number of xenobiotics and trace compounds were detected in the river water, sewage water and groundwater. The substances investigated are indicators of anthropogenic impact and the majority appear to be introduced upstream via the River Weisse Elster which flows through Leipzig.

The fate of the xenobiotics and the metabolic effects remain uncertain; almost no information currently exists with regard to the risk potential for ecosystem and human health.

Using stable isotopes of sulphur and nitrogen, supported by hydrochemical analyses, the sources of the primary contaminants have been identified. Enriched $\delta^{15}N$ nitrate in groundwater is associated with leakage from the urban sewer system. Changes in the sulphur isotopic composition of river water during its passage through the city can be attributed to the diffuse inflow of urban groundwater to the River Saale. In future, the study will be improved through isotopic analysis of other dissolved compounds such as inorganic carbon.

Contaminant loading calculations suggest that the majority of the xenobiotics, notably the endocrine disruptors (BPA and NP) and the antiepileptic (CA) are introduced via the River Weisse Elster which, upstream, passes through the city of Leipzig. Loadings of the fragrances Galaxolide and Tonalide appear to increase as the River Saale passes through the city. The work demonstrates the mobility and potential threat of xenobiotics in the urban environment; however, further work is required to document the potential impacts in detail.

REFERENCES

Asmussen, G. and Strauch, G. 1998. Sulphate reduction in a flooded lignite mining pit by stable sulfur and carbon isotopes. Water, Air and Soil Pollution, 108, 3–4, 271–284.

Balk, F. and Ford, R.A. 1999. Environmental risk assessment for the polycyclic musks AHTN and HHCB in the EU I. Fat and exposure assessment. Toxicol. Lett., 111, 57–79.

BUA. 1997. BUA – Beratergremium für umweltrelevante Altstoffe der Gesellschaft Deutscher Chemiker (ed.) Bisphenol A, BUA-Stoffbericht 203. Stuttgart: Hirzel.

Chahoud, I., Gies, A., Paul, M., Schönfelder, G. and Talsness, C. (ed.). 2001. Bisphenol A: Low Dose Effects–High Dose Effects. Abstracts of a symposium NOV 18–20 2000 in Berlin, Germany.

Chang, C.C.Y., Langstone, J., Riggs, M., Campbell, D.H., Silva, S.R. and Kendall, C. 1999. A method for nitrate collection for $\delta^{15}N$ and $\delta^{18}O$ analysis from water with low nitrate concentrations. Can. J. Fish. Aquat. Sci., 56, 1856–1864.

Dorn, P.B., Chou, C.S. and Gentempo, J.J. 1987. Degradation of Bisphenol A in Natural Waters. Chemosphere, 16, 1501–1507.

Gehre, M., Kowski, P., Städter, W. and Knöller, K. 2000. Guidelines LSI for EA and HTP isotope analyses (HCNOS), Laboratory of Stabile Isotopes, Centre for Environmental Research: Leipzig, 2000.

Heemcken, O.P., Reincke, H., Stachel, B. and Theobald, N. 2001. The occurrence of xenoestrogens in the Elbe river and the Noth sea. Chemosphere, 45, 245–259.

Hennebrüder, K., Wennrich, R., Mattusch, J., Stärk, H.J. and Engewald, W. 2004. Determination of gadolinium in river water by SPE preconcentration and ICP-MS. Talanta, 63, 309–316.

Lobos, J.H., Leib, T.K. and Su, T.M. 1992. Biodegradation of Bisphenol A and other Bisphenols by a Gram-Negative Aerobic Bacterium. Appl. Environ. Microbiol., 58/6, 1823–1831.

Mayer, B., Boyer, E.W., Goodale, C., Jaworski, N.A., Howarth, R.W. and Seitzinger, S. 2002. Sources of nitrate in rivers draining sixteen watersheds in the northeastern US: Isotopic constraints. Biogeochemistry, 57, 171–197.

Mersmann, P. 2003. Transport- und Sorptionsverhalten der Arzneimittelwirkstoffe Carbamazepin, Clofibrinsäure, Diclofenac, Ibuprofen und Propyphenazon in der wassergesättigten und -ungesättigten Zone. Dissertation, Institut für Angewandte Geowissenschaften der Technischen Universität Berlin, Fakultät VI · Bauingenieurwesen und Angewandte Geowissenschaften.

Möller, P., Dulski, P., Bau, M., Knappe, A., Pekdeger, A. and Sommer von Jarmersted, C. 2000. Anthropogenic gadolinium as a conservative tracer in hydrology. Journal of Geochemical Exploration, 69–70, 409–414.

Nakamura, S., Takino, M. and Daishima, S. 2000. Determination of Chlorophenols, Bisphenol A and 17 Beta-estradiol by Gas Chromatography/Negative-Ion Chemical-Ionization Mass Spectrometry. Bunseki Kagaku, 49, 3, 181–187.

Preuß, G., Willme, U. and Zullei-Seibert, N. 2001. Verhalten ausgewählter Arzneimittel bei der künstlichen Grundwasseranreicherung – Eliminierung und Effekte auf die mikrobielle Besiedlung; Acta Hydrochim. Hydrobiol., 29(5), 269–277; Weinheim (Wiley).

Ricking, M., Schwarzbauer, J. and Franke, S. 2003. Molecular markers of anthropogenic activity in sediments of the Havle and Spree rivers (Germany). Water Research, 37, 2607–2617.

Sebilo, M., Billen, G., Grably, M. and Mariotti, A. 2003. Isotopic composition of nitrate-nitrogen as a marker of riparian and benthic denitrification at the scale of the whole Seine River system. Biogeochemistry, 63, 35–51.

Stachel, B., Ehrhorn, U., Heemken, O.P., Lepom, P., Reincke, H., Sawal, G. and Theobald, N. 2003. Xenoestrogens in the River Elbe and its tributaries. Environmental Pollution, 124, 497–507.

Ternes, Th.A. 1998. Occurrence of drugs in german sewage treatment plants and rivers. Water Research, 32(11), 3245–3260; Great Britain (Elsevier).

Trettin, R. 2004. pers. communication, UFZ-Centre for Environmental Research Leipzig-Halle, Dept. Hydrogeology.

Wassenaar, L.I. 1995. Evaluation of the origin and fate of nitrate in the Abbotsford aquifer using the isotopes of ^{15}N and ^{18}O in NO_3^-. Appl. Geochemistry, 10, 391–405.

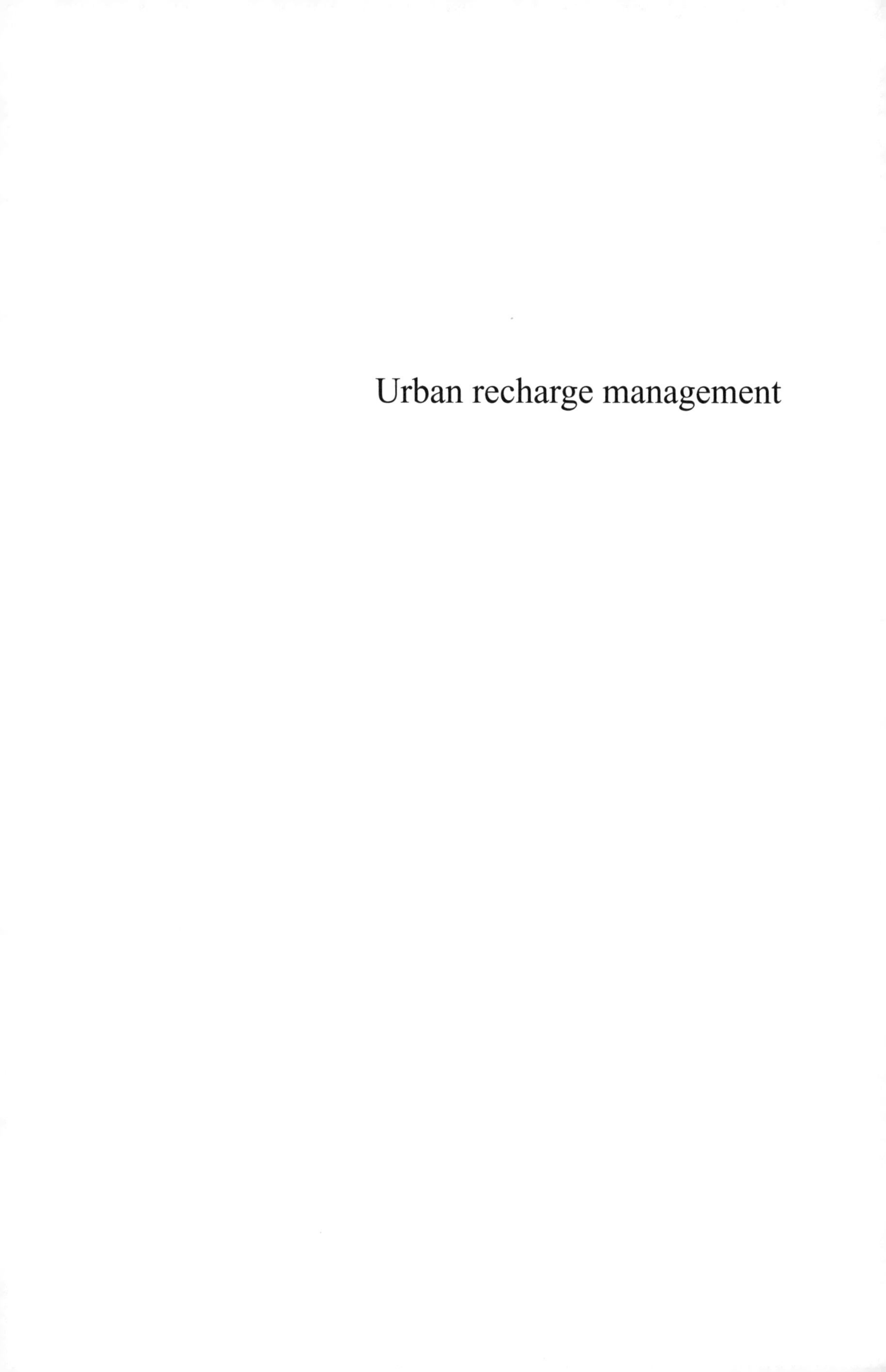

Urban recharge management

CHAPTER 12

Conjunctive use of urban surface water and groundwater for improved urban water security

Peter Dillon, Simon Toze and Paul Pavelic
CSIRO Land and Water, Australia

ABSTRACT: In arid and semi-arid areas, water security for cities can be enhanced by substituting stormwater and reclaimed water (initially) for non-drinking uses of conventional potable supplies drawn from stressed catchments or aquifers. Intentional recharge of aquifers from these urban water sources has been practiced in two Australian cities: Adelaide and Perth. This is analogous to the unintentional recharge that occurs in most cities and demonstrates the capacity for aquifers to store and passively treat water on a sustainable basis. Research projects have been undertaken to define the changes in the quality of stored water and the sustainability of contaminant and pathogen attenuation processes, and thereby identify the highest valued uses of the recovered water. Key findings from two of these projects (Andrews Farm and Halls Head) are presented. With climate change, population growth and limits on yields of catchments and aquifers, sustainable sources of water are needed for cities. Aquifers, even saline ones, can have a role in improving the security of urban water supplies and increasing the value of uses of water from all sources.

1 INTRODUCTION

Intentional groundwater recharge (via recharge wells and infiltration basins) has been undertaken in two Australian cities, Adelaide and Perth, for many years. These experiences have yielded lessons about technologies, economics, planning, hydrogeological mapping and investigations in urban areas, environmental and health regulation, water policies, system integration, training, and the effort required to develop enabling institutional arrangements. This paper shares these experiences with the aim of assisting other water-short cities to implement conjunctive use systems that are appropriate and sustainable. Of necessity, this demands a holistic approach to water resource management, and hydrogeologists have a key role in identifying innovative options for cities and undertaking the necessary risk assessments.

Stormwater runoff generated in urban areas commonly has a high peak flow rate and volume of discharge in comparison with the equivalent storms in similar rural areas. Urban runoff persists for only short periods following storms, contains a variety of contaminants, and is rarely seen as an urban water resource. Where there are separate sewerage and stormwater systems, treated sewage effluent has relatively uniform flow rate throughout the year, and is usually regarded as having poor quality with respect to standards for water supply expected in urban areas. By recharging aquifers with these waters, conjunctive use of groundwater and these urban surface waters is possible making use of an aquifer's capacity to store and passively treat water. However the opportunity to do this is limited by:

- lack of information about the quality of source water,
- lack of information on the sustainable capacity for contaminant attenuation in aquifers, and
- costs of monitoring to ensure compliance with water quality objectives.

However these gaps are not insurmountable and several concepts are becoming an important basis for water resources management and allow some jurisdictions a broader perspective on water supply.

Firstly, a pro-active risk assessment is required which identifies potential public health and environmental risks. Using a Hazard Analysis and Critical Control Points (HACCP) plan such risks can be managed through education, training, and licensing as well as structural solutions including flow control systems and additional water quality treatment (Nadebaum et al., 2004). An HACCP approach builds on an improved understanding of the sources of contaminants, the incidents by which they could potentially enter the surface water system and implementing procedures to avert this. It also deals with contingency plans in the event of the multiple-barriers being breached. Monitoring is needed to demonstrate compliance but is not the primary means of assuring water quality. Treatment within the surface water system and during groundwater recharge and subsurface storage (e.g. Dillon et al., 2005) are also taken into account.

Secondly, a groundwater protection policy is required. For example in Australia, under the National Water Quality Management Strategy (ARMCANZ and ANZECC, 1995), a differential protection policy is adopted to protect aquifers and ensure their initial beneficial uses (environmental values) are preserved or enhanced. Different aquifers may be protected as drinking water supplies, irrigation supplies or for ecosystem support, or various combinations depending on the quality of their native groundwater and taking account of historical use. The water quality requirements for recharge water depend on the environmental values to be protected. In addition, allowing for an attenuation zone in the vicinity of recharge locations based on the sustainable attenuation capacity of aquifers, reduces the need for pre-treatment.

Thirdly, water allocation processes and ownership need to be well-defined. Investments in groundwater recharge and water supply infrastructure can be secured when rights to harvest surface waters are assigned. Similarly for water stored in aquifers, proponents need assurance that they have an entitlement to recover a reasonable proportion of the water they have recharged, and that it will not be extracted by other groundwater users who have not contributed to the costs of recharge (Pitman, 2003).

Finally, the storage capacity of any aquifer is finite and the allocation of storage capacity to recharge projects, while it will take place chronologically, needs also to take into account equity and a long term aquifer management plan that accounts for quality, sustainable yield, and any limitations on hydraulic heads to satisfy geotechnical or environmental requirements.

In the cases that follow, none of these proposed policies have yet to be fully implemented, most are under development, and some have yet to be considered even at a preliminary level by regulatory agencies. In spite of these policy and institutional impediments, projects in Adelaide and Perth have been successfully established due to the imperatives of producing economic water supplies, or of discharging waste waters affordably and in an environmentally benign way. However, without such policies demonstration projects such as these cannot be expected to proliferate.

2 CONJUNCTIVE USE OF SURFACE WATER AND GROUNDWATER IN ADELAIDE

Adelaide is a city of 1 million people situated on a coastal plain in South Australia and has a Mediterranean climate with an average annual rainfall of 500 mm and evaporation of 1800 mm. Its water supply (~200 million m^3/yr) comes from water-stressed catchments dominated by agricultural and pastoral land use, and is treated in water filtration plants before supply to the city. The sewage system discharges (~90 million m^3/yr) via four sewage treatment plants along the coast, which previously discharged all treated effluent into the Gulf of St Vincent. Fifteen percent of this water is now reclaimed and used to irrigate horticulture and viticulture (Radcliffe, 2004). A separate stormwater system discharges approximately160 million m^3/yr to the Gulf, with 3 to 5 million m^3/yr of rainfall being captured and used in the 38% of dwellings that have rainwater tanks (South Australian Government, 2004). Low yielding shallow alluvial aquifers, interbedded with surficial clays, have variable salinity and in the minor areas where groundwater is sufficiently fresh, such as near streams, a small amount is used for irrigation from less than 2000 wells. Recharge to the Quaternary alluvium is thought to be on the order of 10 million m^3/yr, with negligible intentional recharge. In coastal areas saline shallow groundwater ingresses leaky sewage pipes, restricting the potential for productive use of reclaimed water (Bramley et al., 2000). Two deeper confined Tertiary limestone aquifers at depths ranging from 50 to 200 m provide drinking water and irrigation supplies of ~30 million m^3/yr mostly north and south of the city where it is freshest. Usage has been capped by a license system due to historical over-abstraction. Elsewhere this aquifer is brackish, but when recharged with fresh stormwater harvested from urban wetlands, provides good supplies of irrigation quality water.

Twenty two aquifer storage and recovery (ASR) systems have been installed, between 1992 and 2004 and the majority within the last two years, with a combined capacity of 2 million m^3/yr (Table 1). Each of these systems is recharged by stormwater runoff, which is typically detained and passively treated in wetlands prior to injection into an aquifer. The most common storage zone is Tertiary limestone, followed by fractured bedrock and several small-scale ASR projects are in alluvium. Considering the potential for rising water tables and potential contamination in unconfined aquifers, confined aquifers are generally preferred. A further 32 projects are currently under investigation. Most of the ASR projects

Table 1. ASR operational sites in Adelaide in 2004, classified by aquifer type (adapted from Hodgkin, 2005).

Aquifer type	No. of sites	Volume stored ($10^3 m^3/yr$)
Limestone	12	1740
Fractured rock	7	250
Alluvium	3	3

are owned and operated by local government, or private organisations, such as developers and a horse-racing club, to replace reticulated potable water for irrigation of parks and sports fields.

Two state government departments are primarily responsible for the licensing of ASR operations; the Department for Water, Land and Biodiversity Conservation (responsible for water allocation and water quantity), and the Environment Protection Agency (responsible for water quality protection). The Department of Human Services also becomes involved if there are public health issues. Furthermore, four Catchment Water Management Boards have responsibility in the metropolitan and adjacent areas for producing and implementing catchment water management plans, and local government has responsibility for urban stormwater infrastructure and planning regulations. The complications in obtaining approvals for starting ASR projects were considerable, so a South Australian ASR Coordinating Committee was established to improve coordination among regulatory and research agencies, to provide accurate information to proponents and discourage spurious applications, to identify research needs, to oversee demonstration projects, to coordinate inputs to a database on operational ASR sites and applications, and to provide information suitable for the public and for government (e.g. Martin and Dillon, 2002).

The first stormwater ASR project commenced in 1993 at Andrews Farm, situated 25 km north of the Adelaide city centre on a new urban subdivision where a wetland was established in a park for urban stormwater flood mitigation and water quality improvement. The developer, Mr Alan Hickinbotham, recognised the potential of ASR for landscape irrigation and wetland top-up in summer, together with his notion of a "Bordeaux-style" development of villages interspersed between market gardens and vines. He was instrumental in achieving participation of state government in this research and demonstration project, which proved to be highly successful. Prior to becoming an operational site in 2000, Andrews Farm was the focus of a five year research trial to evaluate the technical feasibility and environmental sustainability of ASR using passively treated urban stormwater, and in doing so, to gain an understanding of the key subsurface physical, chemical and microbial processes which operate.

The Tertiary aquifer targeted for storage is comprised of interbedded sequences of variably cemented limestone and sand. This aquifer is moderately transmissive with significant induced heterogeneity close to the ASR well due to enhanced sand recovery. Three observation wells were drilled at distances of 25, 65 and 325 metres down-gradient of the ASR well on a transect to trace the fate of the injected water. Injection occurred each winter/spring for four years between 1993 and 1996 and recovery occurred in 1997 and 1998. A net total of $256 \times 10^3 m^3$ of water was injected at average rates of 1300–1700 m^3/day and $151 \times 10^3 m^3$ was subsequently recovered.

The 55 km^2 surface water catchment supports sheep grazing on non-irrigated pasture, and a relatively small, but expanding residential area. Characteristically, the stormwater is colder, has higher concentrations of dissolved oxygen, suspended solids, nitrogen, organic carbon, faecal coliforms and higher pH and turbidity than the ambient groundwater, but has lower salinity, iron, arsenic and boron and episodic inputs of trace organics such as atrazine and pentachlorophenol (Table 2). The injectant could be readily traced using Cl or electrical conductivity (EC), and during initial events with temperature, but stable isotope variability in source water was too great to be quantitatively useful.

Conduits for flow were created through sand removal during earliest well development which had a significant effect on the hydrologic regime in the vicinity of the ASR well. As a result, direct hydraulic connection between the ASR and 25 m wells resulted in little or

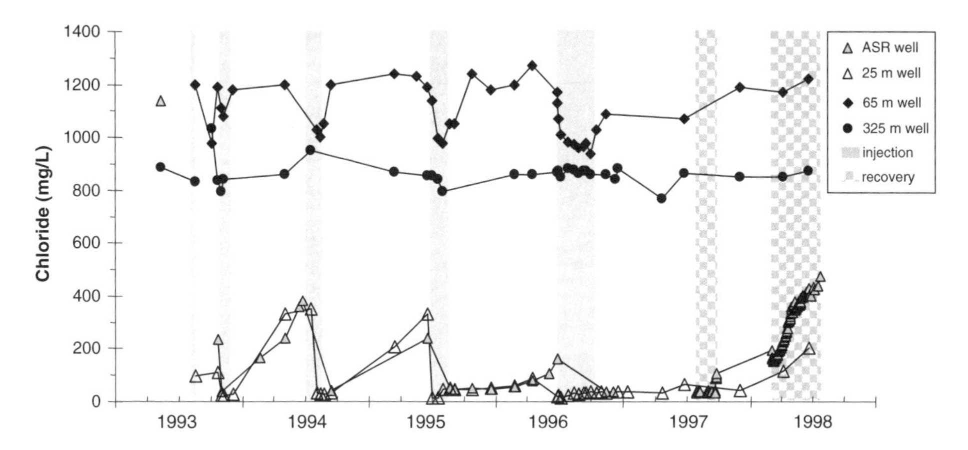

Figure 1. Chloride concentration in injection and observation wells over the duration of the experiment (after Pavelic et al., 2000).

no piezometric head-difference and very rapid solute breakthrough. Only partial breakthrough of fresher water occurred at the 65 m well in each year of injection, with the complete return to ambient water quality during intervening storage periods (Figure 1). No breakthrough was observed at the 325 m well (Pavelic et al., 2000).

The anticipated well clogging problems when injecting turbid stormwater did not materialise, despite the high particulate content in the injectant (yearly average total suspended solids (TSS) levels ranged from 29 to 169 mg/L). Redevelopment by airlifting approximately every $40 \times 10^3 m^3$ was effective in restoring injection rates. Most of the clogging is offset by the increased porosity caused by calcite dissolution (Herczeg et al., 2004) and sand recovery. While well efficiency decreased within each cycle of injection as determined from flow meter and piezometers, redevelopment was effective in restoring well efficiency to at least its initial value on each occasion over the four years of injection.

The main biogeochemical reactions follow a sequence that can be divided into two distinct phases. During or immediately following injection there is an aerobic acid producing phase (organic matter and sulphide oxidation with consequent CO_2 production, carbonate dissolution and cation exchange), following which, further anaerobic organic matter oxidation is accompanied by mobilisation of iron, sulphate reduction and carbonate precipitation (Herczeg et al., 2004).

Batch and column adsorption studies were undertaken using local core material for six reference contaminants: phosphate, cadmium, 2,4-D, atrazine, diuron and parathion (Oliver et al., 1996). Retardation factors were less than five for 2,4-D, atrazine and diuron, approximately 20 for parathion, less than 100 for phosphate and greater than 2000 for cadmium. The additional time that adsorption allows for biodegradation of organics is unlikely to ensure removal, particularly when water is recovered from the injection well.

Microbial quality of the injected water improved significantly during aquifer storage (Table 2). Pathogen studies were performed by monitoring the fate of injected indicator organisms, and by *in-situ* diffusion chamber experiments (Dillon et al., 2005). *In-situ* survival experiments in filter-sterilized water showed that survival of microbial pathogens and indicators could vary widely with 1 log (90%) reduction times ranging from 5 to 122 days

Table 2. Typical composition of injected stormwater, ambient groundwater and recovered water (after Dillon et al., 2005).

Parameter (mg/L)	Injected stormwater[I]	Ambient groundwater[II]	Recovered water[III]
Temperature (°C)	8.7	23.8	18.6
pH	8.2	7.5	7.1
Dissolved oxygen	11.3	0.8	0.8
EC (μS/cm)	277	3500	1420
Total Dissolved Solids (TDS)	150	1900	660
Calcium	21	130	64.2
Magnesium	7.6	66.8	22.9
Sodium	22.2	530	151
Potassium	5.1	11.8	5.6
Bicarbonate	98	382	226
Sulphate	12.2	258	75.1
Chloride	30	794	228
Fluoride	0.24	0.19	0.23
Ammonia-N	0.051	0.022	0.08
TKN-N	0.44	0.12	0.30
Nitrate + nitrite-N	0.27	0.06	<0.005
Total phosphate-P	0.077	0.166	0.032
Arsenic	0.001	0.006	0.008
Boron	0.072	0.278	0.102
Cadmium	<0.0002	0.0002	–
Chromium	<0.005	0.004	–
Copper	0.008	<0.005	<0.005
Iron	0.59	14	0.521
Lead	0.004	<0.001	<0.001
Manganese	0.019	0.109	0.034
Zinc	0.03	1.17	0.076
Suspended solids	38	–	5
Turbidity (NTU)	50	–	2.1
Total organic carbon	4.1	0.8	2.6
Dissolved organic carbon	3.6	0.8	2.6
PAHs (μg/L)	n/d	n/d	n/d
BTEX (μg/L)	n/d	n/d	n/d
Total hal. phenols (μg/L)	0.02	n/d	n/d
Pentachlorophenol	0.02	n/d	n/d
Total insecticides (μg/L)	n/d	n/d	n/d
Total herbicides (μg/L)	5.9	n/d	n/d
Atrazine (μg/L)	5.9	n/d	n/d
Coliforms (cfu/100 mL)	160	0	0
Faecal coliforms (cfu/100 mL)	38	0	0
Faecal streptococci (cfu/100 mL)	26	–	
Het. iron bacteria (mL)	240	–	<10
Col. count (mL)	5200	–	10

[I] Sampled on 15/8/94; [II] 325 m well sampled on 1/11/93; [III] ASR well sampled on 6/4/98 (after 52 ML pumped).

depending on the water type and the presence or absence of aquifer media. Indicator bacteria were not detected in the recovered water.

Injecting fresh water (<200 mg/L TDS) into brackish groundwater (1900–2500 mg/L TDS) constrained the quantity of good quality water recovered (the locally defined limit for agricultural production is 1500 mg/L). However, in spite of the large storage time (between 1 and 3 years), the recovery efficiency (proportion of injected water that was recovered at an acceptable salinity) exceeded 60%. Over half of the injected stormwater was not recovered and would serve to increase recovery efficiencies in subsequent cycles. The cost of recovered water was about half the cost of the reticulated potable water supply (Dillon et al., 1997).

Success at Andrews Farm and subsequent stormwater ASR schemes inspired research on exploring the potential for ASR with reclaimed water from a new irrigation pipeline. A three stage study commenced in 1997 with a desktop and lab study based on a single well with core sampling, geophysics and pump testing. This was followed by the establishment of an ASR trial at Bolivar Sewage Treatment Plant, with intensive monitoring of injectant, groundwater at 16 piezometers and wells and recovered water, and the preparation of plans to scale up from one site to up to 40 sites on the Northern Adelaide Plains. The trial has been successful and the ASR operation with reclaimed water was demonstrated to be more cost-effective than ASR with stormwater owing to the continuous supply of water over the injection period (Dillon et al., 2003).

Evaluation of water quality improvements during ASR at Andrews Farm and Bolivar, supported by information from a number of other sites (Dillon et al., 2005), led to the concept of injecting wetland-treated stormwater at one well with recovery from a separate well to provide drinking water. Aquifer Storage Transfer and Recovery (ASTR) can ensure that residence time of injectant within the aquifer will exceed a specified minimum period. This assures inactivation of any pathogens that may be contained in injectant, and gives time for biodegradation of some trace organics found in stormwater. A research project is commencing to demonstrate the viability of producing drinking water by injecting passively treated stormwater into a brackish aquifer. An important part of that project is to perform a HACCP plan to ensure that recovered water is fit for potable use, taking account of sources and pathways for contamination of stormwater within the catchment of the wetland (Swierc et al., 2005). This will be used to manage the risks and increase confidence in the quality of injected and recovered water.

3 CONJUNCTIVE USE OF SURFACE WATER AND GROUNDWATER IN PERTH

Perth has a population of 1.5 million and is located on a sandy coastal plain in Western Australia. It also has a Mediterranean climate with a mean annual rainfall of 850 mm and evaporation of 1700 mm. Only 15% of Perth's rainfall occurs in the summer compared with 27% for Adelaide, and therefore has a greater need for storage of water than Adelaide. On average almost half of the Perth region's water (570 million m^3/yr) is sourced from nearby forested surface water catchments or from groundwater of the Gnangarra Mound, a thick surficial sand aquifer, that has been protected from development. Water from these sources (270 million m^3/yr) is distributed as a potable supply through the Water Corporation's distribution system. The remaining water is supplied by approximately 100,000 privately owned wells in the surficial aquifer for irrigation of domestic gardens, parks, schools and for industrial use, with a large amount of this (100 million m^3/yr) used for agricultural and horticultural irrigation on the outskirts of the city. Generally this water is fresh but because of the potential for pollution it is not viewed as a drinking water resource.

Sewage effluent is treated at nine plants along the western side of the city, generally on the coast, where most of the 120 million m^3/yr is discharged. Six percent of the water is now in reuse (Radcliffe, 2004), mainly for industrial purposes at Kwinana, and the state has developed a Water Strategy (Gallop, 2003) that articulates plans to reuse 20% of Sewage Treatment Plant (STP) discharge by 2012. Stormwater management gives preference to infiltration through the sandy soils, at domestic and suburban scales. This has been vital in maintaining groundwater levels in spite of the high rates of extraction. Excess urban stormwater discharges to the Swan River or the ocean and, although poorly defined, the annual discharge is thought to be smaller than the volume recharged. A combination of reduced rainfall since the mid 1970s and increasing numbers of wells has resulted in summer groundwater levels declining and the drying of some ecologically significant wetlands and caves in low-lying areas of Perth. Lower rainfall has also had a dramatic effect on streamflows and halved catchment yields over the last 25 years compared with the 60 preceding years. Against this backdrop of declining security of supply due to climate change and urban expansion, a range of options to extend Perth's innovation in conjunctive water management are currently being explored. Among these options is reuse of wastewaters that are currently discharged to the environment. There is potential for the treated sewage effluent and stormwater that Perth currently discharges via ocean outfalls to be returned into the community as a substitute for some uses of potable water. One of the more recent water reuse projects developed in the greater Perth region is the Halls Head Indirect Reuse project.

This indirect reuse project is located at the Halls Head STP in Mandurah, 74 km south of Perth. The activated sludge plant provides secondary treatment to approximately $2.3 \times 10^3 m^3$/day. Undisinfected treated effluent is then discharged via infiltration ponds to the shallow aquifer beneath the treatment plant. The indirect reuse project was initiated by recovering the infiltrated water from the aquifer via two recovery wells, SPB1 and SPB2, located 100 m and 60 m respectively from the infiltration ponds. The recovered water is then piped to a storage tank were it is stored prior to use for the irrigation of green open spaces in the neighbouring residential development. The indirect reuse project was monitored over a 24-month period for potential health and environmental risks from major contaminants, with particular emphasis placed on the potential presence of microbial pathogens in the recovered water and the influence of treated effluent on the local groundwater system (Toze et al., 2004).

A hydrogeological analysis and tracer tests undertaken during the monitoring period indicate that the aquifer beneath the Halls Head STP site consists of karstic limestone which is very heterogenous and strongly influenced by sea level variation at diurnal, seasonal and inter-annual time scale. The pre-existing groundwater table is approximately 2–3 m below ground surface and the infiltration basins are situated in a thin layer of coarse sand overlying the limestone. The limestone above the groundwater table becomes fully saturated when the basin is filled with treated effluent.

The tracer tests indicate that particles travelled at a rate of approximately 2.5 metres/day in the vicinity of tracer bores, suggesting that particles would take a minimum of 24 days to reach the closest recovery well. It was also determined, based on Total Dissolved Solids

(TDS) concentrations, that 80% of the recovered water originated from infiltrated treated effluent, with the remainder being drawn from the background groundwater.

The water quality monitoring undertaken for health risks associated with the recovered water focused on the potential presence and numbers of enteric bacteria and viruses and the concentration of heavy metals in the recovered water. Throughout the 24 month monitoring period Thermotolerant Coliform (TTC) numbers detected in the recovered water never exceeded 1 cell/100 mL and were only detected in the recovered water on two occasions. Coliphage and the human pathogenic enteroviruses group were never detected in the recovered water or the background groundwater despite consistent detection in the treated effluent prior to infiltration. A series of experiments examining the factors responsible for removing the pathogens from the infiltrated treated effluent in the aquifer determined that all of the pathogens had a 1 log attenuation time of less than 16 days and that the major removal process was due to the action of the indigenous groundwater microorganisms in the aquifer (for more details see Toze et al., 2004).

Analysis for heavy metal concentrations in the recovered water determined that the only trace metals measured above detection limits were lead and zinc. The maximum concentrations of lead and zinc were 0.0063 mg/L and 0.03 mg/L, respectively. These concentrations, however, were all lower than the maximum allowable levels defined in the Australian Drinking Water Guidelines of 0.01 mg/L for lead and 3 mg/L for zinc (NWQMS, 2001).

The environmental parameters of major interest were TDS, nitrogen as nitrate, ammonium and Total Kjeldahl Nitrogen (TKN), phosphorous and Total Organic Carbon (TOC). The average TDS concentrations in the recovered water (<1000 mg/L) were consistently lower than in the background groundwater (>2500 mg/L) (Figure 2a) making the reclaimed water more suitable for green open space irrigation than the background groundwater. Average phosphorous concentrations in the recovered water (0.019 mg/L) were consistently less than the concentration in the treated effluent (10.8 mg/L) and similar to that in the background groundwater. It is probable that the remaining small concentrations would be taken up as a nutrient source by irrigated plants and would not be returned to the aquifer. While TOC concentrations of up to 140 mg/L were observed in the treated effluent prior to infiltration, TOC concentrations were never higher than 9 mg/L in the recovered water or the background groundwater.

The average total nitrogen concentrations in the recovered water (2.4 mg/L for SPB1 and 4.2 mg/L for SPB2) were lower than in the treated effluent (6.6 mg/L) and similar to that in the background groundwater (2.5 mg/L) (Figure 2b). TKN and ammonia (NH_3-N) concentrations were always less than 1.2 mg/L and 0.7 mg/L, respectively in the recovered water which were a similar concentration to that observed in the background groundwater. Nitrate (NO_3-N) was the only nitrogen compound which was consistently higher in concentration than the background groundwater (average of 2.2 mg/L) (Figure 2b). The shallow karstic aquifer was always aerobic, thus preventing the removal of the nitrate via biological denitrification, however the concentration of nitrate in the recovered water never exceeded 6 mg/L which was well below the 50 mg/L limit in the Australian Drinking Water Guidelines (NWQMS, 2001). As with the remaining phosphorous in the reclaimed water, it was considered that nitrate present in the concentrations detected in the recovered water would not pose any environmental concerns and would probably be consumed by irrigated plants.

It was concluded from this research project that the recovered water obtained through indirect reuse was of a suitable quality for irrigation purposes and had negligible associated

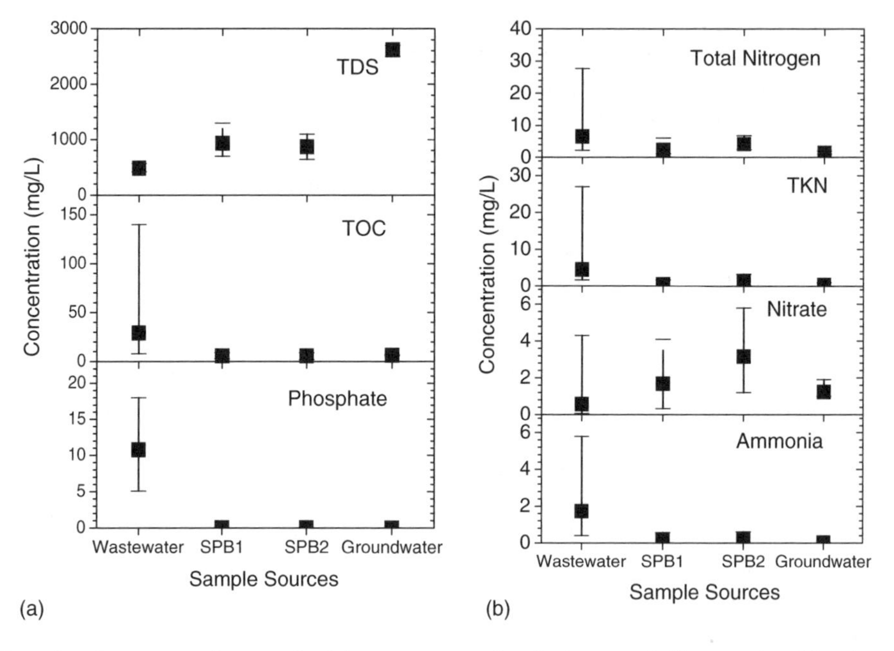

Figure 2. Average, maximum and minimum concentrations in the treated effluent and reclaimed water recovered from the two recovery wells (SPB1 and 2) and the background groundwater over the 24 month monitoring period, of (a) TDS, TOC and phosphate and (b) various forms of nitrogen (Toze et al., 2004).

health or environmental risks. The hydrogeological analysis also determined that the Halls Head Indirect Reuse scheme was capable of being expanded by the establishment of more recovery bores as demand for the reclaimed water increased. To date a second storage tank has been added to the system to cope with the increased demand for the recovered reclaimed water and plans are underway for the establishment of additional recovery wells.

The success of this project has led to increased interest in the potential for managed aquifer recharge to have a major role in the reuse of reclaimed water in the southwest of Western Australia. Scatena and Williamson (1999) identified numerous opportunities to enhance groundwater recharge for increasing supplies of irrigation or potable water. One of these involves injection of potable water into the deep aquifer system for recovery at times of water stress. Studies are continuing, and have included evaluation of biogeochemical changes (Toze et al., 2001). Another reuse option currently being explored is to recharge reclaimed water via subsurface galleries to augment irrigation supplies in established urban areas, without the need for infiltration ponds.

4 CONCLUSIONS

With climate change, population growth and limits on yields of catchments and aquifers, sustainable sources of water are increasingly needed for cities throughout the world. Urban stormwater and reclaimed water are a vast and undervalued resource with potential to provide substantial volumes of non-potable (and potentially potable) supplies if they can be stored inter-seasonally. Intentional groundwater recharge has been undertaken with waters of impaired quality in Adelaide and Perth to provide non-potable supplies for many years. In Adelaide there are currently twenty two ASR schemes with a combined capacity of 2 million m^3/yr and a similar number in the planning stages. In Perth, stormwater infiltration has been practised for many years and is relied on to sustain domestic wells. Reclaimed water infiltration has started to enhance groundwater recharge. Against a backdrop of reduced security of supply and expansion of the city, a range of options are being explored.

Research projects have been undertaken to define the changes in the quality of stored water and the sustainability of contaminant and pathogen attenuation processes, and thereby identify the highest valued uses of the recovered water. At Andrews Farm in Adelaide, a five year stormwater ASR project was highly successful. The anticipated clogging problems when using turbid wetland-treated stormwater did not materialise, water quality improvements during storage were substantial, groundwater protection was achieved, recovery efficiency was good and the cost of recovered water was lower than traditional sources. At Halls Head near Perth, where undisinfected secondary treated effluent is infiltrated into a shallow karstic aquifer and recovered from two nearby production wells, the recovered water over a two year monitoring period was of a suitable quality for irrigation purposes with negligible health or environmental risks.

ACKNOWLEDGEMENTS

The South Australian Department of Water Land and Biodiversity Conservation, SA Water, Hickinbotham Group and Urban Water Research Association of Australia are acknowledged for their contribution to the Andrews Farm ASR research project, and Water Corporation of Western Australia supported the research at Halls Head water reclamation site. Technical assistance was provided by Karen Barry and Jon Hanna, CSIRO Land and Water. The authors thank the two anonymous reviewers for their helpful comments on the manuscript and are grateful to Prof Ken Howard for the opportunity to contribute to this tribute to Dr Matthias Eiswirth, a valued friend and source of inspiration.

REFERENCES

ARMCANZ and ANZECC. 1995. Guidelines for Groundwater Protection in Australia. http://www.mincos.gov.au/pdf/nwqms/guidelines-for-groundwater-protection.pdf (accessed 7 Mar 2005).

Bramley, H., Keane, R. and Dillon, P. 2000. The potential for ingress of saline groundwater to sewers in the Adelaide metropolitan area: an assessment using a geographical information system. Centre for Groundwater Studies Report No. 95, July 2000, 33p.

Dillon, P., Pavelic, P., Sibenaler, X., Gerges, N. and Clark, R. 1997. Aquifer storage and recovery of stormwater runoff. Aust. Water and Wastewater Assoc. J. Water 24(4):7–11.

Dillon, P., Martin, R., Rinck-Pfeiffer, S., Pavelic, P., Barry, K., Vanderzalm, J., Toze, S., Hanna, J., Skjemstad, J., Nicholson, B. and Le Gal La Salle, C. 2003. Aquifer storage and recovery with reclaimed water at Bolivar, South Australia. In: Proc. 2nd Australian Water Recycling Symposium, Brisbane 1–2 Sept 2003. Australian Water Association, Artarmon, NSW, Australia.

Dillon, P., Toze, S., Pavelic, P., Skjemstad, J., Davis, G., Miller, R., Correll, R., Kookana, R., Ying, G.-G., Herczeg, A., Filderbrandt, S., Banning, N., Gordon, C., Wall, K., Nicholson, B., Vanderzalm, J., Le Gal La Salle, C., Gibert, M., Ingrand, V., Guinamant, J.-L., Stuyfzand, P., Prommer, H., Greskowiak, J., Swift, R., Hayes, M., O'Hara, G., Mee, B. and Johnson, I. 2005. Water Quality Improvements

During Aquifer Storage and Recovery, Vol 1. Water Quality Improvement Processes. AWWARF Project 2618, Final Report.

Gallop, G. 2003. Securing Our Future – A State Water Strategy for Western Australia. Govt. of W.Aust.

Herczeg, A.L., Dillon, P.J., Rattray, K.J., Pavelic, P. and Barry, K.E. 2004. Geochemical processes during five years of aquifer storage recovery. Ground Water 42(3):438–445.

Hodgkin, T. 2005. Aquifer storage capacities of the Adelaide region. South Australia Department of Water, Land and Biodiversity Conservation Report 2004/47.

Martin, R. and Dillon, P. 2002. Aquifer storage and recovery – future directions for South Australia. Dept. for Water Land and Biodiversity Conservation Report. DWLBC 2002/04. Aug 2002.

Nadebaum, P., Chapman, M., Morden, R. and Rizak, S. 2004. A guide to hazard identification and risk assessment for drinking water supplies. CRC for Water Quality and Treatment Research Report 11.

NWQMS. 2001. Australian Drinking Water Guidelines. Pub. National Health and Medical Research Council/Agriculture and Resource Management Council of Australia and New Zealand (ISBN 0 642 24462 6).

Oliver, Y.M., Gerritse, R.G., Dillon, P.J. and Smettem, K.R.J. 1996. Fate and mobility of stormwater and wastewater contaminants in aquifers. II. Adsorption studies for carbonate aquifers. Centre for Groundwater Studies Report No. 68.

Pavelic, P., Dillon, P.J. and Gerges, N.Z. 2000. Challenges in evaluating solute transport from a long-term ASR trial in a heterogeneous carbonate aquifer. Proc. IAH Congress, Groundwater: Past Achievements and Future Challenges (Eds. Sillio et al.), Balkema, Rotterdam, ISBN 90 5809 159 7, p.1005–1010.

Pitman, C. 2003. Stormwater harvesting and utilization. In: Proc. Water Recycling Aust. 2nd National Conf., Brisbane, 1–3 Sept, 2003. Australian Water Association, Artarmon, NSW, Australia.

Radcliffe, J. 2004. Water Recycling in Australia. Review for the Australian Academy of Technological Sciences and Engineering. www.atse.org.au (accessed 7 Mar 2005).

Scatena, M.C. and Williamson, D.R. 1999. A potential role for artificial recharge within the Perth Region. Centre for Groundwater Studies Report No 84.

South Australian Government. 2004. Water Proofing Adelaide: A Thirst for Change. http://www.waterproofingadelaide.sa.gov.au/main/publications.htm

Swierc, J., Van Leeuwen, J. and Dillon, P. 2005. Preparation for a Hazard Analysis and Critical Control Points (HACCP) Plan for Stormwater to Drinking Water Aquifer Storage Transfer and Recovery (ASTR) Project. CSIRO Land and Water Tech Report No. 20/05.

Toze, S., Hanna, J. and Ladbrook, M. 2001. Biogeochemical changes during ASR of potable water into the Leederville aquifer. CSIRO Land and Water Report to Water Corporation, Western Australia.

Toze, S., Hanna, J., Smith, T., Edmonds, L. and McCrow, A. 2004. Determination of water quality improvements due to the artificial recharge of treated effluent. *Wastewater Reuse and Groundwater Quality* IAHS Publication 285:53–60.

CHAPTER 13

Stormwater infiltration technologies for augmenting groundwater recharge in urban areas

Ken Howard[1], Stephen Di Biase[1], Joanne Thompson[2], Herb Maier[1] and John Van Egmond[3]

[1]Groundwater Research Group, Earth and Environmental Sciences, University of Toronto at Scarborough, 1265 Military Trail, Toronto, Ontario, M1C 1A4 Canada
[2]W.B. Beatty & Associates Ltd., 18 King St. East, Bolton, Ontario, L7E 1E8 Canada
[3]Egmond Associates Ltd., 27 Hall Rd., Georgetown, Ontario, L7G 5Y7 Canada

ABSTRACT: Although stormwater infiltration has been practiced successfully for many years, some regulatory agencies are hesitant to adopt such technologies due to concerns over water table mounding, aquifer contamination and efficiency losses due to clogging. The onus, in such cases, is on the research community to perform the work necessary to demonstrate the effectiveness and reliability of the approach. In recent years, the Groundwater Research Group at the University of Toronto has worked with industry partners to develop and refine technologies suitable for the infiltration of stormwater in urban areas. The work has involved comprehensive field, laboratory and computer model studies and the primary focus has been the efficiencies and maintenance needs of two contrasting technological approaches to stormwater infiltration. Research to date indicates that both approaches can be effective for a wide range of hydrogeological conditions and that potential water quality concerns are unlikely to be an issue.

1 INTRODUCTION

In rapidly growing population centres, the provision of a safe and adequate water supply often proves difficult. The greatest challenge is to meet an accelerating demand for good quality water through extraction of finite water resources which are being increasingly degraded by urban-sourced contaminants (Howard, 2002; Howard and Gelo, 2002). Proactive solutions are required. Resource conservation, management and protection are all important, but ultimately the demand for water can only be met through additional sources of water. In many urbanised areas, stormwater runoff is often neglected as a potentially valuable water source.

In temperate climates, stormwater runoff is a major component of the urban water balance. Figure 1, for example, shows pre-development and post-development water balance calculations for a proposed urban subdivision located close to Toronto, Canada (Howard et al., 2000). The creation of an impermeable surface across 32% of the developed area reduces direct aquifer recharge (RLAND), but significantly increases the net amount of water available by reducing evapotranspiration (E). In effect, urbanisation "creates" water, in this example by over 500,000 m^3/a or almost 25% of the annual precipitation. The scientific and

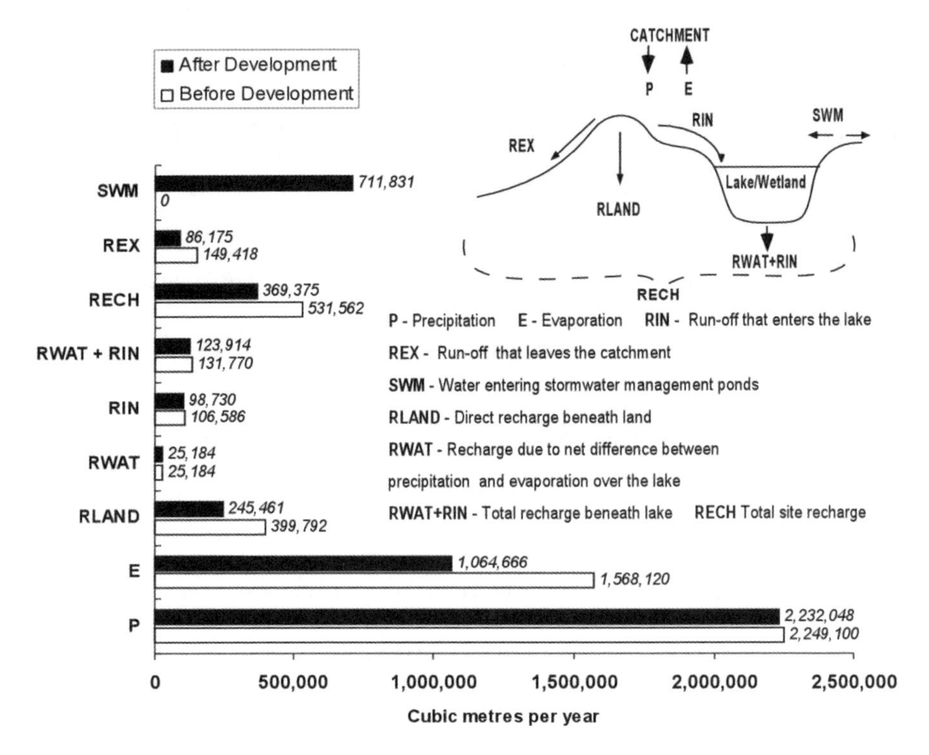

Figure 1. Pre- and post-development water balances for a proposed subdivision near Toronto, Canada, showing the large volume of stormwater (SWM) that can be generated.

engineering challenge is to "harvest" this additional water – manifest as a large component of urban "stormwater runoff" (SWM) – and utilize it to augment the water resource.

Typically, the most effective approach to "harvesting" stormwater is to direct excess into the subsurface thereby taking advantage of aquifer storage. This technique is generally referred to as artificial recharge or "recharge management" (Jacenkow, 1984; Asano, 1986; Li et al., 1987; Dillon et al., 1994; Watkins, 1997). This approach has proved extremely successful in many parts of the world; however, some regulatory agencies are hesitant to adopt stormwater infiltration technologies due to concerns over aquifer contamination and efficiency losses due to clogging (Pitt et al., 1996). This concern is especially prevalent in the Province of Ontario, Canada where groundwater contamination in the town of Walkerton (O'Connor, 2000; Howard, 2001) and fears over urban development along the Oak Ridges Moraine (Howard, 1997) have heightened sensitivity.

During the past five years, the Groundwater Research Group at the University of Toronto, in collaboration with industry, has been conducting research on the hydrogeological design of urban subdivisions. The overall goal is to develop new technologies and novel approaches to urban water management which can reduce the potentially detrimental impacts of urban development on the quality and quantity of groundwater. The ultimate goal is the development of a versatile infiltration technology that could adequately treat and infiltrate large volumes of stormwater, cost-effectively and with minimal maintenance. In effect, such a technology would transform a serious urban problem (management of stormwater) into a major resource asset. Special considerations in Canada are the winter conditions which can freeze soil and the fact that stormwater often contains elevated

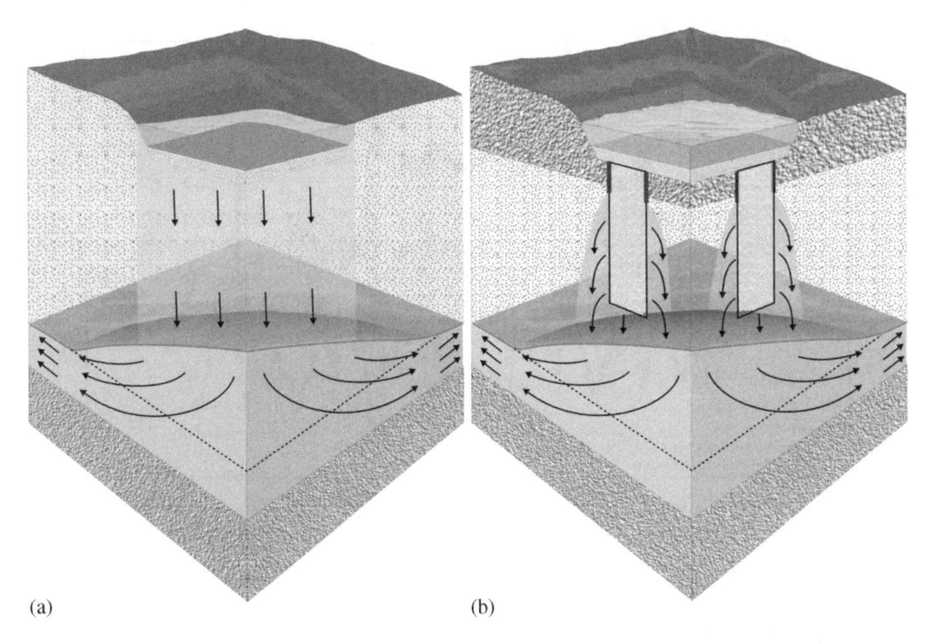

Figure 2. Application of the low maintenance filter-basin technology (a) using a sand filter blanket for aquifers overlain by permeable sediments; (b) using drainage columns to allow stormwater to penetrate low permeability sediments such as glacial tills.

concentrations of sodium chloride (NaCl) road de-icing chemicals (Howard et al., 1993; Howard and Livingstone, 2000).

In work conducted to date, field, laboratory and computer model studies have been used to investigate the functionality, efficiencies and maintenance needs of two contrasting technological approaches to stormwater infiltration. The first approach (the primary subject of this paper), uses a permanently installed, low maintenance filter basin (LMFB) to infiltrate stormwater released from a pre-treatment settling pond. The second technological approach (SAGES™) is somewhat less conventional and uses replaceable filter cartridges, installed in wells, through which stormwater infiltrates directly to the water table. Work on the latter began just recently, and is discussed only briefly here.

2 THE LOW MAINTENANCE FILTER-BASIN TECHNOLOGY

Most of the work conducted during the past five years has focused on the low maintenance filter-basin technology. It began with field pilot studies and later progressed to the laboratory where a series of experiments were conducted to study infiltration rates, clogging, and optimal filter design. Subsequently, computer modeling work was carried out to investigate water table mounding and water quality issues.

2.1 *Pilot field studies*

The basic design of the LMFB is shown in Figure 2a. A filter blanket, comprising fine to medium grained sand, permits water to drain vertically through it to the aquifer with minimal

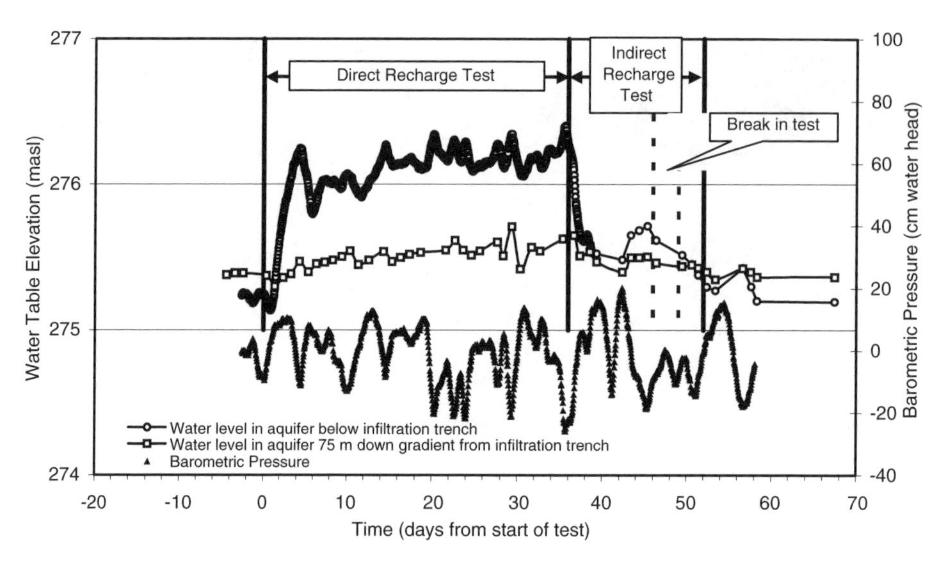

Figure 3. Pilot field study data showing the response of the water table immediately beneath the site and at an observation well 75 m from the site. During the direct recharge test (Phase 1) water was introduced to the columns directly; during the indirect test (Phase 2) the basin was flooded such that flow initially took place via the filter blanket.

resistance to flow. Since aquifers throughout much of Canada are overlain by low permeability sediments such as glacial tills, a variant on the design includes one or more vertical drainage columns, filled with gravel or coarse sand to provide direct access to the underlying aquifer (Beatty and Thompson, 2000) (Figure 2b). Pilot field studies were conducted for the latter scenario, thus allowing both the filter blanket and the drainage column design to be evaluated.

At the pilot field study site, three 18 m deep drainage columns were constructed, 0.61 m in diameter to penetrate 6 m of glacial till. To accommodate these columns, a triangular filter basin was constructed 40 m^2 in area, and a network of 6 observation wells (15 to 56 m deep) was installed in and around the basin to monitor the test results.

The pilot test was conducted under two operating conditions:

- Phase 1. Direct application of water to all three columns, thus testing the hydraulic capacity of the individual columns and of the receiving aquifer;
- Phase 2. Flooding the basin to achieve column infiltration following drainage of water through the filter blanket.

Tests began in 1999. Figure 3 shows the response of the water table directly beneath the study site and at a well 75 m down gradient. Significantly, the water levels respond both to recharge and changes in atmospheric pressure. The maximum infiltration capacity of each column was about 1 L/s and all observation wells responded within hours of the test start. With all three columns replenishing the aquifer at 1 L/s during Phase 1, a 1 m high mound developed immediately beneath the basin but the zone of influence extended no further than 300 m.

During Phase 2 (the indirect recharge test) (Figure 4), the flow rate was reduced to accommodate the comparatively lower infiltration capacity of the filter blanket. As shown in Figure 3, the reduced water supply caused the water table mound to decline rapidly.

Figure 4. Pilot field study site during Phase 2 infiltration studies (basin flooding).

Under the flooded-basin condition, the effects of freeze-thaw cycles and variable water levels in the basin were also investigated. A week of unusually cold air temperature (~ minus 30°C) allowed the basin performance to be observed under frozen conditions. Initially, water flow to the basin was shut off for 3 days to allow the filter blanket to drain and freeze. When water flow to the basin resumed, the pre-frozen recharge rates were immediately re-established. This demonstrates that the permeability of a naturally drained filter blanket is not adversely affected by freezing (Figure 5).

During the 52-day pilot study test period, the aquifer was replenished with 10,000 m^3 of water at rates ranging between 1.5 and 2.6 L/s. The results confirmed the technology as an effective means of artificially replenishing aquifers capped by low permeability sediments. Based on the pilot study, it was concluded that infiltration rates as high as 1000 m^3/day could be achieved with sub-basin water table mounds of no more than 4 to 5 m. In practice this height of mounding will rarely be achieved since the infiltration of stormwater will occur intermittently, thus allowing ample opportunity for water table mounds to decay.

2.2 *Laboratory studies*

Laboratory studies were conducted primarily to investigate the extent to which physical retention of suspended solids, chemical precipitation, sorption of dissolved solids and bacterial or algae growth would "clog" the filter materials and reduce their efficiency over time. Experiments were carried out using 10 sand columns, 4 cm in diameter and 1.25 m in length (Figure 6). Five columns were filled with a well characterized sand from Borden, Ontario (Cherry et al., 1983); for the remaining 5 columns, a slightly less well sorted sand

Figure 5. Field studies conducted during the winter months demonstrate that the filter blanket functions well in freezing conditions.

from the Oak Ridges Moraine (ORM) was used (Figure 7). The tops of the columns were sealed with removable rubber caps and two holes were drilled into each cap for connection to a water supply source and a discharge manometer. The manometer allowed entrapped air to escape from the columns and helped maintain consistent discharge rates. The experiments were conducted over a period of 100 days. Locally collected stormwater was passed through 8 columns, either intermittently or continuously; distilled de-ionized water was passed continuously through two columns used for experimental control. At the closure of the experiment, each column was dissected into five 10 cm sections and grain-size analyses were performed to determine where fines and other material had accumulated.

Flow results for the Borden sand columns are shown in Figure 8. Columns B2, B3 and B4 showed a significant reduction in discharge rate compared to B1 (the clean water control) and B5 (which was vegetated with grass). Occasional scraping of the sand at the top of column B4 (e.g. Day 60) improved flow rates temporarily.

When the columns were dissected, no significant accumulation of fine grained material was found below a depth of 10 cm. Field experiments were subsequently conducted and confirmed that the sand filter blanket efficiencies could be maintained by regular scraping (raking) of the surface and rare replacement of the top 10 cm of the blanket material.

2.3 *Mounding studies*

An important constraint on the design of infiltration facilities is the degree to which underlying aquifers are able to accommodate and distribute the infiltrated water. Highly permeable aquifers raise very few problems; however, significant recharge mounds can develop beneath infiltration basins where aquifer materials are poorly to moderately permeable,

Figure 6. Experimental apparatus included a series of 10 drainage columns and a supply of both stormwater and distilled de-ionized water.

thus limiting the amount of water that can be safely introduced. Mounding studies were conducted using a combination of field study and finite element numerical modeling.

As an example of the modeling studies, Figure 9 shows the height to which a mound would develop beneath an infiltration basin 15 m by 2 m in dimension, receiving 45 m^3/day over a period of two days. The aquifer thickness is 15 m. Where high mounding heights are anticipated, the only option is to distribute the infiltration, in effect exchanging a single high mound for a series of much smaller ones (Figure 10).

2.4 *Model studies of water quality concerns*

While most of the technical problems of replenishing aquifers with stormwater can be overcome, water quality issues persist. In Canada, a particular concern is the potential contamination of groundwater by stormwater containing elevated concentrations of sodium chloride derived from road salt. Currently, there are no cost-effective means of removing chloride from water and dilution and dispersion provide the only practical means of ameliorating impacts.

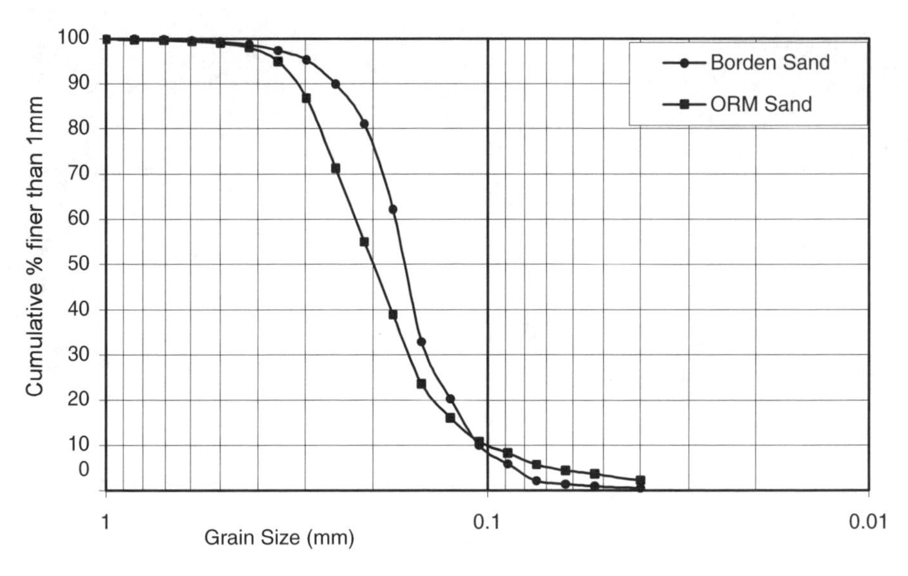

Figure 7. The experiments were conducted using two types of sand – Borden sand and Oak Ridges Moraine (ORM) sand.

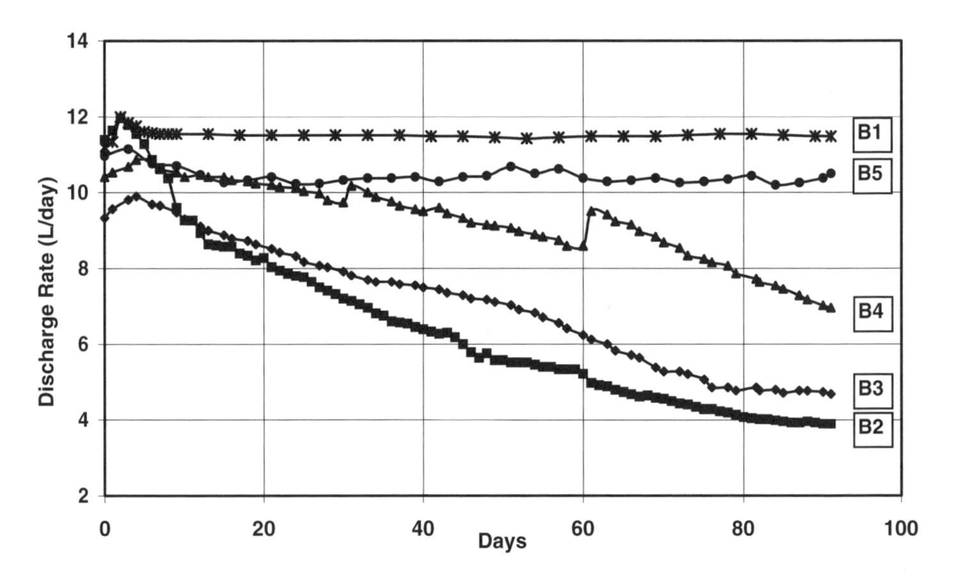

Figure 8. Discharge rates for columns containing Borden sand.

To investigate the role of dilution and dispersion in the aquifer, sub-basin plume behaviour was investigated using AT123D (a transient multi-dimensional analytical groundwater transport model developed in 1981 at the Oak Ridge National Laboratory, USA). The model was run for an aquifer 40 m thick with a hydraulic conductivity of 2×10^{-2} cm/s and an effective porosity of 20%. Dispersivity was set to 20 m in the direction of groundwater flow and 2 m transverse to this direction. The basin was assumed to be 40 m by 50 m

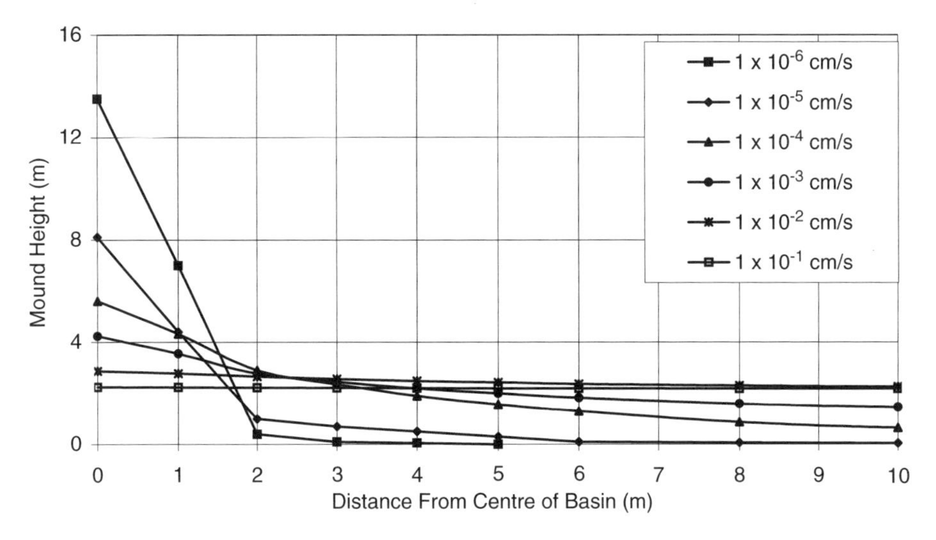

Figure 9. Mound height as a function of the hydraulic conductivity.

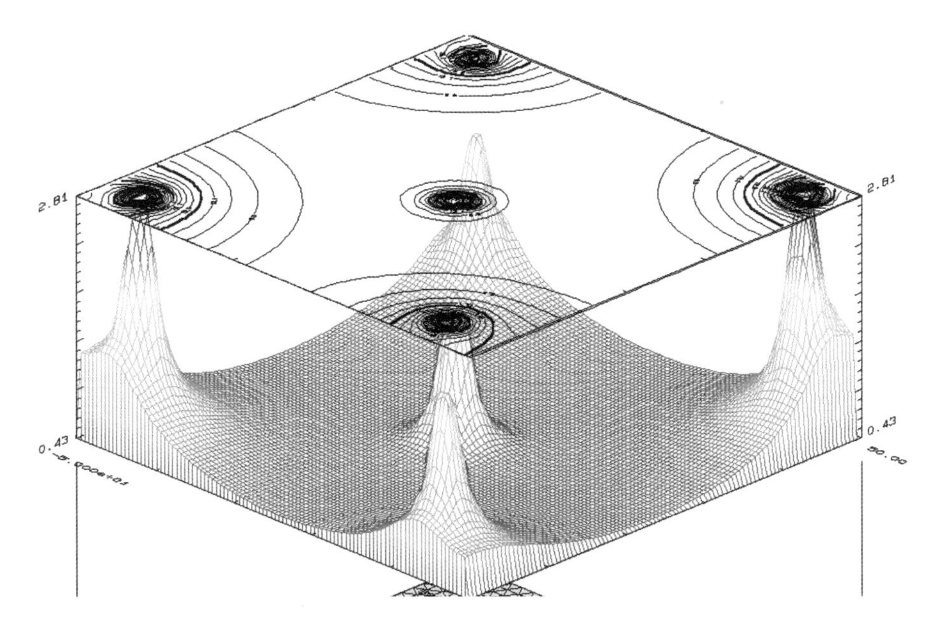

Figure 10. Multiple recharge mounds formed by distributing infiltration across five locations.

in size and recharged the aquifer at a rate of 120,000 m^3 per year. During 4 months of the year (i.e., the winter road salt application period), the chloride concentration of the infiltration water was assumed to average 220 mg/L; during the remaining 8 months, the chloride concentration was set at 0 mg/L. Natural flow in the underlying aquifer was assumed to be low such that no significant mixing occurs with groundwater immediately beneath the site i.e. dilution occurs in response to dispersion only. It was also conservatively assumed that no dilution occurs as a result of aquifer recharge down-gradient of the site.

The simulation was carried out for a period of 58 months so as to include five, 4-month road-salting seasons. The chloride input function is shown on Figure 11a. Figures 11b, 11c and 11d show the predicted chloride concentration at the top of the aquifer, along the centre line of flow, 100 m, 300 m and 600 m respectively, down-gradient from the basin area.

The results of the simulations demonstrate that dispersion will cause significant smoothing of the original chloride input function at very short distances from the infiltration basin. Within 300 m, concentrations of chloride reduce to less than 40 mg/L chloride and seasonal fluctuations are less than 15 mg/L. At 600 m, no response is seen until midway through the second year when concentrations begin to rise gradually to a steady state concentration of just over 20 mg/L. This represents a 10-fold dilution compared to the input concentration during the winter season. Chloride concentrations were also found to decrease significantly with depth, although this effect becomes less pronounced with distance along the flow line.

3 SAGES™

SAGES™ (Stormwater and Groundwater Enhancement System) is an infiltration technology designed to treat stormwater during the artificial replenishment of aquifers. It is patented in the U.S.A. and Canada. The device consists of three filter materials (Figures 12 and 13) installed as replaceable cartridges in an array of vertical recharge wells (Figure 14). In combination these filters remove suspended solids and a wide range of pollutants

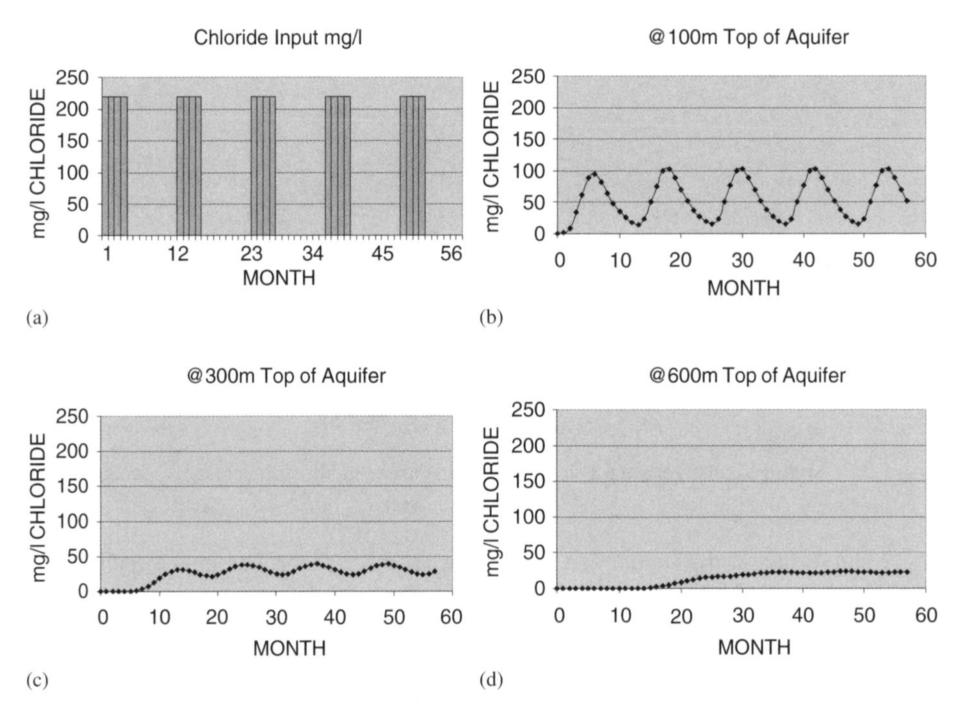

Figure 11. AT123D model results showing the effects of dispersion on the chloride plume as it migrates away from the basin.

Figure 12. SAGES™ filter materials (a) coarse sand; (b) activated carbon; (c) gravel. Each image is approximately 50 mm square.

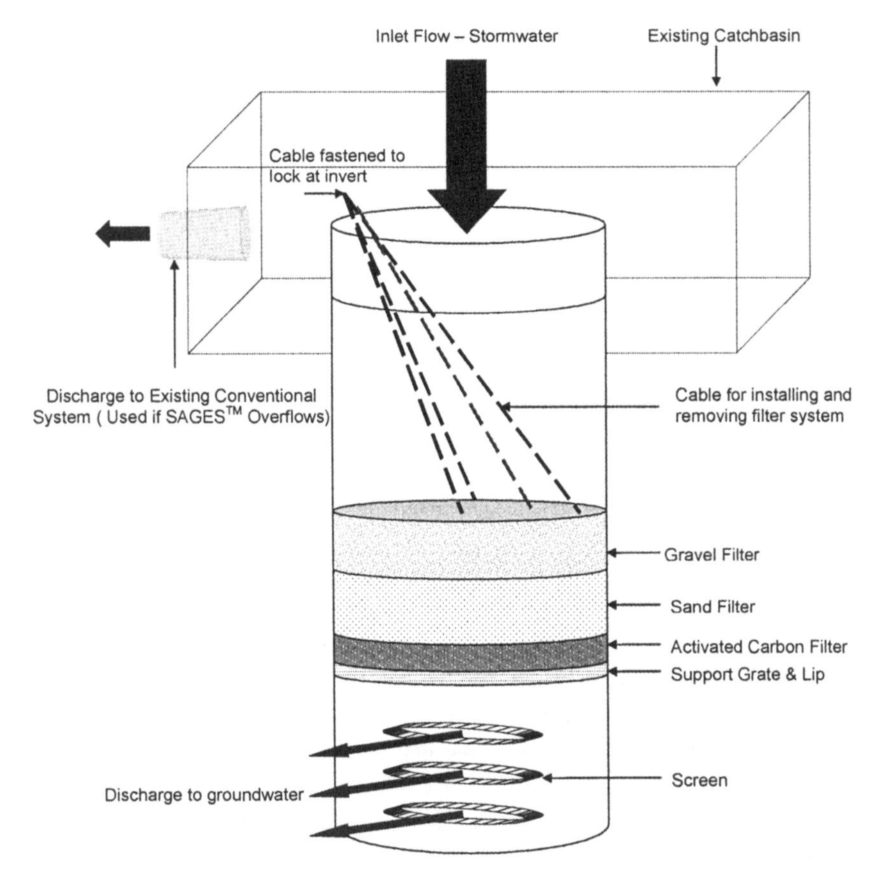

Figure 13. SAGES™ consists of three filter materials installed within a replaceable cartridge in a vertical recharge well.

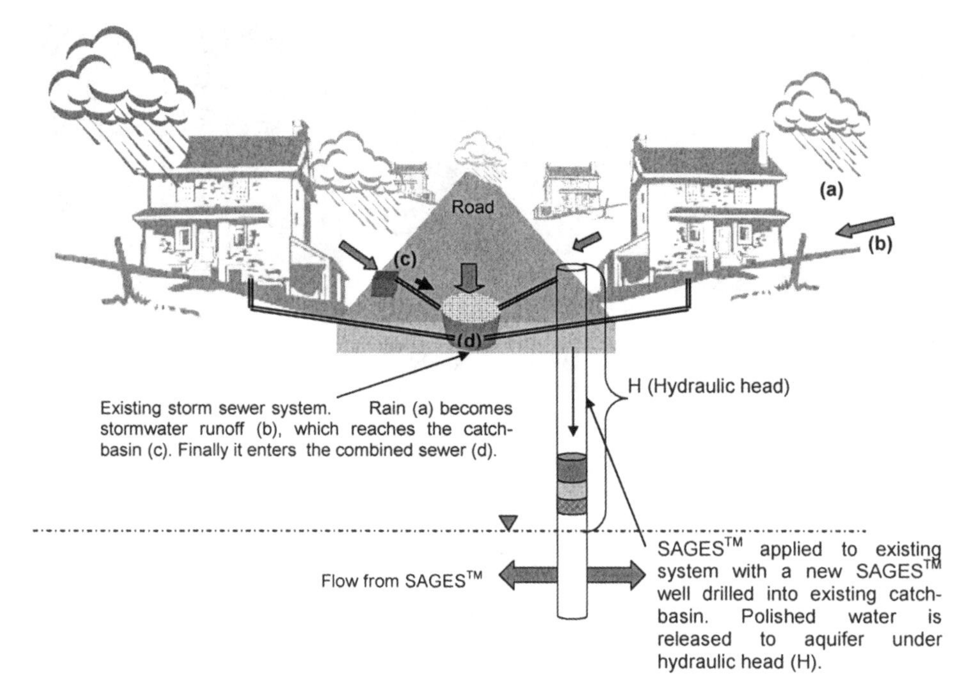

Figure 14. SAGES™ is designed for installation in existing catch-basins.

including, but not limited to, trace metals and organics. Especially appealing aspects of the SAGES™ technology are:

- Minimal maintenance – the cartridges are simply removed and replaced when their efficiency becomes impaired.
- Minimal space impact – infiltrated stormwater does not require pre-treatment and thus large, expensive and often controversial settling ponds are unnecessary.
- The development of recharge mounds is minimized since stormwater infiltration is distributed through an array of wells over a very large area.

Research is currently in progress at the University of Toronto that aims to:

1. Characterize and document the chemical behaviour of the filter system and quantitatively assess the removal efficiencies for sediment and key chemical species;
2. Characterize the flow characteristics of the filter system, e.g. clogging rates and the effects of clogging on system efficiency;
3. Investigate the functional life span of the system components; and
4. Identify those components of the system that would benefit from modifications and refinement.

Much of this work is being conducted under carefully controlled laboratory conditions and involves column experiments on scaled-down versions of the filter system. At a later stage, finite element flow modeling, chemical reaction path modeling, tracer studies and field tests will be carried out to investigate the potential merits of alternative filter materials. In the longer term, the overall goal is to instill consumer/regulatory confidence in the SAGES™ technology and make it readily marketable.

4 CONCLUSIONS

As a technique for managing urban runoff and augmenting groundwater supplies, stormwater infiltration has been practiced successfully for many years. Nevertheless, some regulatory agencies are hesitant to adopt stormwater infiltration technologies due to concerns over aquifer contamination and efficiency losses due to clogging. The onus, in such cases, is on the research community to perform the work necessary to demonstrate the effectiveness and reliability of the approach.

In recent years, the Groundwater Research Group at the University of Toronto has worked with industry partners to develop and refine technologies suitable for the infiltration of stormwater. The work has involved comprehensive field, laboratory and computer model studies, and the primary focus has been the efficiencies and maintenance needs of two contrasting technological approaches to stormwater infiltration. Most of the work has been conducted on low maintenance filter basins (LMFBs) that infiltrate stormwater released from pre-treatment settling ponds. This versatile design appears to be very effective for a broad range of hydrogeological conditions and is easily maintained by regular raking of the sand filter blanket. Mounding problems can be addressed by distributing infiltration amongst several sites, and the presence of low permeability layers above the water table is readily overcome by the installation of vertical drainage columns.

Work on SAGES™, a less conventional technique that uses replaceable filter cartridges to infiltrate untreated stormwater, remains in progress. Preliminary work on the physical attributes of the device suggests this method has considerable potential, particularly in major urban areas where there is limited space for stormwater settling ponds.

ACKNOWLEDGEMENTS

The authors gratefully acknowledge Karina Gelo, Susie Mattson, Massimiliano Pizzato and Patrizia Gatti and for their help in collecting and analyzing the data, preparing figures and reviewing the manuscript. Initial work was supported by research grants to Howard from the Great Lakes University Research Fund (GLURF) (sponsored by Environment Canada and the Natural Sciences and Engineering Research Council (NSERC)) and the Ontario Ministry of the Environment.

More recently, significant financial contributions have been made by NSERC (Strategic Grant to Dr. Emil Frind and others), and the Centre for Research in Earth and Space Technology (CRESTech). The views presented here are those of the authors and are not necessarily endorsed by colleagues or by the funding agencies.

REFERENCES

Asano, T. 1986. *Artificial recharge on groundwater*. Butterworth Publishers: 767p.

Beatty, B. and Thompson, J. 2000. Maintaining groundwater recharge in urban areas via stormwater infiltration. *Proceedings of the 1st Joint IAH-CNC and CGS Groundwater Specialty Conference, 53rd Canadian Geotechnical Conference*, Montreal, October 2000.

Cherry, J.A., Gillham, R.W., Anderson, E.G. and Johnson, P.E. 1983. Migration of Contaminants in Groundwater at a Landfill: A Case Study 2: Groundwater Monitoring Devices. *Journal of Hydrology* Vol. 63, pp. 31–49.

Dillon, P.J., Hickinbotham, M.R. and Pavelic, P. 1994. Review of international experience in injecting water into aquifers for storage and reuse. *25th Congress of the International Association of Hydrogeologists/International Hydrology and Water Resources Symposium of the Institution of Engineers, Australia.* Adelaide, 21–25 November 1994. Preprints of Papers NCPN No. 94/14, pp. 13–19.

Howard, K.W.F., Boyce, J.I., Livingstone, S. and Salvatori, S.L. 1993. Road salt impacts on groundwater quality – the worst is yet to come! *GSA Today* Vol. 3, No. 12, pp. 301–321.

Howard, K.W.F. 1997. Incorporating policies for groundwater protection into the urban planning process. In Chilton, J. et al. (eds.), *Groundwater in the Urban Environment.* A.A. Balkema, Rotterdam Vol. 1. pp. 31–40.

Howard, K.W.F. and Livingstone, S.J. 2000. Transport of urban contaminants into Lake Ontario via sub-surface flow. *Urban Water* Vol. 2, pp. 183–195.

Howard, K.W.F., Beatty, B., Thompson, M.J. and Motkaluk, S.D. 2000. Advancing technologies in the hydrogeological design of urban subdivisions. In Sililo, O. et al. (eds.), *Groundwater – Past Achievements and Future Challenges. XXX Congress of the International Association of Hydrogeologists*, Cape Town, South Africa, November 2000, pp. 947–952.

Howard, K.W.F. 2001. Polluted groundwater – deadly lessons from Walkerton, Ontario, Canada. *Proceedings of the XXXI Congress of the International Association of Hydrogeologists*, Munich, Germany, September 2001, 4p.

Howard, K.W.F. 2002. Urban groundwater issues – an introduction. In Howard, K.W.F. and Israfilov, R. (eds.), *Current problems of hydrogeology in urban areas, urban agglomerates and industrial centres. NATO Science Series IV Earth and Environmental Sciences 8*, pp. 1–15.

Howard, K.W.F. and Gelo, K.K. 2002. Intensive groundwater use in urban areas: the case of megacities. In Llamas, R. and Custodio, E. (eds.), *Intensive use of groundwater: Challenges and Opportunities*, Balkema, pp. 35–58.

Jacenkow, O.B. 1984. Artificial recharge of groundwater resources in semi-arid regions. *Proceedings of the Harare Symposium, IAHS publ. No. 144*, pp. 111–119.

Li, C., Bahr, J.M., Reichard, E.G., Butler, J.J. Jr. and Remson, I. 1987. Optimal siting of artificial recharge: An analysis of the objective functions. *Ground Water* Vol. 25, pp. 141–150.

O'Connor, D.R. 2000. *Report of the Walkerton Inquiry – Part 1 – The events of May 2000 and related issues.* Queens Printer for Ontario, 504p.

Pitt, R., Clark, S., Parmer, K. and Field, R. 1996. *Groundwater Contamination from Stormwater Infiltration*, Ann Arbor Press, Inc., Chelsea, Michigan.

Watkins, D.C. 1997. International practice for the disposal of urban runoff using infiltration drainage systems. In Chilton, J. et al. (eds.), *Groundwater in the Urban Environment: Volume 1: Problems, Processes and Management*; *Proc. of the XXV11 IAH Congress on Groundwater in the Urban Environment*, Nottingham, UK, 21–27 September 1997. Rotterdam: Balkema, pp. 205–213.

CHAPTER 14

Aspects of urban groundwater management and use in India

Shrikant Daji Limaye

Director, Ground Water Institute, 2050, Sadashiv Peth, Pune 411030, India

ABSTRACT: While urban groundwater in India is often highly polluted, it remains a valuable resource. The municipal piped water supply in most major and mid-sized cities is far from satisfactory and people have to use and treat groundwater for domestic and industrial purposes with proper precautions. Large, extensive urban areas are an integral part of the hydrologic basin and affect the quality and quantity of both surface water and groundwater. City planners are becoming increasingly aware of their responsibility for efficient management of water resources in river basins and are now managing the demand for water through water conservation, recirculation and reuse. Another strategy being tested by city planners is the artificial recharge of urban aquifers.

1 INTRODUCTION

In India, civic authorities (municipalities or municipal corporations) provide all municipalities, ranging from large cities to small towns, with piped water for domestic supply. The main source of water for those areas underlain by fractured rock is from perennial rivers or reservoirs created by dams built across seasonal rivers. Groundwater, other than that from infiltration galleries in river beds, is only occasionally used as an auxiliary or emergency source of supply. However, in cities located on thick alluvial aquifers, deep tube wells drilled and maintained by municipalities or municipal corporations form the source for urban water supply. There is no private sector involvement in this field.

In most cities, citizens experience shortages of drinking water or piped water, especially during the summer season. This shortages is due to inadequate availability of raw water, lack of funding available to civic authorities, old designs of tube wells, insufficient capacity of water treatment plants, lack of adequate pressure due to insufficient numbers of high level storage reservoirs, small diameters of water mains and leakage from the distribution systems.

The cost of domestic water supply is highly subsidised and the income received from water bills is insufficient to cover the system's operating and maintenance costs, let alone new investments for improvements. In many cities, the piped water supply is available only for about 3 hours in the morning and 3 hours in the evening. During periods of drought when the sources of raw water are depleting every day, the piped water supply is often restricted to a few hours per week. The water supply service cannot improve because of a lack of funding and citizens do not wish to pay more because the service is poor. This is a vicious circle.

The increased pressure on the urban water supply in India is a result of:

- the unplanned growth of cities,
- the ever increasing number of slums due to the influx of rural population towards urban areas,
- the political necessity to provide "free" water to slum dwellers,
- the high percentage of "unaccounted for water" (UFW) due to pilfering from pipelines,
- political interference in the distribution of available surface water between farmers and citizens, and
- corruption at various levels of government.

The larger urban centres have financial and political clout and so their water supply problems receive attention, while the smaller cities have equally serious problems related to their water supply but receive much less attention.

2 URBAN GROUNDWATER

Investment in the sanitation sector generally lags behind that in the water supply sector. The sewer lines in many cities are old, congested and leaking. Additional pollution of groundwater takes place through the percolation of highly polluted water in streams and rivers flowing through urban areas. In urban areas, leaks from stormwater drains, petrol pumps and chemical storage facilities contaminate groundwater. Newly extended suburban areas are without sewers and leaks from septic tanks, industrial waste landfills and nearby agricultural activities contribute to groundwater contamination both locally and regionally. Where urban growth is up-gradient from the main city, polluted groundwater flows towards the city and becomes further degraded beneath the city. On the other hand, if urban growth takes place down-gradient of the main city, it receives groundwater already highly polluted by the city.

Shallow groundwater in urban and suburban areas is virtually always polluted. Upper-class, educated people as well as poor slum dwellers are aware of this situation. The civic authorities responsible for providing drinking water supply seldom monitor the quality of raw water coming into their water treatment plants or the quality of the outgoing treated water. The task of checking the quality of shallow urban groundwater totally falls outside their responsibility and scope of work. Only a few non-governmental organizations or research scholars from universities are interested in the pollutants appearing in shallow urban groundwater.

The depth to which urban pollutants reach in aquifers depends upon the local hydrogeological setting. Four typical settings include:

- high rainfall areas in hard (fractured) rock aquifers,
- low rainfall areas in hard (fractured) rock aquifers,
- high rainfall areas in alluvial aquifers, and
- low rainfall areas in alluvial aquifers.

In high rainfall areas, the pollution in shallow aquifers builds up in the summer season and is diluted in monsoon rains from June to September. In fractured hard rock aquifers, it is rare to find water below 100 m depth irrespective of the amount of rainfall. Pollutants can travel fast in these aquifers without much attenuation. Old dug wells of 3 to 5 m diameter and 10 to 15 m depth in city areas provide highly polluted water which may only be used for gardens or toilets. Bore wells of 150 mm diameter and 60 to 100 m depth offer a better alternative. If the first 10 to 15 m depth of the bore well is cased with PVC pipe, groundwater of better quality can be obtained in the lower aquifers. In alluvial areas, if the water table is deep due to low rainfall, attenuation of some pollutants takes place in the unsaturated zone, and prevents contamination reaching the water table. However, if the water table is shallow, such attenuation is not possible. Deeper aquifers in a thick alluvial sequence often provide a reliable source of good quality water, especially if the hydraulic gradient is upwards. In many cities located on alluvial aquifers, water from deep tube wells, up to 300 m in depth, is distributed without treatment, even if the deep tube wells are located within an urban area.

Exploration for suitable sites for bore wells in urban areas is very difficult in hard rock terrain. It is important to acquire information generated during other construction projects regarding the nature of strata; the inventory of existing dug wells and bore wells is also helpful. DC Resistivity exploration is almost impossible due to stray currents but low frequency AC Resistivity techniques can be used. However, if the area is paved or asphalted, it is necessary to use a jack hammer to drill a series of 40 mm diameter holes up to 50 cm depth and insert electrodes for Wenner profiling.

In areas with many buildings, access is often a problem for the large "down-the-hole hammer" drilling trucks capable of drilling 150 mm diameter bores up to 100 m depth. In such cases, the compressor is located on a main road and compressed air is carried through a hosepipe to a smaller drilling unit erected on site. These units can drill a 100 mm diameter hole up to 60 to 70 m depth. In alluvial aquifers, small, low-cost tube wells for external household use are drilled with locally assembled "water jet drilling units". The strainer or screen at the well bottom is usually a PVC pipe with slots or holes around which a coir rope is wound.

3 DOMESTIC USE

Many urban residents, especially those living in private bungalows and apartment complexes of cooperative housing societies, prefer to augment the piped water supply available from the municipal pipeline with groundwater obtained from bore wells, tube wells or shallow dug wells. Of these alternatives, only deep tube wells in alluvial aquifers normally provide good quality drinking water. Dug wells and shallow bore wells generally yield polluted groundwater.

Despite a daily consumption of about 120 to 150 liters per capita, only about 10 to 15 liters are necessary for drinking and cooking purposes. If this quantity of drinking water supply were available from municipal pipelines, the groundwater supply from the bore wells could be used exclusively for gardening and similar, non-potable use. Unfortunately, builders do not provide homes with a dual piping system. It is, therefore, a common practice to mix the municipal water (which is chlorinated and is supposed to be safe for drinking) and bore well water in a ground level tank and pump the mixture to a high level tank on the terrace for distribution. Individual owners often install small UV purification units for use in the kitchen. Alternatively, some people prefer to treat water for drinking by boiling.

In urban areas on hard rock terrain, civic authorities do not normally invest in infrastructure to obtain groundwater for drinking. However, in some cities the civic authorities have

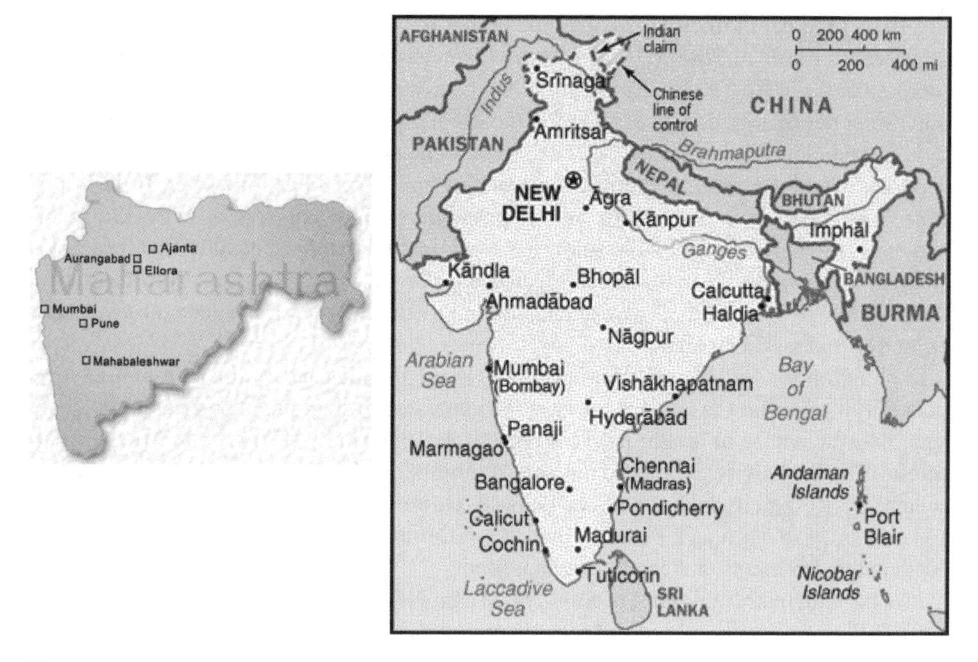

Figure 1. Location of Maharashtra in India.

drilled bore wells by the road side and installed hand pumps, so that people may obtain water for non-potable use. In the basaltic hard rocks of Maharashtra state in western India (Figure 1), many such wells have been drilled by municipal authorities in mid-sized cities like Kolhapur, Satara and Solapur. Mumbai (Bombay) Municipal Corporation once planned a scheme for drilling bore wells in developing suburban areas for emergency use, but in the basaltic terrain the yields from the first few bores were meager and the project was abandoned.

In Kolkata (Calcutta), located in the alluvial deltaic region of the Ganges, raw water pumped from one of the branches of the Ganges is highly polluted. Even after treatment, many people prefer not to drink it. However, deep tube wells yield better quality water and people in some localities transport large quantities of water from such tube wells to use for drinking water in their homes. Municipal piped water is available only for a few hours per day. The municipal corporation has, therefore, drilled shallow tube wells along road-sides and installed hand pumps so that people can obtain additional water for non-potable use at any time.

4 INDUSTRIAL USE

Many industries receive much less than their daily requirement for water from the municipal pipelines. The price set by the municipality for industrial use of water is also much higher than that for domestic use. These two factors are responsible for increasing industrial demand for groundwater supply. If the hydrogeological conditions are favorable, industries prefer to be self-sufficient in terms of their industrial water supply. Because urban groundwater is normally polluted, however, industries often use municipal connections for their drinking water needs.

When constructing new buildings builders often prefer to drill bore wells rather than obtain temporary municipal connections at a high price. Any additional requirements at the building site are met through tanker water supply. An entire industry has developed to supply water in tanker trucks to building sites and other industries facing water shortages. Depending on the proposed use, these tankers either take highly polluted water from shallow wells in the city area or better quality water from bore wells in suburban locations.

In urban areas on thick alluvial aquifers, the shallow aquifer is the first to become polluted. Industries go to the deeper second or third aquifer to find better quality water. Over the years, excessive pumping from these aquifers has created a high downward hydraulic gradient and pollution from the shallow aquifer is now reaching lower levels. Industries must then drill even deeper to reach unpolluted groundwater. Companies manufacturing bottled water or mineral water also depend on deep tube wells. The water pumped from deep tube wells by these companies is good quality raw water which is subjected to micro-filtration, ozonation and UV treatment to ensure potability and is often treated with calcium and sodium salts to produce mineral water.

5 WATER HARVESTING IN URBAN AREAS

During the last decade or so, many international organizations such as UNESCO, the World Water Council, and the Global Water Partnership have been advocating "integrated water resources management" (IWRM) and "integrated river basin management". Urban areas sprawling over many thousands of hectares form a major part of many river basins and have a major influence on the quality and quantity of the regimes of surface water and groundwater. Urban areas consume large quantities of good quality water in the basin and produce comparably large quantities of waste water which flows into nearby streams after receiving little or no treatment. Roads and buildings in urban areas reduce natural infiltration from rainfall to the groundwater table but this is often more than compensated by leakage from water supply pipes and sewers and by the water added directly to lawns and gardens by homeowners. City planners have recently examined ways to make water management in the basins more efficient. One approach is to control demand for water through conservation, recirculation and reuse. Another is to augment replenishment of recharge to urban aquifers using artificial recharge techniques.

In urban areas with an undulating landscape hard rock terrain, groundwater pumping from wells and bores in low lying areas often causes lowering of the water table and a reduction of the yield from wells. In alluvial areas, water table and piezometric decline due to over-exploitation, may also cause land subsidence. In coastal areas, saline water intrusion may take place. In order to counter the effects of the depletion of urban groundwater, city dwellers are now being asked to assist with collecting water for artificial aquifer recharge. During the monsoon season, rain water is collected from terraces and roofs, passed through sand filters and directed into dug wells or bore wells. Some cities are piping water from terraces into sand-filled pits around wells that function as filter jackets.

Some municipal corporations have made the collection for storage of rain water and its storage in cisterns or its recharge to groundwater mandatory for builders seeking permission for new construction projects. Although this works well in alluvial areas, areas with hard rock support low yielding wells (up to 2.0 m^3 per hour) and cannot accept high rates of artificial rainwater recharge. Excess water spills out as rejected recharge. Despite this constraint, the quality of water recharged into the aquifer is normally equal to or better than the native water in the aquifer, because the rain water collected from a roof or terrace is typically very clean.

6 CONCLUSIONS

1. The water supply condition in most of the major cities in India is critical with regards to the quantity and quality of water. However, the worst problems are encountered in smaller towns that have little financial or political power.
2. Urban groundwater is usually polluted, but in the absence of any alternative it is used to augment the municipal piped water supply.
3. Urban groundwater is used for domestic and industrial supply. Due to the absence of dual potable/non-potable plumbing systems in many homes, domestic users are forced to mix the municipal water with groundwater pumped from wells and then use UV purification units for water used in the kitchen. Alternatively water is boiled.
4. Roof or terrace collection of rain water for recharging wells and bore wells is a relatively recent strategy for augmenting water supply, but is gaining popularity. However, local hydrogeological conditions must be taken into account before mandating such techniques in the construction of new buildings.

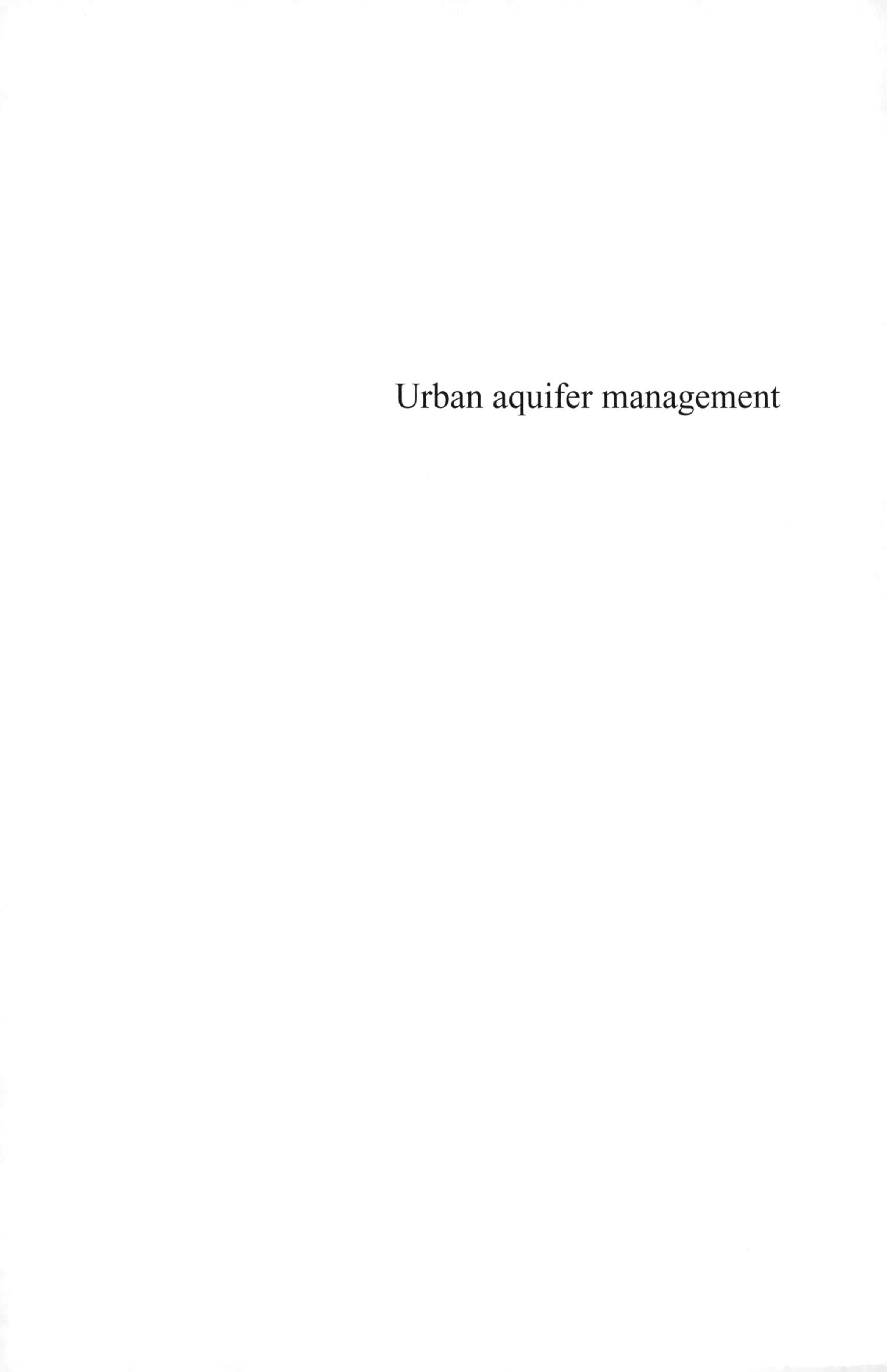

Urban aquifer management

CHAPTER 15

Evaluating groundwater allocation alternatives in an urban setting using a geographic information system data model and economic valuation techniques

Suzanne A. Pierce, John M. Sharp, Jr. and Beatriz Garcia-Fresca
John A. and Katherine G. Jackson School of Geosciences, The University of Texas at Austin

ABSTRACT: In order to meet future water demand scenarios, aquifers must be managed sustainably. The over-abstraction of aquifer systems may occur if methods used to determine groundwater allocation and management strategies fail to be equitable and efficient.

To analyse allocation scenarios, a data model linking hydrogeologic conditions with the economic attributes of individual wells is presented. Input data derive from a combination of hydrogeologic models for water balance calculations loosely coupled with economic valuation attributes and simplified urbanisation parameters. Well locations that experience extreme drawdown under variable pumping scenarios can be identified through comparison with the spatial distribution of saturated thickness within an aquifer. Economic attributes assigned to well data are used to determine the relative costs and benefits incurred to maintain or replace water supplies for affected wells.

This groundwater decision support system (GWDSS) data model allows the analysis and presentation of groundwater simulation information in a manner that is valuable within a decision support context and provides a method for the relative ranking of various water allocation strategies. Resulting calculations can be readily included in an interactive spatial decision support system for sustainable yield calculations. In rapidly urbanising areas spatial decision support systems can provide flexible and adaptable mechanisms for improving resource management.

1 INTRODUCTION

Groundwater sources, or aquifers, are valuable stores of fresh water. Therefore, it is not surprising that as demand grows, the management of groundwater and problems associated with its use and development (e.g., subsidence, contamination, etc.) become an important focal point for the success of communities worldwide. As demands for reliable freshwater sources continue to increase, groundwater resources will become critical to sustaining the world's population, and for maintaining existing ecosystems. Groundwater is the predominant source of water supply for rural areas in the United States, primarily for agriculture and domestic use, and this trend is expected to expand into urban areas in the future (National Research Council, 1997).

Science can contribute to water resource allocation by providing information about the workings of an aquifer system and creating tools to quantify the amount of available water. To address issues related to allocation, decision makers need tools that can assist with the evaluation and presentation of water management alternatives.

This paper presents a preliminary decision tool that provides proof of concept for a method to link simulation and optimisation models with stakeholder preferences to calculate an aquifer yield. Initial components are simplified, but allow flexibility to expand and aggregate additional elements to calculate sustainable yield for urban aquifers. The work presented herein provides the prototype that will be incorporated into a groundwater decision support system (GWDSS) to be applied to the Barton Springs segment of the Edwards aquifer in Austin, Texas, which is a rapidly urbanising area.

2 SUSTAINABILITY AND AQUIFER YIELD

Sustainability describes the rates of use for a resource that are considered appropriate for the current generation's benefit offset by preserving the viability of that same resource for future generations (WCED, 1987). As demands for water resources outpace supplies, the development of methodologies and tools that can be used to define clear paths to reach sustainable groundwater yields becomes increasingly important.

The ubiquitous nature of groundwater in certain areas makes it one of the most valuable and accessible natural resources. Groundwater can provide a reliable source of usable fresh water when an aquifer can "yield", or provide, supplies to water wells. The concept of a "safe yield" was first defined for groundwater resources by Lee (1915) as "… the limit to the quantity of water which can be withdrawn regularly and permanently without dangerous depletion of the storage reserve."

Despite a wide array of papers regarding the topic, clarity with regard to a definition for safe yield remains elusive. Considerations have primarily included natural recharge, together with environmental, economic, and legal considerations (Meinzer, 1923; Conkling, 1945; Todd, 1959; Kazmann, 1968; Loáiciga and Leipnik, 2001; NRC, 1997; Alley et al., 1999; Sophocleous, 2000). Despite much debate in the technical literature, in practice safe yield continues to be calculated as an extraction rate equivalent to long-run annual average recharge (NRC, 1997; Mace et al., 2001).

From its inception, the term "safe yield" has been controversial and is often the pivotal element when calculating acceptable pumping rates for an aquifer system. Bredehoeft et al. (1982) examine misconceptions about the implementation of safe yield practices and points out that safe extraction rates cannot be equal to annual recharge without expected negative impacts. Sophocleous (2000) and Mace et al. (2001) also conclude that a true sustainable yield of an aquifer must be less than the annual recharge rate. In addition, Bredehoeft et al. (1982) aptly point out that pumping inherently modifies a system's equilibrium state and the lag times needed to determine when over-abstraction has occurred may be too long and too late. Considering that the recharge rate for an aquifer is not an adequate metric for calculating acceptable extraction rates, an alternative method for quantifying a water budget needs to be used to determine if over-abstraction has occurred.

As the modern management paradigm shifts towards concepts of sustainability that emphasise society's ability to meet current resource demands while still preserving long-term resource viability for future demands, the technical practices used to calculate yield are coming under scrutiny. An alternative term for a more inclusive calculation is sustainable yield. A sustainable yield is considered to be the volume of water that can be removed from an aquifer that: 1) does not exceed natural and urban-induced recharge rates; 2) avoids negative water quality impacts; 3) preserves economic viability; 4) complies with existing permit constraints; 5) maintains environmental flows; and 6) protects inter-generational equity.

Using the broad guidelines of sustainability, a successful methodology requires linking physical system behaviour with socioeconomic needs to create a more realistic model that manages various scenarios. Figure 1 presents a conceptual diagram of the broad elements that should be included within a sustainable yield calculation. In addition, the diagram reflects the iterative approach needed to implement sustainable yield calculations successfully into active aquifer management programs.

The conceptual elements shown in Figure 1 generally reflect the input and output variables of the water balance equation.

$$I(t) + O(t) = \partial S/\partial t \tag{1}$$

where $I(t)$ = lumped input or recharge components as a function of time; $O(t)$ = lumped output components, or discharge, as a function of time; and $\partial S/\partial t$ = the change in storage relative to time. A key difference in the diagrammatic representation and the hydrologic equation comes about in the distinction between natural and artificial flows. The distinction is necessary because in order to calculate accurate sustainable yields the overall water budget must consider both natural and human-induced flows.

After recognising the necessity of quantifying sustainable yield, a further delineation must be created that links the mathematical equation to the spatial distribution of recharge and discharge sources. As detailed in the following section, some work to link these two systems has already been completed within urban settings. One possible mechanism for addressing the subsequent step of linking urban and natural flows within a decision support context is presented in the methodology section of this paper.

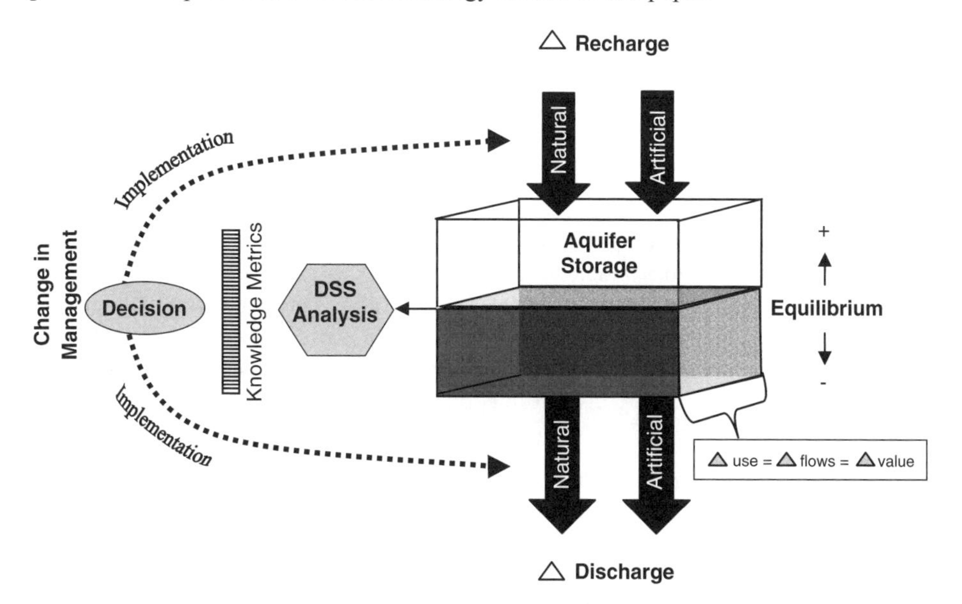

Figure 1. Conceptual components of sustainable yield calculations.

3 SUSTAINABLE YIELD IN AN URBAN SETTING

According to the United Nations (2002), 50% of populations worldwide, live in urban areas and growth rates for urban populations are expected to continue at accelerated rates. From a resource allocation perspective, the influence of urban activity on natural systems has become an important area for research to preserve the standards of living at healthy levels (Howard, 2001). The impact of urban populations is exacerbated by urban sprawl, or the tendency of a human population to spread into low-density, decentralised configurations resulting in the rapid extension of municipal infrastructure (Johnson, 2001). The combination of increased freshwater demands, primarily from groundwater reserves, and urban-induced modifications to the groundwater flow regime adds complexity to the problem of sustainable yield calculations. At the same time, due to rapidly increasing water demands in these areas, urban centers are likely to be areas where stresses on the groundwater system result in policy and allocation conflicts.

Loss rates for cities worldwide compiled by Garcia-Fresca and Sharp (2003) indicate that recharge due to urban water infrastructure leakage can be expected to increase overall recharge to aquifer rates significantly. Yang et al. (1999) concluded that the two main volumetric sources of urban recharge are natural precipitation and urban water infrastructure, for potable and waste flows in most cities. Leakage from these waterlines is a significant source of recharge to urban aquifers and cannot be overlooked when developing groundwater allocation and valuation scenarios. Incorporating the spatial distribution of recharge from urban water distribution systems, together with volumetric estimates, will be an important aspect for calculating a sustainable yield for urban aquifers.

Computerised decision support systems that integrate urban aspects in a flow system, particularly recharge, may demonstrate that urban leakage can provide a valuable increase in urban water resources. Various urban water balance methodologies are proposed in existing literature; two examples include papers by Yang et al. (1999) and Eiswirth et al. (2004). Each focuses specifically on methods for linking the spatial distribution of urban water infrastructure together with estimates for the quantity and quality impact of leakage rates. These studies provide the primary platform upon which the present methodology builds.

4 METHODS

The methods set out for this proposed project are geared to achieve a usable spatial decision support tool that provides a logical, science-based decision sequence for determining groundwater allocation strategies and provides a connection between community preferences, simulation, and optimisation models. The GWDSS is useful for comparing alternative management scenarios, can aid decision-makers when conceptualising a problem, and provides a mechanism for displaying information regarding aquifer system status.

4.1 *A data model and link to hydrologic modeling programs*

To build the initial GWDSS prototype a data model was constructed and a preliminary schema for implementation was applied to an ArcGIS 9.0 geodatabase (Zeiler, 1999; ESRI, 2000) and subsequently versions of the more openly accessible MYSQL database program. A spatially indexed database provides the storage capacity for both structured and unstructured data attributes, while the schema defines principle relationship classes within an urban groundwater context. Data model elements can be divided by functional and data format categories as detailed in Table 1 below. It should be noted that the GWDSS data model presented here is not comprehensive, but it presents an initial integration strategy for point and line features common to urban groundwater problems within a spatial distribution framework, or a loose construction and logical format. Specific data model elements are selected with the intent of proving the potential for linking modeling systems and proposing a functional link with mathematical optimisation equations.

Once developed, the data model schema may then be applied to a database and populated for use in simulation modeling and analysis. A preliminary schema outline for the data model is presented in Figure 2.

The use of geographic information systems (GIS) within a water allocation decision support system provides an important means for integrating spatial data together with numerical models and empirical calculations. Various studies have developed the use of spatial data or GIS within a larger decision support system applied to groundwater (Belmonte et al., 1999; Chowdary et al., 2003; Brown, 1999) but the limitations of GIS to simple displays of surface features has reduced the effectiveness of subsurface applications to date. Developing a relational database link between a GIS and groundwater models, such as MODFLOW (Harbaugh and McDonald, 1996) that can be presented and analysed in the GWDSS are potentially more effective and architecturally the system is more robust than embedded simulation models.

Using raster calculations, the aquifer polygon features (see Table 1, CellPoly) can be generated to correspond to the hydrologic simulation model grid. In effect, the polygon mesh acts as the geometric representation of the hydrogeologic flow model and attributes can be assigned on a cell-by-cell basis. This delineation of spatial features within GIS is the property that makes spatial decision tools unique because the physical system can be modeled while retaining data in its relevant spatial layout.

As the simulation model is run under various management scenarios the output files may then be linked by grid cell back into the geodatabase. This enables analyses such as the creation of either raster or vector based files that represent potentiometric surfaces at different times and conditions. Iterative model runs and analyses can then be collected and compared.

Some traditional hydrogeological methods for estimating the parameters include: water budgeting, numerical modeling, optimisation simulation, chemical tracing, chemical mixing models, flow-net construction, pump testing, slug testing, and geophysical methods (Weight and Sonderegger, 2001). With the ability to link the model results to a spatially indexed database, water budget estimates may now be completed via GWDSS applications. This approach is very similar to the Urban Value Quality (UVQ) approach defined by Eiswirth (2001) with the added ability to evaluate simultaneous well performance. The next step includes linking management objectives and stakeholder weighting mechanisms into the data model, such that a multiobjective optimisation system of equations can be evaluated.

To link the valuation, management, and stakeholder attributes into the database, the object classes and their respective feature attributes are added to the existing data model. As shown in Table 2, these feature attributes span a range of calculated values and user defined input values. For example, the *WellCost* and *PipeCost* object classes are generated from spatial analyses completed within the database, methods may be programmed independently or spatial analysis tools available in some proprietary software products may be used. These object classes are used to calculate the costs associated with either drilling an

Table 1. GIS Data model elements and key attributes.

Feature attributes/ Field name	Data type	Description
Object Class – HydroFeature*		
HydroID	Auto-generated ID	Primary key for all hydrological components of system
HydroCode	Text string	Indicates type of hydrologic feature in ArcHydro format
Object Class – Cell		
HydroID	Auto-generated ID	Foreign key for all hydrological components of system
CellID	Auto-generated ID	Primary key to link fishnet cell with hydrologic elements
CellNam	Text string	Cell row and column name used in groundwater model
CellPoly	Polygon Geometry	Fishnet mesh generated to match finite difference model cells
Object Class – Aquifer		
HydroID	Auto-generated ID	Foreign key for all hydrological components of system
AQID	Auto-generated ID	Primary key to link aquifer with hydrologic elements
AQNam	Text string	Aquifer name used for groundwater availability model
AQtype	Drop-down option	Aquifer behaviour (confined = linear/ unconfined = non-linear)
Object Class – Well		
HydroID	Auto-generated ID	Foreign key for all hydrological components of system
WellID	Auto-generated ID	Primary key to link well objects and hydrologic elements
WellNam	Text string	State designated well identification number
WellType	Text string	Description of reason for well construction & use
DDLatitude	Number	Latitude for well site
DDLongitude	Number	Longitude for well site
LSDElevati	Number	Land surface elevation
WellDep	Number	Well completion depth
AQNam	Text string	Aquifer unit well completed within
WLMeasure	Number	Water level measurement (*Stored as time series)
WLMDate	Date	Date water level collected (*Stored as time series)
Discharge rate	Number	Rate of discharge
Well_life	Short	Anticipated life of well
WellUse	Text	Well use category
Object Class – Pipeline		
HydroID	Auto-generated ID	Foreign key for all hydrological components of system
PipeID	Auto-generated ID	Primary key generated for each pipeline segment
LPoly	Long Integer	Length of pipeline segment
INode	Long Integer	Initial start point for pipeline segment
ENode	Long Integer	End point for pipeline segment
Length	Double	Length of pipe segment
Diameter	Float	Measure of inner diameter for pipeline segment
Year_Inst	Short Integer/Date	Date of original pipe installation
Up_Elev	Float	Upper elevation along segment
Down_Elev	Float	Lowest elevation along segment
Material	Text	Material type construction
Object Class – Pump		
HydroID	Auto-generated ID	Foreign key for all hydrological components of system
PumpID	Auto-generated ID	Primary key generated for each pump
Setdepth	Double	Depth of pump installation within well
Effic	Short	Pump operation efficiency (in kilowatt/hour)

*The Hydrofeature class is a key link into the pre-existing ArcHydro data model framework as developed by the Center for Research in Water Resources at The University of Texas at Austin in conjunction with ESRI (Maidment, 2002).

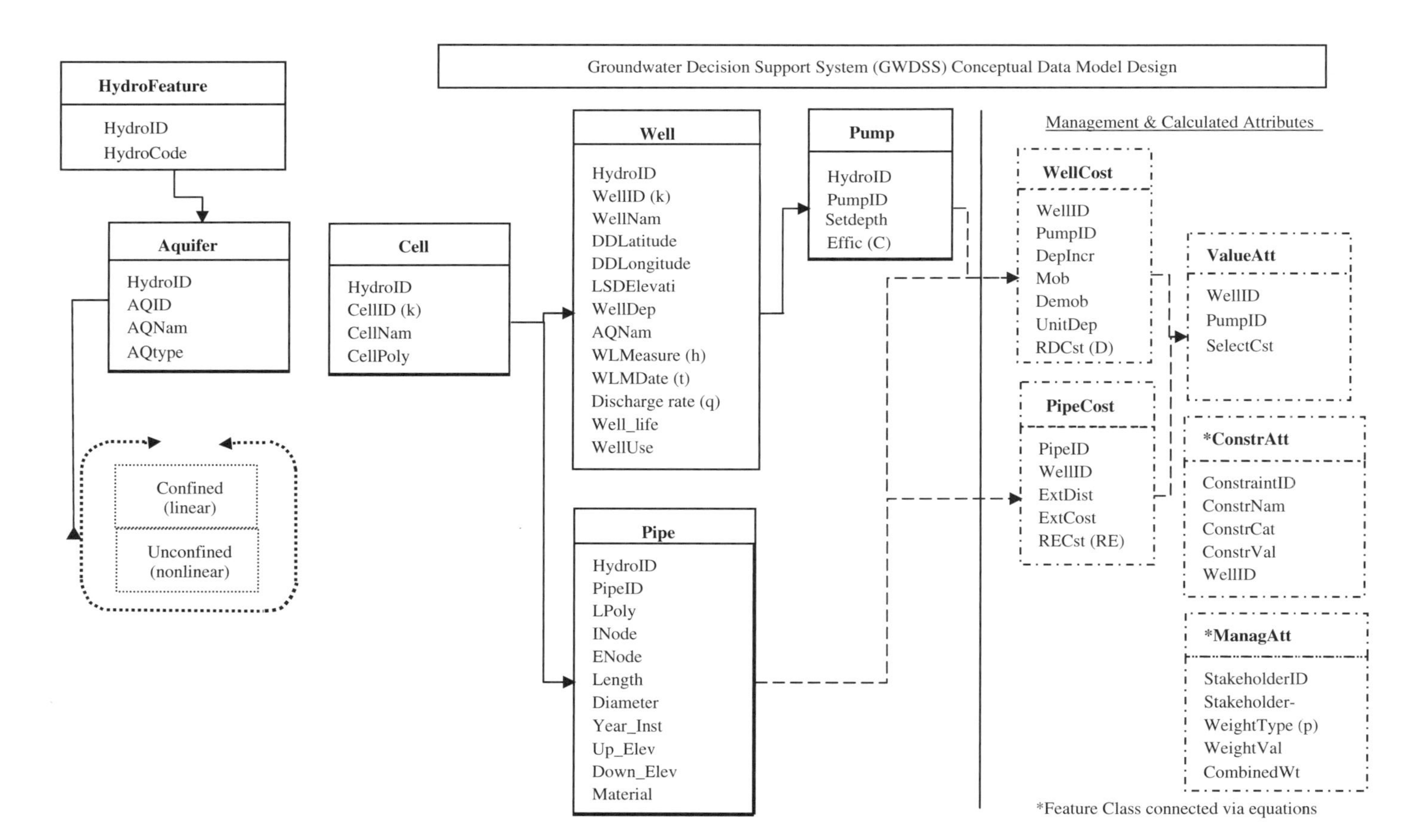

Figure 2. Conceptual and logical Groundwater Decision Support System (GWDSS) data model design to support a database linked to simulation and management models for groundwater. Note that either the Cell or Well may be treated as the atomic unit within the model. Variables noted with () are used in optimisation system of equations.

Table 2. GIS data model calculated fields for management, optimisation, and multiobjective delineation.

Feature attributes/ Field name	Data type	Description
Object Class – WellCost		
WellID	Auto-generated ID	Foreign key for all wells in system
PumpID	Auto-generated ID	Foreign key generated for pumps in wells for system
DepIncr	Double	Calculated depth to reach acceptable saturated thickness
Mob	Double	Mobilisation costs for drilling equipment
Demob	Double	De-mobilisation costs for drilling equipment
UnitDep	Double	Cost per length to extend well depth
RDCst	Double	Cost calculated for deepening well
Object Class – PipeCost		
PipeID	Auto-generated ID	Foreign key generated for each pipeline segment
WellID	Auto-generated ID	Foreign key for all wells in system
ExtDist	Double	Calculated distance from well to nearest pipeline
ExtCost	Double	Unit cost per length to extend pipeline to well location
RECst	Double	Cost calculated to extend infrastructure to affected well
Object Class – ValueAtt		
WellID	Auto-generated ID	Foreign key for all wells in system
PumpID	Auto-generated ID	Foreign key for all pumps in wells for system
SelectCst	Double	Calculated cost per affected well, least expensive option
Object Class – ConstrAtt		
ConstraintID	Auto-generated ID	Primary key for all hydrologic constraints
ConstrNam	Text	Constraint name
ConstrCat	Text	Constraint category description
ConstrVal	Double	Constraint value input by stakeholder for each category
CombinedWt	Double	Combined weights calculated from WeightVal (above)
Object Class – ManagAtt		
StakeholderID	Auto-generated ID	Foreign key for all stakeholder participants
StakeholderNam	Text	Stakeholder group name
WeightCat	Text	Categories established by stakeholders for weighting
WeightVal	Double	Weighted value input by stakeholder for each category
CombinedWt	Double	Combined weights calculated from WeightVal (above)

affected well deeper, resulting in the *RDCst* attribute field, or extend the pre-existing infrastructure lines to the well location, *RECst* attribute field. After creating new cost function fields in the data model the new object class, ValueAtt is generated. The *ValueAtt* object class includes calculated value fields by selecting the least expensive option between the *RDCst* and *RECst* fields, this determination is controlled by a binary variable which is retained in the active memory by the GWDSS.

The importance of spatial data analysis and calculated fields generation is that an original potentiometric surface, either measured or simulated, may be input into the database and management constraints may then be used to calculate a new management scenario. These newly generated value attributes are then input back into the groundwater simulation model and a new potentiometric surface using identified wells can be completed. Steps might include the incorporation of management constraints, via the *ConstrAtt* object class, such as springflow limits or potentiometric minimums, to further improve the groundwater management application. As a final step, the data model includes the *ManagAtt* object class to input weighting preferences for stakeholder groups.

After developing the basic structure for the relational database files and schema, management scenarios can be designed. Figure 3 depicts a hypothetical urban groundwater analysis that might begin with the calculation of distributed urban recharge, linking back into the groundwater simulation model. The next step might include the calculation of saturated thickness levels to identify wells that could be considered at risk for extreme drawdown.

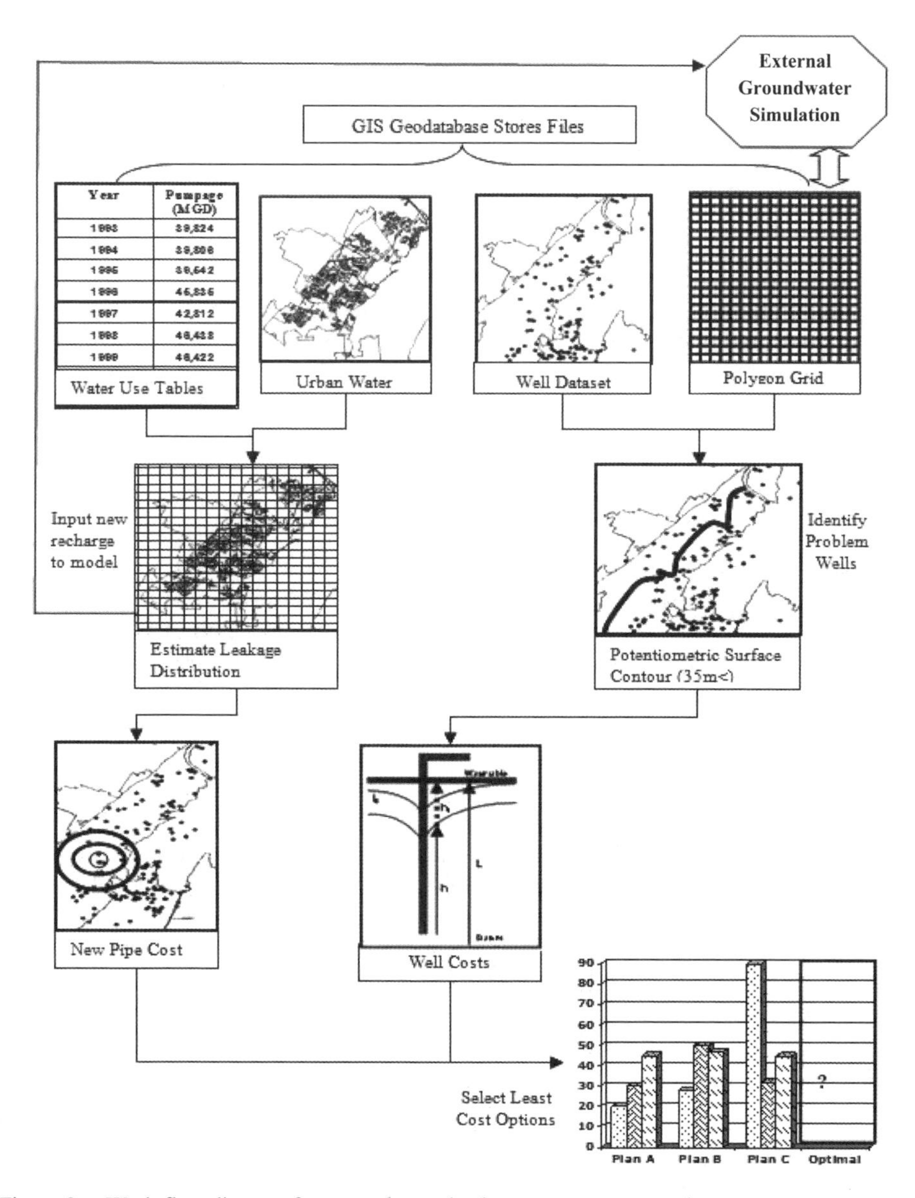

Figure 3. Work flow diagram for example geodatabase management analyses.

Replacement options for wells identified in the simulated potentiometric surface analysis could be quantified using the spatial analyst feature to identify buffer zones of water main infrastructure or drilling costs for increasing the total depth of affected wells. Once a suite of affected wells have been identified and a replacement option selected, then a present worth value for a given management scenario could be calculated.

4.2 *Optimisation modeling*

Using the newly created data model and hydrologic simulation model link it is possible to consider possible multiobjective optimisation scenarios. Multiobjective optimisation can be defined as a weighting and hierarchal approach to project or for strategic ranking analyses that usually requires subjective components and attempts to optimise the objective function while minimising the trade-off for other constraints (Yang et al., 2001).

Multi-objective optimisation models have been proven as effective groundwater management analysis tools by as far back as Gorelick (1983) which gives a precise review of early computer modeling techniques for groundwater. More recently Das and Datta (2001) present a review of optimisation models specifically with the inclusion of multi-objective methods applied to groundwater. Pairing these recognised optimisation techniques, the following sections present two simple objective functions and constraints that are coupled with the groundwater data model presented in previous sections and for use with stakeholder weighting functions for multi-objective analysis.

Figure 4 shows a schematic diagram for a pumping well and the variables that are used in two forms of a cost minimisation objective. One considers variable lift costs only and the second includes capital cost considerations. Lift cost minimisation is a classic agricultural economics approach that allows for the overall cost to lift water to be calculated using the unit cost (C), which is a function of the pump efficiency, set depth, and energy price, multiplied by the well discharge rate, (q), and the difference in the average water table head (L), and the head (h) at a given well, for any well location (k). Additional realism can be added into the set of equations with an objective function that considers both variable costs and capital costs. The system of equations is developed as a cost minimisation problem accounting for both capital and operational costs on a well-by-well (k) basis (or if the user determines that it is appropriate, a cell-to-cell basis can be used as the atomic unit for grid computing). In this setup the overall costs are minimised (Z) with operational costs for pumping (C) at extraction rate (q), and capital costs for either well deepening (DC) or infrastructure extension (EC) are included in the calculation when appropriate through the use of binary variables (M and G respectively). Constraining functions for upper (u) and lower (l) hydraulic head (h) boundaries can be set and output fluxes are constrained such that the difference between flow (F), and extraction rates (q), at any given time (t), with an initial flow value cannot be less than a regulated lower limit (D) and may not exceed maximum total pumping limits (N). Each of the constraining equation limits can be contained within the ConstrAtt table of the data model. This system of constraints touches on the issue of ecosystem management within the context of water resource management.

Classic lift cost minimisation:

$$\text{Min} \sum_{k=1}^{n} W_k = C_k q_k ((L + h_L)_k - h_k) \tag{2}$$

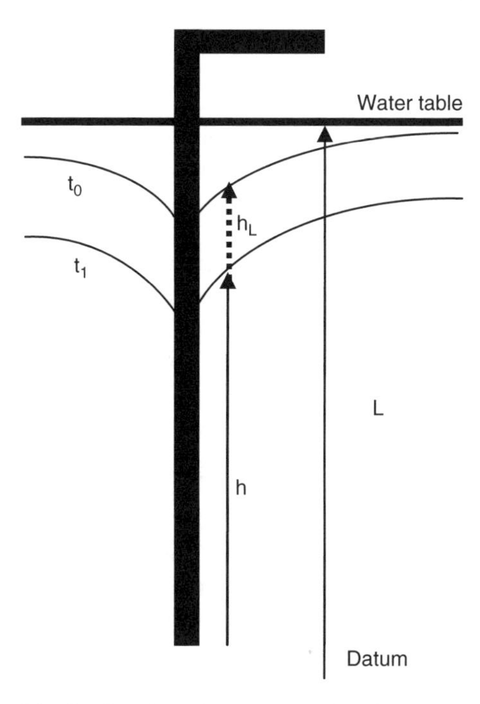

Figure 4. Two possible objective functions that can be used within a callable data model format for linked simulation-optimisation. Classic pump lift minimisation schematic showing the relationship between lift cost (W), unit cost (C), well discharge rate (q), and the difference in the average water table head (L), and the head (h), for any well location (k) at an initial time (t_0) and a subsequent time (t_1). An alternative objective function includes variable costs (C) and capital costs from either well deepening (DC) or infrastructure extension (EC) as controlled by binary variables (M) and (G). Additional constraints for upper (u) and lower (l) head (h) elevations, changes in flow (F) conditions and the maximum pumping limits (N) for a set of candidate wells (X) (modified from Ahlfeld and Mulligan, 2000).

Minimise Capital and Operational Costs:

$$\text{Min } Z = \sum_{k} \sum_{t=1}^{T} [C_k q_{k,t} + (DC_{k,t} \cdot M_{k,t}) + (EC \cdot G_{k,t})] \tag{3}$$

Subject to

$$h^l_{k,t} \leq h_{k,t} \geq h^u_{k,t} \tag{4}$$

$$F^0_{k,t} - F_{k,t}(q) \geq D^l_{k,t} \tag{5}$$

$$\sum_{k} q_k \leq N^u \tag{6}$$

Let $M_{k,t} = \{1 \text{ if well deepened at site k, else } 0\}$
$G_{k,t} = \{1 \text{ if pipeline extended to site k, else } 0\}$

Finally, modifying the classic approaches from above and adding the data model weights, WeightVal (w) for specific uses a generalised multi-objective optimisation approach. To assess the problem in a multi-objective format we can re-write equation 3 such that, extraction locations (k) are also assigned a use type index, NamWeightVal (p) that may be defined within the ManagAtt table of the GWDSS data model.

$$\text{Minimise } Z(x) = w_1 Z_1(x_1) + w_2 Z_2(x_2) + \cdots + w_p Z_p(x_p) \quad (7)$$

where

$$x = \sum_k \sum_{t=1}^{T} [C_k q_{k,t} + (DC_{k,t} \cdot M_{k,t}) + (EC \cdot G_{k,t})] \quad (8)$$

Each of the mathematical system of equations can be incorporated into a groundwater decision support tool using the data model elements and GIS geodatabase techniques described previously. One of the primary goals of this research is the development of a database architecture that can support the use of groundwater simulation-optimisation techniques in a management context, such that the ultimate results are readily explained and used by both managers and stakeholder groups. Linking optimisation techniques, simulation modeling, and more intuitive visualisation provided by a geographic information system data model may prove to be a successful avenue for achieving real-world implementation of science-based resource management.

4.3 *Ranking integrated results*

Using the applications and framework presented above, a ranking tool can be applied to a hypothetical groundwater management scenario. Operational cost estimates for well suites may be determined using traditional cost of lift calculations, while capital costs associated with either infrastructure extension or increasing well depth can be calculated as well. Non-market elements for an aquifer can be represented as either hard or soft constraint functions and, with the development of a user-interface, it is possible that stakeholder groups could potentially assign preference weights.

While the methodologies here do not reflect an absolute valuation of water, they provide a mechanism for comparing disparate management scenarios and ranking the relative outcomes for the simulation, valuation, and optimisation techniques.

5 PROPOSED FUTURE APPLICATIONS

With the completion of the prototype data model, schema application, and mathematical system of equations, a test case can be generated. Future work will involve the implementation of this evaluation scenario for use on an urban aquifer system. A test case with an existing distributed parameter model (e.g., a calibrated MODFLOW model) will be used to define the initial hydrogeologic characteristics and numerical model aquifer conditions under various scenarios. Groundwater demand will be calculated by using pumping information for the specific test case from GIS shapefiles available through municipal agencies. Tools for evaluating the economic cost of water replacement has been developed using standard engineering cost estimating techniques. These are appropriate for initial monetary valuations, but subsequent studies should include new methodologies for incorporating non-market valuation techniques for ecosystem elements. The optimised behaviour of the aquifer system can then be determined through the use of either classical optimisation formulations, such as the objective functions presented here (Ahlfeld and Mulligan, 2000), or non-classical search algorithms, such as TABU search (Glover, 1986).

The philosophies of ecosystem and adaptive management (Christensen et al., 1996; Reichman, 1996) fit well with the calculation of sustainable yields for groundwater resources as approached in the context of this project. According to an ecosystem management approach, the optimisation of a single resource parameter can be detrimental to other related elements within the ecological system (Kessler, 1992). For the purposes of an initial evaluation, ecosystem constraints may be embedded into aquifer management functions as constraining minimum flow levels for spring discharges and potentiometric surface minimums as established by regulating agencies. Using the methodology outlined above, a simplified sustainable yield calculation based on real world policy observations regarding the relationship between saturated thickness and well levels can be converted into a simplified management plan ranking system.

6 CONCLUSIONS

A truly sustainable groundwater resource management model will not optimise a single indicator to define a long-term groundwater management regime, but will take into account the various hydrogeologic, economic, legal, environmental, and other factors to estimate the most appropriate yield for all parties concerned. Currently observed general practice continues to result in the implementation of some version of safe yield (Mace et al., 2001). However, new applications using groundwater decision support systems are capable of providing alternative means for approaching water resource management operations.

English (1999) recognised that future research and work in the area of decision support development can be expected to flourish in areas that 1) develop new tools that are increasingly transparent to the user groups; 2) improve the integration of tools into daily use by decision makers (in other words, keeping the tools off the shelf and in use); and 3) continued collection of input parameter data and improve data measurement.

Initial steps for creating a flexible, user-friendly GWDSS that can provide an avenue for improved communication with regard to water resource conflicts and management are presented. Important advances include the successful development of a data model for integrating urban, hydrogeologic, and economic parameters within a management framework. The application of a data model improves the total description of the hydrologic system by allowing assignment and storage of otherwise disparate data, of both quantitative and qualitative nature, within the same data warehouse.

Additionally, the application of the data model schema to a database provides a link between a spatial geodatabase, groundwater simulation models, and optimisation algorithms. In rapidly changing urban environments, this adaptability is very useful and can prove invaluable to decision-makers.

The increased ability to analyse through the spatial decision tool allows for a comparison and relative ranking of management scenarios in a decision support context. Ultimately, a systems optimisation and multi-objective approach can be evaluated using the new GWDSS data model providing avenues for increased understanding of human influences on aquifer systems and potential for direct stakeholder involvement in the evaluation process.

REFERENCES

Ahlfeld, D.P. and Mulligan, A.E. 2000. Optimal Management of Flow in Groundwater Systems, Academic Press, San Diego.

Alley, W.M., Reilly, T.E. and Franke, L.O. 1999. Sustainability of Ground-water Resources. U.S. Geological Survey Circular 1186, U.S. Geological Survey, Denver, Colorado.

Belmonte, A.C., González, J.M., Mayorga, A.V. and Fernández, S.C. 1999. GIS tools applied to the sustainable management of water resources: Application to the aquifer system 08-29. *Agricultural Water Management*, 40(223): 207–220.

Bredehoeft, J.D., Papadopulos, S.S. and Cooper, H.H., Jr. 1982. Groundwater: The Water-Budget Myth, National Academy Press, Washington, DC.

Brown, M.C. 1999. MODFLOW data reader: A processor for importing MODFLOW data sets into ArcView. HSI GeoTrans, Inc., Sterling, VA.

Chowdary, V.M., Rao, N.H. and Sarma, P.B.S. 2003. GIS-based decision support system for groundwater assessment in large irrigation project areas. *Agricultural Water Management*, 62: 229–252.

Christensen, N.L., Bartuska, A.M., Brown, J.H., Carpenter, S., D'Antonio, C., Francis, R., Franklin, J.F., MacMahon, J.A., Noss, R.F., Parsons, D.J., Peterson, C.H., Turner, M.G. and Woodmansee, R.G. 1996. The report of the Ecological Society of America committee on the scientific basis for ecosystem management. *Ecological Applications*, 6(3): 665–691.

Conkling, H. 1945. Utilization of groundwater storage in stream system development. Proceedings of the American Society of Civil Engineers, New York, NY: 33–62.

Das, A. and Datta, B. 2001. Application of optimisation techniques in groundwater quantity and quality management. *Sadhana*, 26(4): 293–316.

Eiswirth, M. 2001. Hydrogeological Factors for Sustainable Urban Water Systems. Proceedings of the NATO Advanced Research Workshop on Current Problems of Hydrogeology in Urban Areas, Urban Agglomerates and Industrial Centres, Baku, Azerbaijan: 159–183.

Eiswirth, M., Wolf, L. and Hötzl, H. 2004. Balancing the contaminant input into urban water resources. *Environmental Geology*, 46: 246–256.

English, L.P. 1999. Improving data warehouse and business information quality methods for reducing costs and increasing profits. New York.

ESRI. 2000. CASE tools tutorial: Creating custom features and geodatabase schemas. Environmental Systems Research Institute.

Garcia-Fresca, B. and Sharp, J.M. 2003. Hydrogeologic considerations of urban development-Urban-induced recharge. Humans as geologic agents, Geological Society of America Reviews in Engineering Geology.

Glover, F. 1986. Future paths for integer programming and links to artificial intelligence. *Computers and Operations Research*, 13(5): 533–549.

Gorelick, S.M. 1983. A review of distributed parameter groundwater management models and methodologies. *Water Resources Research*, 19(2): 305–319.

Harbaugh, A.W. and McDonald, M.G. 1996. User's document for MODFLOW-96, an update to the US. Geological Survey modular finite-difference ground-water flow model.

Howard, K.W.F. 2001. Urban Groundwater Issues – An Introduction. Proceedings of the NATO Advanced Research Workshop on Current Problems of Hydrogeology in Urban Areas, Urban Agglomerates and Industrial Centres, Baku, Azerbaijan: 1–13.

Johnson, M.P. 2001. Environmental impacts of urban sprawl: a survey of the literature and proposed research agenda. *Environment and Planning*, 33: 717–735.

Kazmann, R.G. 1968. From Water Mining to Water Management. *Ground Water*, 6(1): 26–28.
Kessler, W.B. 1992. New perspectives for sustainable natural resources management. *Ecological Applications*, 2: 221–225.
Lee, C.H. 1915. The determination of safe yield of underground reservoirs of the closed basin type. *Trans. Amer. Soc. Civil Engrs.*, 78: 148–151.
Loáiciga, H.A. and Leipnik, R.B. 2001. Theory of sustainable groundwater management: an urban case study. *Urban Water*, 3: 217–228.
Mace, R.E., Mullican, W.F. and Way, T.S. 2001. Estimating groundwater availability in Texas. 1st annual Texas Rural Water Association and Texas Water Conservation Association Water Law Seminar: Water Allocation in Texas, The Legal Issues, Austin, Texas, 16.
Maidment, D.R. 2002. ArcHydro: GIS for Water Resources, ESRI Press.
Meinzer, O.E. 1923. Outline of groundwater hydrology, with definitions.
National Research Council. 1997. Valuing Ground Water: Economic Concepts and Approaches, National Academy Press, Washington: DC.
Reichman, O.J. 1996. The scientific basis for ecosystem management. *Ecological Applications*, 6(3): 694–696.
Sophocleous, M. 2000. From safe yield to sustainable development of water resources – the Kansas experience. *Journal of Hydrology*, 235(1–2): 27–43.
Todd, D.K. 1959. Groundwater Hydrology, John Wiley & Sons, Inc., New York.
United Nations. 2002. World Urbanization Prospects – 2001 Revision. United Nations Population Division, Department of Economic and Social Affairs, ESA/P/WP.173, New York.
WCED. 1987. Our Common Future. World Commission on Environment and Development, Oxford (England).
Weight, W.D. and Sonderegger, J.L. 2001. Manual of Applied Field Hydrogeology, McGraw-Hill, New York.
Yang, Y., Lerner, D.N., Barrett, M.H. and Tellam, J.H. 1999. Quantification of groundwater recharge in the city of Nottingham, UK. *Environmental Geology*, 38(3): 183–198.
Yang, Y.S., Kalin, R.M., Zhang, Y., Lin, X. and Zou, L. 2001. Multi-objective optimization for sustainable groundwater resource management in a semiarid catchment. *Hydrological Sciences Journal*, 46(1): 55–72.
Zeiler, M. 1999. Modeling Our World: The ESRI guide to geodatabase design, Environmental Systems Research Institute, Inc., Redlands, California.

CHAPTER 16

Can urban groundwater problems be transformed into new water resources?

Eric van Griensven[1], Martine Verhagen[1], Frank van Swol[2], Jan Eerhart[3], Erik Hendrickx[4], Suzan Krook[5], Dana Kooistra[6] and Jos Peters[6]

[1]*Brabant Water, Postbus 1068, Shertogenbosch, the Netherlands*
[2]*Eindhoven Municipality, Eindhoven, the Netherlands*
[3]*Province of Noord-Brabant, the Netherlands*
[4]*Water Board 'De Dommel', Bosscheweg 56, 5283WB, Boxtel, the Netherlands*
[5]*Brabantse Milieu Federatie (Provincial Environmental Federation of Brabant – NGO), the Netherlands*
[6]*DHV Water, Postbus 484, 3800 AL Amersfoort, the Netherlands*

ABSTRACT: In the Netherlands the loss of wetlands has led to a water management policy that intends to stabilize or decrease the use of rural groundwater. As a consequence, future increase in demand for water must be met from urban groundwater or available surface water resources. Cities commonly face the opposite problem with high groundwater levels leading to problems such as flooded gardens, cellars and basements. This not only causes an inconvenience to residents but represents a threat to public health. Remedial measures are expensive, and require installation of drainage systems that place a heavy additional load on sewers and wastewater treatment plants. Refurbishment of sewers further aggravates the groundwater problems. Current remedial measures appear to be inefficient for both environment and society.

Increased abstraction of groundwater can alleviate high water table problems in urban areas but can this solve the problem? Urban groundwater has very little protection against contamination and in many cities has been degraded over centuries of development. Treatment of this water can be complex and expensive. Pumping of urban groundwater potentially spreads contaminants in the environment and raises the concern for cleanup. Water companies are aware of this potential risk but do not feel it is their responsibility to manage the groundwater table.

During the last decade, high groundwater levels have become a problem in Eindhoven, one of the larger cities in the south of the Netherlands. Causes are various but are also attributed to a reduction in pumping both by industry and the local water company. Solutions are limited, but one option is to restore the rate of pumping. This approach features both technical and political challenges since only a proportion of the pumped water will be suitable for supply and could require expensive treatment. The water company together with the municipality, the water board and the province, is faced with a difficult decision that has to consider both water supply needs and the issues of cost and sustainability.

1 INTRODUCTION

The Netherlands is a densely populated country (pop. 16 million) with an area of 41,500 km^2. Large parts of the country are below sea level and the country has experienced a long and challenging history with water issues. One of these challenges relates to the groundwater table which is very shallow and occurs at a depth of less than 4 m throughout 90% of the country (Koreimann et al., 1996). In many areas, the water table is less than half a metre below surface during winter periods.

The use of groundwater as a source for potable supply began in the 19th century. Towards the end of the 20th century, total annual abstraction of groundwater had reached 900 million m^3 for both domestic and industrial use with much of the water abstracted close to the cities where demand is greatest. Much of this increase has taken place in recent decades with urban growth accelerating as a consequence of economic development. Unfortunately, much of this urban development has occurred in groundwater protection zones and has compromised raw water quality.

In rural areas, the loss of wetlands has been an issue of growing concern. In response, the provincial authorities have developed water management policies that prevent any increase in the volume of groundwater abstracted for industrial and domestic use, requiring that abstractions affecting wetlands be reduced and either shifted to urban areas or be replaced by water production from surface water sources. This has created a dilemma for the water companies. For reasons of cost efficiency, water companies generally prefer to use rural groundwater resources and have been closing down pumping stations in urban areas where raw water quality is notably inferior. The downside to this policy is that many cities report problems with rising groundwater tables with flooding of basements, cellars and gardens common during wet winter periods. The water companies argue that it is not their responsibility to manage groundwater levels and that part of the problem is caused by industries that have shifted their activities to peri-urban and rural areas causing a further reduction of urban groundwater abstraction. Eindhoven is one such city affected by rising groundwater levels and angry citizens have called upon local politicians to take action. The solution is not easy and requires the co-operation of many parties.

2 DESCRIPTION OF PROBLEM

The city of Eindhoven has grown from a small village to the fifth largest city in the Netherlands with slightly over 200,000 inhabitants in 2001. Today, Eindhoven serves as a focal point of technology in the south of the Netherlands and plays an important role in the economic growth of the region.

Eindhoven is situated in the valley of the Dommel River (Figure 1), a transboundary river that begins its journey across to the south in Belgium. In the mid-1990s, a growing number of citizens complained about flooded cellars and basements and submerged gardens and the municipality began a review to determine the cause and magnitude of the problem and the legal responsibilities. The problem was found to be extensive with high water table conditions encountered in 18 districts of the city (Figure 2). The cause was attributed to a 50% reduction in groundwater abstraction by both industry and the local water company (from over 20 to less than 10 million m^3 per year) (Figure 3), exacerbated by:

- low permeability sediments (fine sands and silt in large parts of the city) that hinder natural drainage, and
- the removal of natural streams and old drainage systems.

Due to the Netherlands' long struggle with water related problems, water management tends to be well organized, especially in rural areas; however, the situation is less satisfactory in

Figure 1. Location of Eindhoven and extent of the Dommel River catchment (after Pieterse et al., 1998).

urban areas where water management is still not fully regulated. As a result, the various parties involved have no legal obligation to take action to solve the groundwater problem.

Studies were conducted by the city to identify and understand the rising groundwater problem and to develop solutions that would both solve the issue and prevent any re-occurrence. The costs involved with resolving these problems were assessed and work commenced in a number of districts. The first solution involved dredging ditches and installing subsurface drains, but this remedial task had proved onerous and less than ideal due to the expense and considerable advanced planning required. Dealing with all the city's known problem areas would take over 10 years, moreover, the remediation work is not problem free. Disposal of drained water places a large additional burden on the existing sewer system and wastewater treatment plant. Capacity, costs and sustainability are the primary concerns. As a result, new, more suitable methods of lowering the city's groundwater levels are required. One option is to restore previous rates of groundwater abstraction in the city and

Figure 2. Areas of shallow water table in the city of Eindhoven where the water problems mainly occur:
- basements, which are situated more than 1 metre below ground level;
- crawl spaces, basements and convector heaters, which are situated less than 1 metre below ground level;
- gardens on ground level.

utilise a proportion of this water for drinking water and industrial use, but the solution is more challenging than it seems.

3 PARTNERS INVOLVED

The responsibility for managing the groundwater table in urban areas is not only poorly defined but also difficult to allocate. Recently, a national committee tried to address the

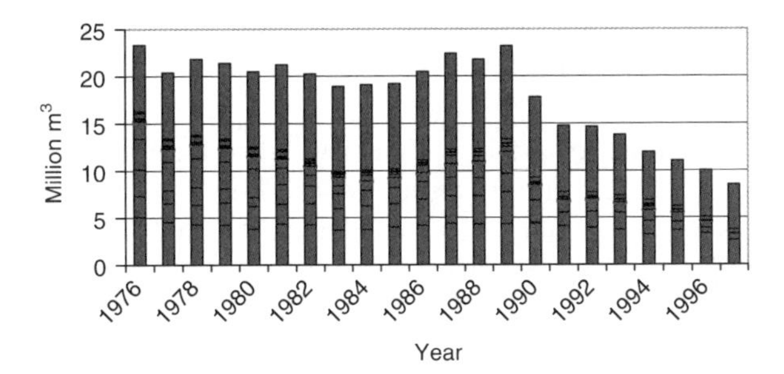

Figure 3. Total annual groundwater abstraction in Eindhoven since 1976 showing a significant decline in pumping during the 1990s.

issue but, disappointingly, was unable to make much progress. In the Eindhoven case the key players are numerous. They include:

1. The Eindhoven municipality. The city of Eindhoven is central to the issue and has embarked upon the long and expensive work program described above that plans to drain problem areas and release discharge into the sewers. The program involves refurbishment and upgrading of the urban sewer system and is heavily criticised by a vociferous group of impatient citizens who have been affected by the problems and feel the work is too costly and too labour intensive.
2. The Water Board De Dommel takes its name from the local river. The Water Board's responsibility is water management with a focus on the quantity and quality of surface water in rural areas. The Water Board is also responsible for sewage treatment and disposal of treated effluent.
3. The Province of Noord Brabant. As the regional government, the province supervises the water authorities, is a shareholder of the water companies, and is responsible for groundwater management (and protection) and for issuing licenses to water companies and industries for the abstraction of groundwater. It is the provincial authority that introduced the legislation requiring wetlands to be protected through a moratorium on rural groundwater development (Province of Noord Brabant, 2002).
4. The Brabant Water company is in charge of public water supply for the region. Although Dutch water companies decline to take any responsibility for managing the groundwater table, they see themselves as actors who might influence it positively. Brabant Water operates 35 drinking water production stations distributed all over the province. From a quality point of view the company prefers either pumping stations located in rural areas or those that draw water from deep aquifers.
5. The Environmental Federation of Brabant represents all ecological and environmental organisations in the province.

The water company, together with the municipality, the water board and the province are faced with a difficult decision that has to consider both water supply needs and the issues of cost and sustainability. The five parties involved have agreed to proceed with a joint initiative that focuses on a potential solution but explicitly avoids any discussion of who is responsible for the groundwater problems or who should take responsibility for action. They have agreed to work together in recognition of their experience, stakeholder interest and potential to contribute constructively to a meaningful resolution.

4 RESEARCH INTO THE FEASIBILITY OF PUMPING TO RESOLVE GROUNDWATER PROBLEMS

The first question that must be asked is: "Can elevated groundwater levels in urban areas be resolved by increased abstraction?" To address this question a study was carried out to determine how existing abstraction wells could be utilised, and to identify an appropriate pumping strategy (DHV, 2003a). Specifically, there was a need to determine:

- how much water needs to be pumped and when (time of the year),
- can the water be used beneficially (e.g. for drinking water supply, to meet industrial demand, or perhaps as a source of artificial recharge) or would the pumped water simply be discharged as waste.

During the study, the willingness of partners to co-operate in the execution of the various management options was also tentatively explored. It was recommended that two pilot studies be initiated in districts seriously affected by the elevated groundwater levels. Locations reporting problems caused by stagnating groundwater were avoided, as too were areas where corrective drainage in combination with sewer renewal had already been implemented. Provisional calculations showed that in the order of 300,000 m^3 per month would need to be abstracted and that some existing abstraction and treatment facilities could be used if required, at short notice (DHV, 2003a).

Finding a beneficial use for the water is a particularly difficult challenge. Ideally the water should be used as a drinking water supply or to meet industrial needs. However, although the water can be treated using advanced techniques, this may be insufficient or cost prohibitive. The quality of urban groundwater has not been well protected in the past and the degree of protection in the future is not known. Moreover, existing sources of water readily meet the present regional demand for drinking water, and the additional urban groundwater supplies are surplus. One interesting option is to treat the abstracted water and transport it to rural areas just outside the city where it would be re-introduced to the groundwater system artificially (Peters, 1998). This could be carried out using gravity-flow artificial recharge facilities, or alternatively using an induced recharge approach by releasing the abstracted groundwater to surface water systems that are under the influence of groundwater pumping. In either case, the urban groundwater would help mitigate the long-term effects of pumping in peri-urban and rural areas and the loss of wetlands.

The first pilot study was implemented during the winter of 2003/2004, primarily to determine:

- the effectiveness of using existing wells,
- the degree to which groundwater levels could be lowered, and
- the most suitable use for the water.

The second pilot study began as a "desk study" in the autumn of 2004. This work focused on the most appropriate way to use the water such that the solution would be sustainable for many years to come.

5 URBAN GROUNDWATER CONTAMINANTS

While increased pumping of shallow groundwater can readily resolve the problem of high groundwater tables in urban areas, it is the quality of urban groundwater that has become

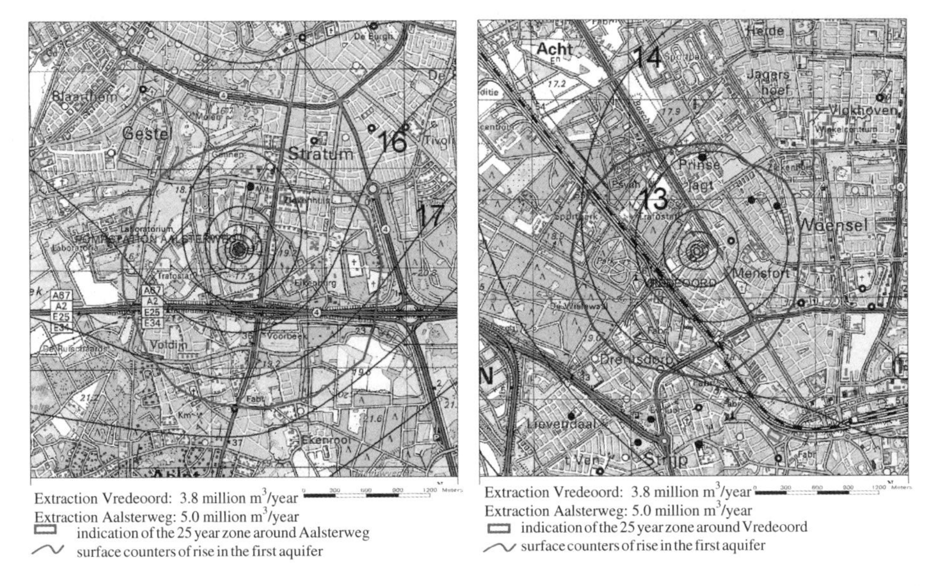

Figure 4. The 25-year time-of-groundwater-travel for pumping wells Aalsterweg and Vredeoord in the city of Eindhoven.

the primary obstacle in establishing sustainable and cost-effective solutions. A major concern is that existing raw water treatment facilities are simply inadequate to deal with the wide range of contaminants commonly encountered in urban areas.

In Eindhoven, the approach adopted involves a thorough review of contaminant sources registered and reported in the immediate vicinity of abstraction sites. Given that, the following important factors include:

- contaminant location and depth,
- contaminant load,
- groundwater flow velocity and travel time, and
- contaminant transport behaviour (mobility, decay, retardation).

Detailed consideration was given to zones established around the Aalsterweg and Vredeoord pumping wells, representing a 25-year groundwater travel time (Figure 4).

From this work it was determined that at a minimum, the water treatment facilities should be enhanced with carbon filters to target reported contaminants such as mineral oils, BTEX, naphthalene and chlorinated hydrocarbons. Tables 1 and 2 show the known contaminants and their current status are given for the Aalsterweg and Vredeoord pumping wells.

6 CONCLUSION – THE DILEMMA

The ultimate question concerns the issue of sustainability i.e. the task is achievable but can abstraction of urban groundwater on a large scale (either for beneficial use or for discharging as waste) be considered a sustainable long-term solution? Urban groundwater is rarely

Table 1. Ground and groundwater contamination within the 25-year time-of-travel zone for Aalsterweg.

Address	Kind of contamination*	Source	Clean up status
Eikenburg	MO	Old tank	No clean up, not an urgent case
Fuchsiastraat 1/ Ranonkelstraat 57	VOCl	Chemical washery	Clean up before 2015
Rubbish dump de Roosten	BTEXN, VOCl	Old rubbish dump	Control measures in place and more than sufficient
Antoon Coolenlaan 1a	MO/naphthalene	Old tank	No clean up, clean up plan is available
prof. Holstlaan 4 (nr. 2360)	MO/BETX		Cleaned up
prof. Holstlaan 4 (nr. 2361)	VOCl	Solvents depot Philips lab. 1962–1982	No clean up
Hippocrateslaan 23	MO + xylene	Leakage garage	Under construction
Gestelsestraat/Laarstraat (Waalre)	MO, BTEXN	Former petrol station	Source of contamination is gone, some contamination of groundwater left (probably control in near future)
Petunialaan 16 (Waalre)	Heavy metals, VOCl	Galvanotechnical company	No clean up
Gestelsestraat (Waalre)	VOCl	Metal protection company	No clean up
Schoenoordstraat (Waalre)	MO, VOCl	Old oil tank	No clean up
Emmastraat 1a (Waalre)	VOCl	Chemical washery	No clean up

*VOCl = volatile chlorinated hydrocarbons; BTEXN = benzene, toluene, ethylbenzene, xylenes, naphthalene; MO = mineral oil.

Table 2. Ground and groundwater contamination within the 25-year time-of-travel zone for Vredeoord.

Address	Kind of contamination*	Source	Clean up status
former Etos ground	MO/BTEXN	Tanks former	No clean up
dr. Cuyperslaan 38	MO/BTEX	bakery	Cleaned up (remainder of contamination is present, risk of spreading is very low)
Joris Minnestraat 2	MO/BTEX	Petrol station	Cleaned up
Boschdijk 263	MO/BTEX	Petrol station/ workplace	Clean up before 2015
Frankrijkstraat 7	MO/BTEX	Petrol station	Unknown, probably not cleaned up yet, there is a clean up plan available
Marconiplein	MO/BTEX	Petrol station	Only the ground is cleaned up, a strong contamination of the groundwater is present
Glaslaan 2, Philips Strijp-S-complex	Heavy metals, PAH, MO, BTEX, VOCl	Philips factories	No clean up, clean up plan is available

*VOCl = volatile chlorinated hydrocarbons; BTEXN = benzene, toluene, ethylbenzene, xylenes, naphthalene; MO = mineral oil; PAH = polycyclic aromatic hydrocarbon (DHV, 2003b).

adequately protected from contamination, and in many European cities has been degraded over centuries. This situation is unlikely to change in the future. Is it wise to consider this water as a suitable source of drinking water when in rural areas sources of far superior quality are available? This is only part of the dilemma. The use of sources in rural areas readily impact wetlands and the natural environment, and the urban groundwater problem quite clearly, cannot be left unsolved. At the very least we need to install an adequate sub-surface drainage system and discharge relatively clean water into the sewer system and subsequently to the wastewater treatment plant.

To address these questions, ongoing work is examining cost and sustainability issues for a number of options. This is a difficult task as sustainability is not an objective parameter. Subjective aspects include the future risk of flooding, the effect on client-trust (the clients' perceptions of the water company with respect to drinking water quality), and the perceived impacts on wetlands. Such concerns are not easily evaluated and compared. To deal with this problem, we have created a process in which each of the parties involved is asked to identify issues of relevance to its organisation and develop appropriate criteria for decision-making. Co-operatively we intend to develop a plan that can solve the urban groundwater problem, make good use of the urban groundwater and is politically acceptable.

REFERENCES

DHV. 2003a. Feasibility study: Suppress the groundwater problems by abstraction of urban groundwater in Eindhoven (in Dutch).

DHV. 2003b. Costs of sustaining the abstraction of Vredeoord (in Dutch).

Koreimann, C., Grath, J., Winkler, G., Nagy, W. and Vogel, W.R. 1996. Groundwater Monitoring in Europe. European Environment Agency. 141pp.

Peters, J.H., ed. 1998. Artificial Recharge of Ground Water. Proceedings of the Third International Symposium, Amsterdam, Netherlands, A.A. Balkema, Rotterdam, Netherlands and Brookfield, VT.

Pieterse, N.M., Schot, P.P. and Verkroost, A.W.M. 1998. Demonstration Project for the Development of Integrated Management Plans for Catchment Areas of Small Trans-Border Lowland Rivers: the River Dommel. 3. Simulation of the Regional Hydrology of the Dommel Catchment (in Dutch, with English Summary), Department of Environmental Science, Utrecht University. 63pp.

Province of North Brabant. 2002. Provincial policy paper on the protection of groundwater (in Dutch).

CHAPTER 17

Hydraulic interactions between aquifers in the Viterbo area (Central Italy)

A. Baiocchi[1], F. Lotti[1], V. Piscopo[1], U. Chiocchini[2], S. Madonna[2] and F. Manna[3]

[1]*Dipartimento di Ecologia e Sviluppo Economico Sostenibile, Università degli Studi della Tuscia, Viterbo, Italy*
[2]*Dipartimento di Geologia e Ingegneria Meccanica, Idraulica e Naturalistica, Università degli Studi della Tuscia, Viterbo, Italy*
[3]*Dipartimento di Studi di Chimica e Tecnologia delle Sostanze Biologicamente Attive, Università degli Studi di Roma "La Sapienza", Rome, Italy*

ABSTRACT: Anthropogenic impacts on groundwater flow in the area of Viterbo have been investigated. The study area, approximately 20 km^2, includes both an urbanised and a rural area, each with a different hydrogeological setting and groundwater usage patterns.

In the urbanised area the impacts on a shallow volcanic aquifer are typical of a growing town, and include increase in runoff and recharge from sewer leakage and urban streams. The hydrogeological setting underlying the rural area is more complex in that there is hydraulic interaction between two aquifers, each used for distinctly different purposes: the shallow volcanic aquifer, characterised by fresh waters heavily used for irrigation, and deep hot waters (up to 64°C) which supply thermal spa centres. In this less urbanised area the impacts on groundwater quality are a result of the hydrodynamic response of the two aquifers. Pumping from the shallow volcanic aquifer can cause deep thermal waters to invade the shallow volcanic aquifer, degrading the quality of fresh waters; on the other hand, over-exploitation of deep groundwater can cause, at least in the longer term, a reversal of the existing vertical gradient, exposing thermal waters to surface pollutants.

1 INTRODUCTION

The town of Viterbo, capital of the Viterbo Province is located within the Lazio region of Italy. Viterbo is located approximately 70 km to the north of Rome and is situated within Central Italy's volcanic region (Figure 1). The town consists of approximately 60,000 inhabitants and boasts the largest population within the province. Viterbo has a small medieval town centre, enclosed by XI-XII century walls and during the 1960s and 1970s urbanisation spread beyond the walls toward the north, east and south of the city centre (Figure 2).

Approximately 3 km west of the town centre, urbanisation is less intense. This more rural area hosts the thermal springs (temperatures between 55 and 64°C) of Bullicame, Carletti and Zitelle. Historically, these springs were used by the Romans for hydrotherapy who in turn named the town "Vita Erbo" (City of Life).

Viterbo is a flourishing agricultural and commercial centre. Urbanisation has been confined to the town centre leaving the rural areas, which host the thermal springs used mainly

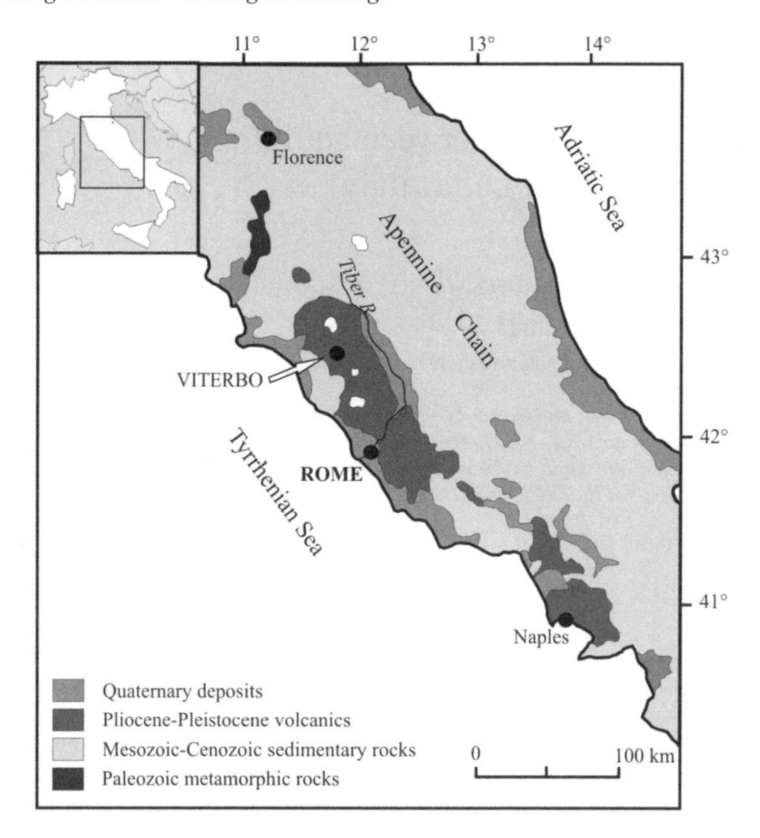

Figure 1. Location of Viterbo Town.

for tourism, less developed. However, further residential areas are planned for the rural area in the future.

The study area is approximately 20 km^2 and includes the town of Viterbo and its surroundings (Figure 2). The aim of this paper is to define the hydrogeological setting of the area and to provide a preliminarily assessment of anthropogenic impacts on groundwater quality and quantity. This in turn can be used for planning and regulating groundwater during future urban developments within the area.

2 GEOLOGICAL AND HYDROGEOLOGICAL FRAMEWORKS

The study area lies within the north-western portion of the Pleistocene Cimino-Vico volcanic district (Figure 3). The area is underlain by volcanic rocks derived from two distinct magmatic cycles, including a felsic cycle of the Tuscan–Latium Province (Cimino Volcanic District) and an alkaline-potassic cycle of the Roman Comagmatic Province (Vico Volcanic District). The Cimino activity (1.35–0.8 My) developed through NW-SE and NE-SW fracture systems and gave rise to lava domes, quartz-latitic surge products and pyroclastic flows (Typical Peperino *Auct.*), quartz-latitic, latitic and olivine-latitic lava

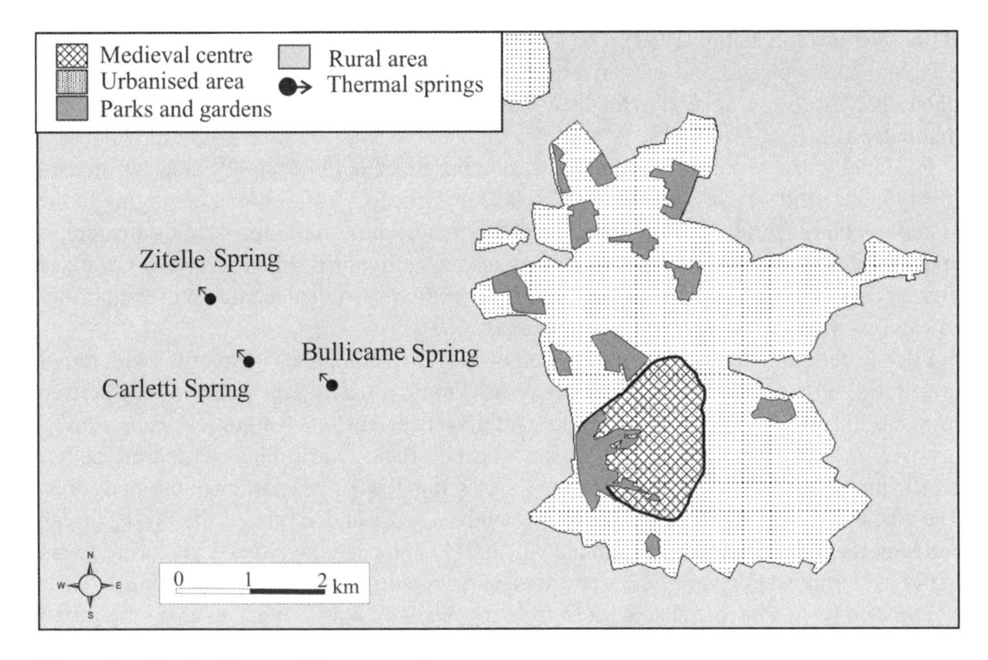

Figure 2. Viterbo Town and its surroundings.

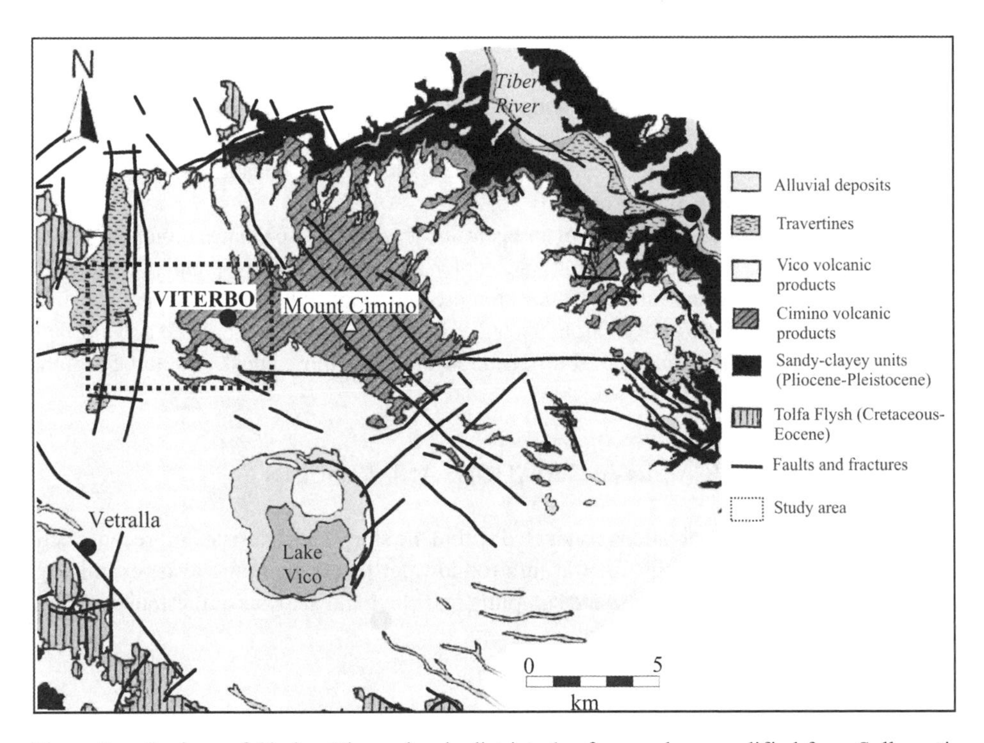

Figure 3. Geology of Cimino-Vico volcanic districts (surface geology modified from Sollevanti, 1983; structural lineaments modified from Mattias and Ventriglia, 1970).

flows (Lardini and Nappi, 1987). The Vico activity (0.7–0.25 My) led to the formation of a typical stratovolcano and gave rise to explosive products, from trachytic to tephritic-phonolitic lava flows, several pyroclastic flows (Ignimbrite A, B, C and D) and hydromagmatic deposits (Locardi, 1965).

Regionally, the volcanic substratum is composed of a Pliocene–Pleistocene, neoautochthonous sandy-clayey unit, a Late Cretaceous–Eocene Tolfa Flysch, belonging to the allochthonous Ligurian units and include marls, sandstones, marly limestone, calcarenites and argillites (Baldi et al., 1974) which was subsequently thrust upon the Meso-Cenozoic Tuscan and Umbria-Marche units, composed primarily of calcareous and evaporitic-calcareous-marly successions (Boccaletti et al., 1980).

Locally, the bedrock geology of the Viterbo study area comprises ignimbrites with minor lava flows, tuffs and loose pyroclastic deposits (approximately 100 metres thick) derived from the Cimino and Vico volcanic events. A thin veneer (up to 6–8 m) of Holocene alluvial deposits and in-fill materials covers these volcanic rocks. Early Pliocene marine pelites ($>$100 metres thick) underlie the volcanic rocks within the eastern portion of the study area. The volcanic rocks in the western part of the study area are underlain by Tolfa Flysh, Tuscan and Umbria-Marche units (Chiocchini et al., 2001). Holocene travertine deposits are found in the westernmost portion of the study area, surrounding the main thermal springs.

The Cimino and Vico volcanic rocks form an extensive aquifer with an area of approximately 900 km^2 and an average yield ranging from 6 to 8 l/s per km^2 (Boni et al., 1986). Recharge of this volcanic aquifer mainly occurs between the Cimino domes and Vico caldera (Figure 3). Groundwater flow is radial in the direction of streams and small springs (discharge less than 10 l/s).

To the west of the town of Viterbo, thermal waters outflow along a stretch coinciding with the uplift of a deep calcareous structure and include springs and deep wells characterised by sulphate-alkaline-earthy water with high temperature (up to 65°C), salinity and gas content (mainly CO_2 and H_2S). The origin of the thermal waters is from deep fluid circulation within the Meso-Cenozoic carbonate rocks, which constitutes a geothermal system at low enthalpy (Calamai et al., 1976).

The climate of the area is sub-coastal and is influenced by the Apennines mountain chain, with maximum precipitation in autumn and winter and minimum precipitation during the summer months. The mean annual values for precipitation and temperature recorded in Viterbo (SIMN 1950–1997) are 821 mm and 13.9°C, respectively. Potential and actual evapotranspiration, estimated using the Thornthwaite-Mather method, equal 761 and 568 mm, respectively (Pistoni, 2004).

3 HYDROGEOLOGICAL INVESTIGATIONS AND RESULTS

The hydrogeological investigations conducted within the study area consisted of: reconstructing the local hydrostratigraphy, measuring groundwater levels, pumping tests, examining surface-groundwater interactions, and sampling and chemical analysis of thermal outflows and groundwater.

3.1 *Hydrostratigraphy*

Three hydrogeological complexes can be distinguished based on rock outcrops identified within the study area and include; pyroclastic, alluvial and travertine rocks (Figure 4).

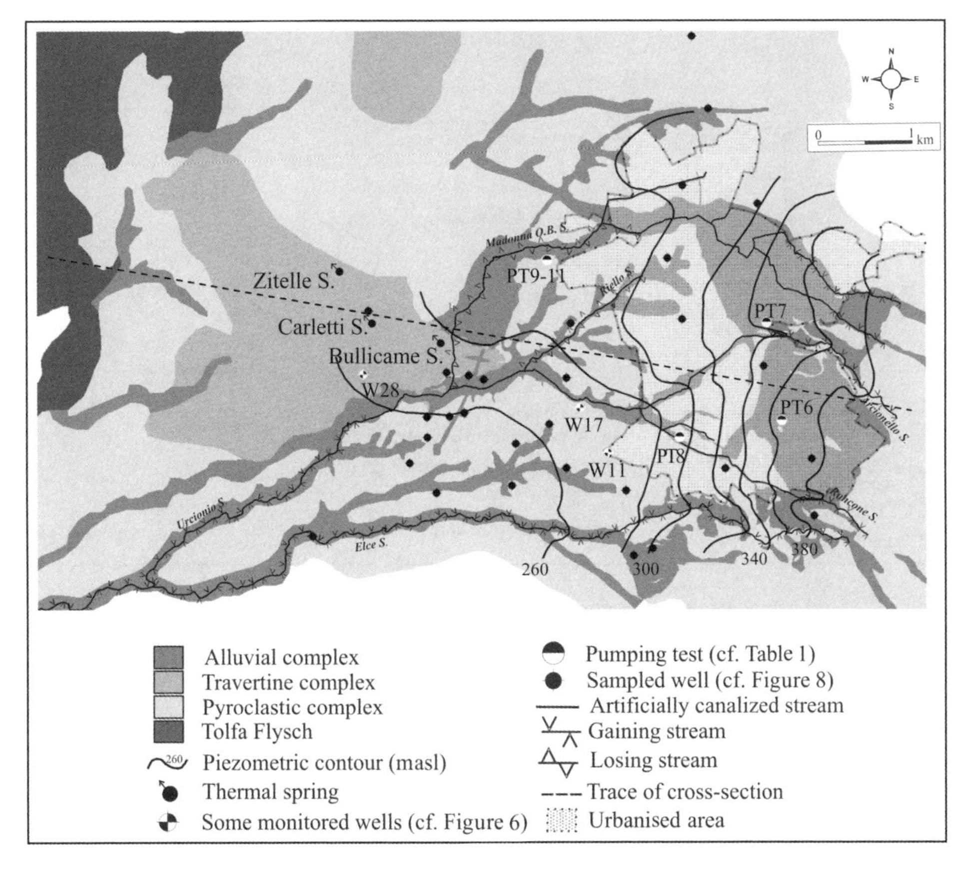

Figure 4. Hydrogeology of the study area.

The study area is predominantly underlain by pyroclastic deposits, composed chiefly of ignimbrites, tuffs and airfall deposits (Figure 4). These deposits perform an important hydrogeological function; they determine the infiltration capacity and define the geometry of the aquifer. Loose pyroclastic layers, are interbedded with ignimbrites that are fissured, which results in a complex porosity (both primary and fractured) and permeability.

A travertine complex is found in the western portion of the study area (Figure 4), and was deposited by the thermal springs.

Alluvial material, found adjacent to rivers and streams (Figure 4) include thin layers of extremely heterogeneous and loose river deposits. These deposits have a limited influence on groundwater flow, yet increase direct infiltration within the more permeable areas of town.

In order to reconstruct the subsoil hydrogeological outline, 50 stratigraphic records were completed. The majority of these data were collected from boreholes and wells penetrating the first hundred metres of subsoil, while data from deeper wells was obtained from historical well records drilled primarily for geothermal and hydrothermal exploration. The cross-section in Figure 5 shows the general hydrogeological profile of the study area. In the urban area the volcanic aquifer is clearly limited by low permeable early Pliocene pelites. The unconfined or leaky aquifer is approximately ten metres thick. To the

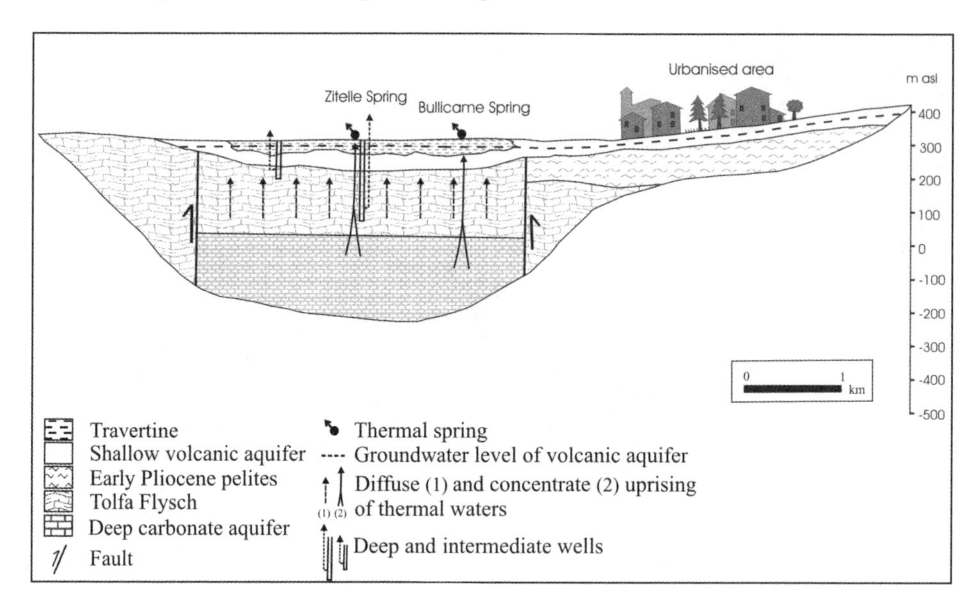

Figure 5. Hydrogeological cross-section.

west, the volcanic aquifer is thicker and includes a low permeable lens of lacustrine deposits. The shallow volcanic aquifer overlies the Tolfa Flysch which hosts the main thermal water-bearing levels tapped by deep wells. Therefore the Tolfa Flysch acts as an aquitard between the shallow volcanic aquifer and the deep carbonate reservoir. The uprising of fluids from the carbonate reservoir can occur owing to the relatively thin layer of calcareous-marly aquitard, which is locally fractured and uplifted (Figure 5).

3.2 *Groundwater level measurements*

During 2001–2002, groundwater levels were measured in approximately 90 wells and piezometers (depths ranged from 10 to 150 metres deep) situated within the shallow volcanic aquifer. Figure 4 shows that groundwater flows from east to west in the volcanic aquifer according to the large scale circulation of the Cimino-Vico hydrostructure. The direction of groundwater flow is coincident with some streams. The hydraulic gradient varies from 0.02 to 0.08 in the eastern part. Lower values of hydraulic gradient and a local recharge area characterise the western sector.

Groundwater level fluctuations were measured monthly in 18 wells and revealed that they generally follow the rainfall regime with a few months delay (Figure 6). The seasonal groundwater fluctuations vary from 0.5 to 2 m, with higher values along groundwater divides (e.g. W28 in Figure 6) and lower values in groundwater drainage areas (e.g. W11 in Figure 6). However, there are some exceptions with certain wells showing groundwater level fluctuations of more than 2 m (up to 11 m, e.g. W17 in Figure 6), even though they are located in groundwater drainage areas. These anomalous fluctuations are most likely due to the wells being situated on irrigated farmlands.

The piezometric levels of the thermal waters were collected from records of deep wells drilled for geothermal exploration and hydrothermal tappings. Results show that the

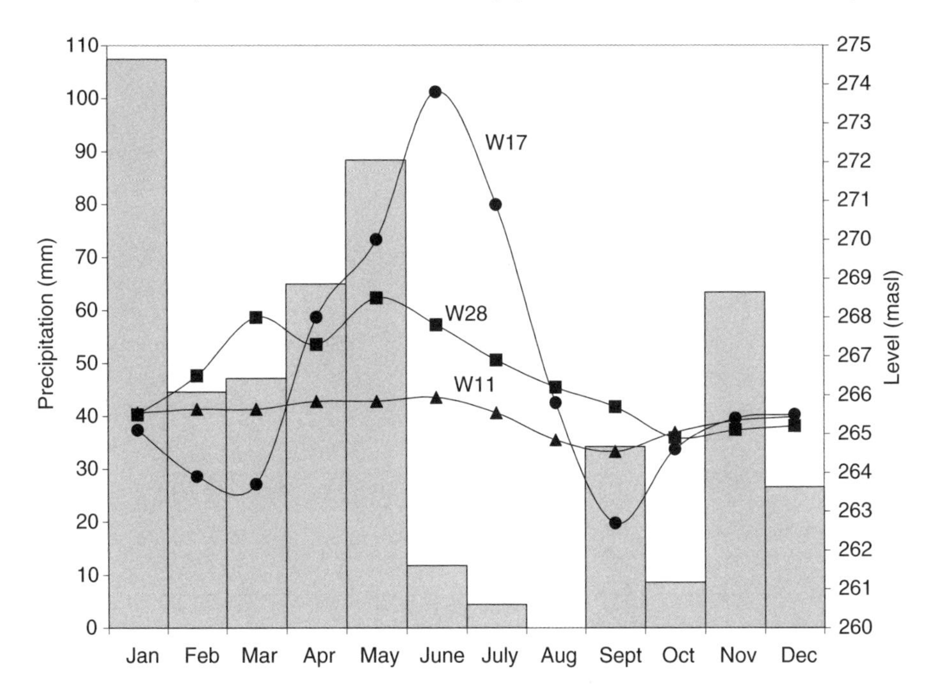

Figure 6. Piezometric level of some wells and rainfall during 2001.

piezometric levels for thermal waters are higher (10–20 m) than those observed within the shallow volcanic aquifer. Additionally, the deep wells often exist under artesian conditions. Therefore within the rural area, there is a vertical flow component from the deep reservoir to the shallow volcanic aquifer via the calcareous-marly aquitard (Figure 5). This rising of water can influence the piezometric surface within the shallow volcanic aquifer in the western sector, creating an area of local recharge and a decrease in the horizontal gradient as illustrated by the piezometric contours in Figure 4.

3.3 *Pumping tests*

Pumping tests were conducted at six of the shallow wells within the study area. Measured and collected data included: hydrostratigraphy of the test site, measuring drawdown during pumping and recovery both in the wells and at times in the observation piezometers.

In the urban area, wells PT6, PT7 and PT8 penetrate the shallow volcanic aquifer which is bounded along its base by impermeable early Pliocene pelites (Figure 4). The saturated thickness varies from 17 to 34 m (prior to pumping) and until pumping ceases when levels approached the level of the submersible pump. The tests, conducted at a low constant rate (less than 2 l/s) created a continuous drawdown in the well (Figure 7a) until a few metres above the submersible pump. Despite the limited pumping time, the tests allow the transmissivity and storativity of the aquifer to be quantified through the analysis of data at time-drawdown observation piezometers, and specific capacity data in the pumping well after 30 minutes of pumping (Table 1).

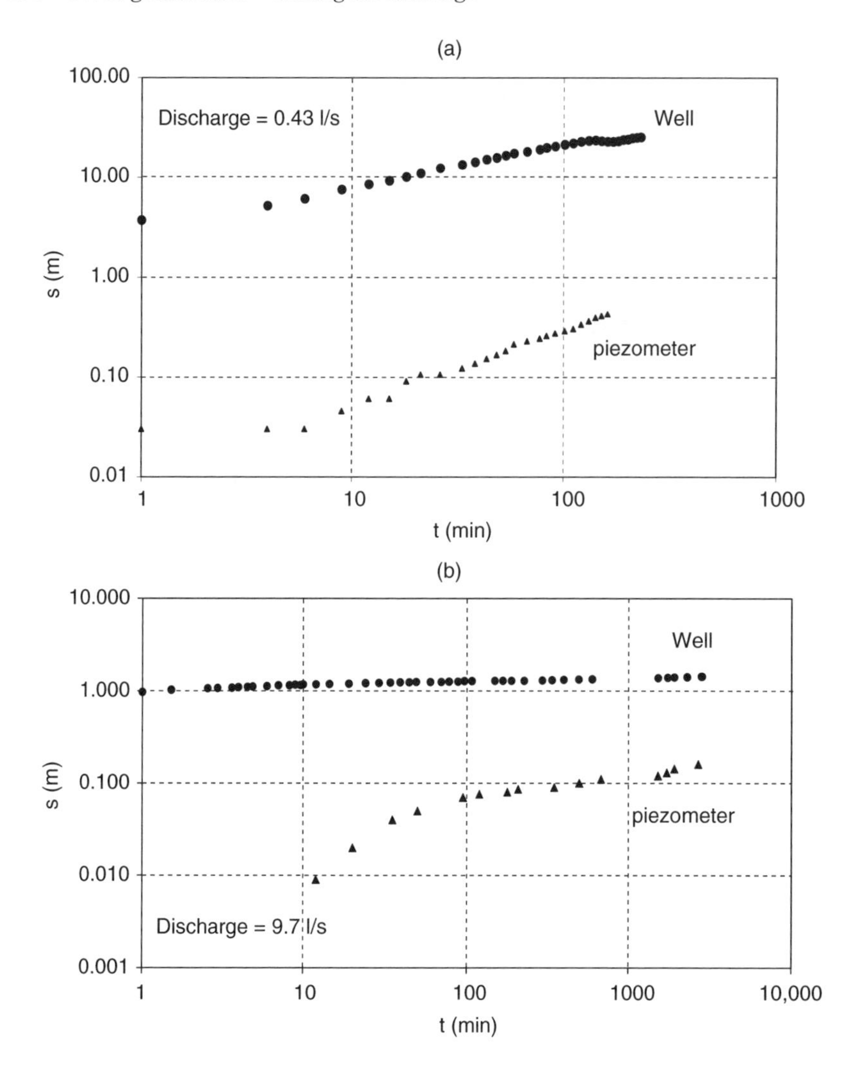

Figure 7. Drawdown (s) vs. time (t) plots of PT6 (a) and PT11 (b) pumping tests.

The other three tested wells (PT9, PT10 and PT11) are located within the rural area, approximately 1 km northeast of the thermal springs (Figure 4). Well screens intersect a saturated layer, ranging between 14 and 34 m deep within the shallow volcanic aquifer. This aquifer includes layers of low permeability lacustrine deposits.

A step-drawdown test was conducted on the PT9 well for 190 min (up to 6 l/s). Results show high value for specific capacity ($1.6 \times 10^{-2} m^2/s$) in comparison to wells PT6, PT7 and PT8 (Table 1).

PT10 and PT11 tests were conducted for 50–60 hours, at a constant rate, and show the highest level of discharge (10–14 l/s) for the whole study area. High values of transmissivity were observed (Table 1) and the drawdown-time follows a pseudo-steady state trend

Table 1. Results of pumping tests.

Well	Q	D	Hsat	SC	T	S
PT6	0.43	241	17	3.3×10^{-5}	1.3×10^{-4}	7.8×10^{-3}
PT7	0.35	33	26	2.0×10^{-5}	2.6×10^{-4}	1.2×10^{-3}
PT8	1.25	305	34	3.2×10^{-3}	3.0×10^{-3}	7.9×10^{-3}
PT9	1.7–6.3	108	30	1.5×10^{-2}	–	–
PT10	14.0	3600	14	1.2×10^{-2}	1.6×10^{-2}	–
PT11	9.7	2830	20	7.9×10^{-3}	3.0×10^{-2}	1.9×10^{-3}

Q = discharge rate (l/s); D = pumping time (min); Hsat = saturated thickness of aquifer (m); SC = specific capacity after 30 min (m^2/s); T = transmissivity (m^2/s); S = storativity.

(Figure 7b). Therefore in this area, the transmissivity of the shallow volcanic aquifer is higher even though it is near a groundwater divide.

3.4 *Interactions between surface and groundwater*

Two main watercourses, the Urcionio and Elce streams (Figure 4) drain the study area. Stream flow rates were measured on several occasions during dry periods in 2001–2003. The Roncone and Arcionello streams, two eastern tributaries of Urcionio, are fed by groundwater (about 10 l/s) in stretches located to the east of the urban area (Figure 4). Within the urban area these streams are artificially canalised above the water table, so that interactions between surface water and groundwater are limited to leakage.

The Riello and Madonna degli Occhi Bianchi streams are two additional tributaries of the Urcionio. Both streams showed a 5 to 10 l/s flow rate to the north of the urban area, mainly feeding from effluent water derived from the outskirts of town. The streams, crossing the north-western area in natural and artificial channels, have water levels higher than the local piezometric head. At the confluence with the Urcionio stream, the two tributaries show a decreased flow rate (around 5 l/s) due to loss of water to the aquifer (Figure 4).

Within the town of Viterbo, many stretches of the Urcionio stream are artificially canalised (Figure 4). After passing the urban area, the Urcionio turns into a gaining stream, based on the relative levels of surface and groundwater. Nevertheless the high flow rate (from 100 to 200 l/s) measured in the western rural area must also be related to the effluent water discharged into the watercourse.

The Elce stream drains the southern part of the study area and is characterised by a natural riverbed. Comparing the measured flow rate of 5 to 10 l/s to local surface and groundwater levels shows that the Elce contributes to groundwater recharge.

3.5 *Thermal springs and wells*

To the west of the urban area, where the travertine complex outcrops, thermal waters outflow from springs and wells; Bullicame, Carletti and Zitelle springs are the most famous outflows (Figure 4). Combined, these three springs have total discharge of about 20 l/s and a temperature ranging from 55 to 64°C.

The Bullicame Spring has an average flow rate of approximately 15 l/s, much of which supplies the Terme dei Papi (a local thermal centre) for therapeutic and heating purposes (warm pools, mud baths, etc.). The Carletti Spring has a flow rate of about 3 l/s and is used for public warm pools. The Zitelle Spring is currently not in use however, it originally had a flow rate of about 2 l/s, but increased to 5 l/s during the 1960s after repairs were made to a 220 m deep perforated well. During the drilling of the well several hydrothermal fissures were encountered, producing a flowing well with a discharge rate of 65 l/s for several days (Figure 5). During the same period the flow rate of the surrounding springs decreased (Conforto, 1954a, b).

Drilling additional wells of various depths within the area has intercepted thermal waters with temperatures ranging from 25 to 60°C (Ruggi, 2004). In these cases the levels of thermal waters are higher than those of the shallow volcanic aquifer. All thermal water has been intercepted within the Tolfa Flysch or at its boundary with the volcanic rocks.

3.6 *Chemical analyses of groundwater*

Water samples collected during 2001 from 30 wells ranging from depths between 10 and 40 metres and water samples collected from the Bullicame, Carletti and Zitelle thermal springs have been analysed using standard Istituto di Ricerca sulle Acque (IRSA) analytical methods (IRSA, 1992). The parameters measured include: temperature, pH, hardness, total dissolved solids (TDS), major and some minor ions (Ca^{+2}, Mg^{+2}, Na^{+}, K^{+}, HCO_3^-, SO_4^{-2}, Cl^-, NH_4^+, NO_2^-, NO_3^-, PO_4^{-3}, SiO_2, Cu^{+2}, total Fe, Mn^{+2}, Cr^{+3}, Zn^{+2}, Pb^{+2}, Co^{+2}, Ni^{+2}, Cd^{+2}, Li^{+}), and dissolved gases (O_2, CO_2 and H_2S).

Many shallow wells had temperatures in the range of 12 to 19°C. Spring waters had temperatures varying from 55 to 64°C, while one sampled well had a temperature of 26°C, which falls between the two ranges mentioned above. During the drilling of wells in the thermal area, waters with temperatures from 25 to 35°C (hypothermal waters) have been intercepted before reaching the hot waters (Ruggi, 2004).

Three geochemical profiles were distinguished based on the measured TDS values and major ion concentrations (Figure 8):

- Group A – bicarbonate-alkaline-earthy or bicarbonate-alkaline waters with low temperature and TDS from 350 to 810 mg/l;
- Group B – sulphate-alkaline-earthy thermal waters with high values of TDS (from 2600 to 3100 mg/l) and gas content (CO_2 = 101–154 mg/l and H_2S = 7–11 mg/l);
- Group C (represented by one well) – hypothermal waters with a chemical formula similar to Group A but with high TDS values (2900 mg/l) and lower temperatures (26°C).

The geochemical features associated with Group A coincide with an active groundwater flow within the shallow volcanic aquifer. Group B on the other hand, coincides with deep circulation influenced by an anomalous geothermic gradient and interaction with Triassic anhydrite of the evaporitic-calcareous-marly succession, while Group C may be caused by mixing of the Group A and Group B waters, however caution should be exercised as Group C is represented by only one well.

No well shows a clear impact of leaking sewers, as indicated by the concentrations of NH_4^+ (mean value = 0.048 mg/l), NO_2^- (mean value = 0.17 mg/l) and O_2 (mean value = 9.17 mg/l). High nitrate concentrations characterise waters of the shallow volcanic aquifer as shown in Figure 9 with values even higher than 50 mg/l in the rural areas and near losing streams.

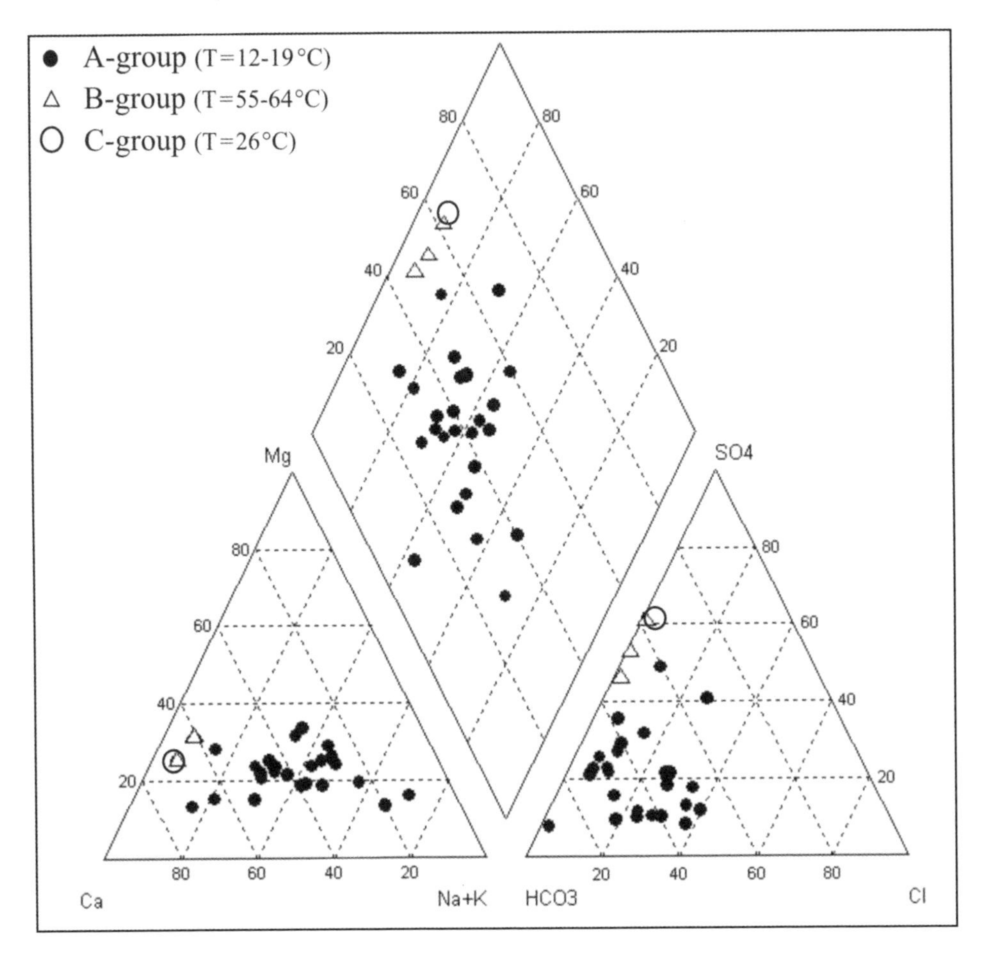

Figure 8. Piper trilinear plot for Viterbo waters.

4 IMPACTS ON GROUNDWATER

The hydrogeological settings within the rural and urban portions of the study area have been identified above and summarised on Figure 10. The potential impacts on groundwater within the rural and urban areas are largely controlled by the different groundwater usage and land use patterns identified for each of these areas (Figure 10).

4.1 *Urban area*

Within the east-west boundaries of the urbanised area, the shallow volcanic aquifer has a saturated thickness which increases from a few metres to tens of metres to the west and is bounded on the bottom by the impermeable early Pliocene pelites. The shallow volcanic aquifer is an unconfined, leaky aquifer and the groundwater flow net forms a subdued replica of the topography, with a high hydraulic gradient of up to 0.02. Transmissivity varies from 10^{-3} to 10^{-4} m^2/s and storativity is about 10^{-3}. The Group A, bicarbonate-alkaline-earthy or bicarbonate-alkaline waters characterised by moderate salinity (<900 mg/l) and low temperatures (<20°C) differentiate this shallow volcanic aquifer.

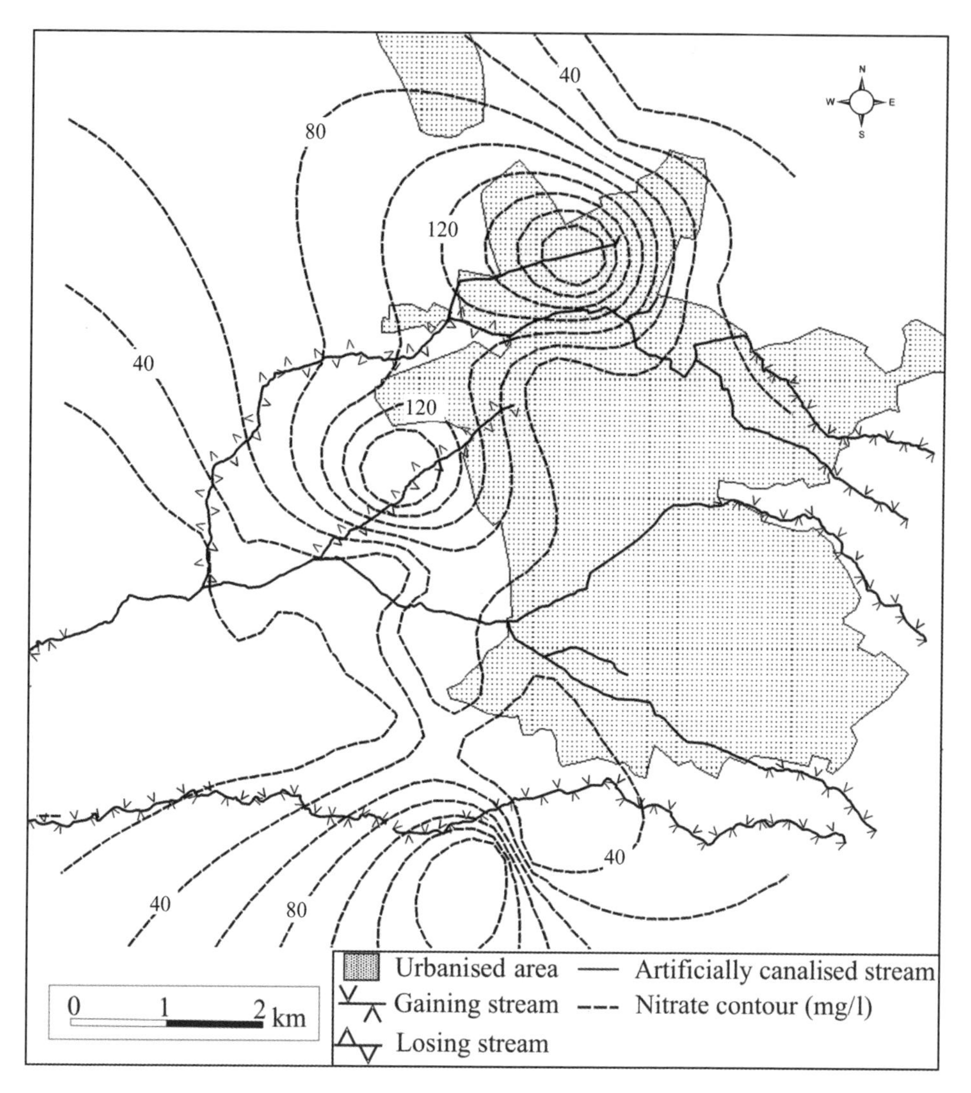

Figure 9. Nitrate contents of shallow groundwater.

The main effects on groundwater quality within the urbanised area are related to precipitation recharge, the public water supply system and pumping from the shallow volcanic aquifer (Figure 10). Recharge from precipitation within the urban area is influenced by the high percentage of gardens, parks, orchards (about 40% of the total area) and by the more impermeable road surfaces (composed of porphyry and thachytic lava pieces) found within the medieval centre. Runoff and evaporation from roofs, roads and other pavements were used to make crude estimates for direct recharge and runoff in this area (about 11 km^2) by using the mean annual precipitation and actual evapotranspiration values (821 and 568 mm, respectively). Considering the relative extent of green and impermeable areas and neglecting the effect of evaporation on the latter, gives a mean annual direct infiltration of approximately 128 mm and runoff of about 284 mm. Runoff is discharged towards the sewer network and then toward the streams. The town's public water is supplied mainly

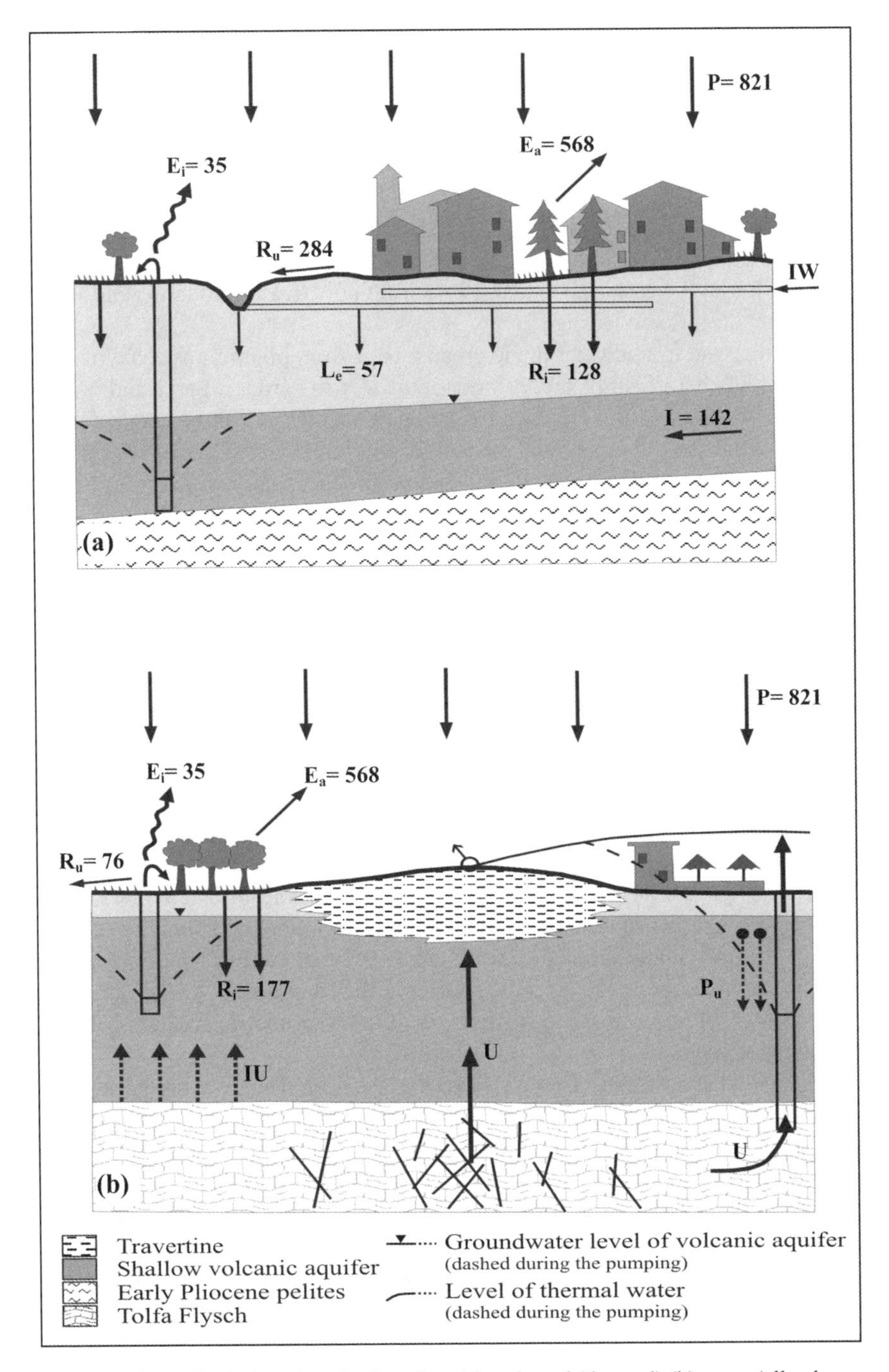

Figure 10. Hydrogeological settings in the urban (a) and rural (thermal) (b) areas (all values are in mm/yr). P = precipitation; E_a = actual evapotranspiration; R_i = direct infiltration; R_u = runoff; IW = imported water; L_e = sewer leakage and streams losses; I = groundwater inflow; E_i = irrigation losses; U = uprising of thermal waters; IU = increase of thermal waters uprising during pumping from shallow well; P_u = downward flow of pollutants during pumping from deep well. NB. Black vertical and sub-vertical lines seen within the Tolfa Flysch unit represent fractures and faults.

from a well field located 5 km to the south-east. Imported water for this densely urbanised area equals about 3 million m^3/yr and is eventually discharged to the sewer network. The recharge rate from sewer leakage and aqueduct losses can be estimated as at least 57 mm/yr, taking into account the mean percentage of losses of the urban network and losing streams. The rate of groundwater inflow from eastern boundaries must be added to the urban recharge of the shallow volcanic aquifer and can be estimated as 142 mm/yr from transmissivity data and piezometric surfaces for the eastern section of the urban aquifer (about 4.5 km long). Therefore, the total recharge of the urban shallow volcanic aquifer comprises 43% from inflow at the eastern boundary, 39% from effective precipitation and 18% coming from sewer and stream leakage.

Another effect on groundwater in the urban area is from pumping the shallow volcanic aquifer. The wells here mainly supply irrigation water to gardens, parks and orchards. A preliminary evaluation of the effective irrigation consumption can be conducted by subtracting the actual from the potential evapotranspiration in green areas, considering that any over-irrigation returns to the aquifer. These estimates were calculated to be equal to 35 mm/yr which is equivalent to 19% of the urban recharge (direct infiltration and sewer leakage).

Precipitation recharge, pumping, sewer and stream leakage alter the natural hydrogeological cycle within the study area (Figure 10). The groundwater chemical analyses do not provide any clear indication of pollution to the shallow volcanic aquifer in the urban area, with the exception of nitrate. Nitrate contamination spreads within the shallow volcanic aquifer and mainly occurs in the suburban rural areas or where groundwater is fed by polluted surface and wastewater.

4.2 *Rural (thermal) area*

To the west of the urban area, where the travertine complex outcrops, the geological setting is more complex and contains two aquifers: a shallow volcanic aquifer and a deep carbonate reservoir. The two systems are separated by a thin, fractured Tolfa Flysch aquitard, which can potentially allow hydraulic interactions between these two aquifers (Figure 10).

Unlike in the urban area, the volcanic aquifer in the rural area is thicker, the hydraulic gradient is lower, the transmissivity is higher and the piezometric surfaces bear no relationship to the topogaphy.

The sulphate-alkaline-earthy (Group B) facies of the deep rising water suggests hydraulic interactions with the carbonate reservoir, including the Triassic anhydrite rocks. Local uplift of the carbonate reservoir, high hydraulic head, high water temperature (up to 64°C) and the presence of dissolved gases contribute to the rising of thermal waters through the faulted and fractured Tolfa Flysch aquitard. The fractured Tolfa Flysch and wells that extend deep into the volcanic aquifer contribute to the rising thermal waters (Figure 10). The hypothermal (Group C) water, with a relatively low temperature (between 25 and 35°C) but high TDS, are located above the deep, thermal waters and below the water in the shallow volcanic aquifer. Therefore unlike in the urban area, the shallow volcanic aquifer underlying the rural area is not well isolated at the bottom and the system is open to a vertical flow that can explain the local piezometric surfaces and the high values of transmissivity.

In this rural area, natural pathways for precipitation are similar to those in the urban area; however, the main difference is a higher rate of direct infiltration, which causes a lower rate of runoff (Figure 10). The effects on groundwater are mainly related to pumping from the shallow volcanic aquifer and to the extraction of thermal waters. The pumping of moderately saline low temperature water (Group A) from the shallow volcanic aquifer for irrigation purposes increases the evapotranspiration losses. In addition to natural springs, thermal and hypothermal waters are also tapped by flowing and pumping wells.

With no pumping, the existing vertical head gradient from deep (>100 m) and shallow (<50 m) wells would likely result in a vertical flow of thermal water towards the volcanic aquifer, especially where the calcareous-marly aquitard is more fractured. This would safeguard the thermal waters from pollution coming from the shallow volcanic aquifer, whose waters often have high nitrate contents due to agricultural activities.

When pumping for irrigation occurs (especially during the dry season), the drawdown of the shallow wells can result in a higher vertical gradient, which causes increased vertical flow towards the volcanic aquifer (Figure 10). Due to the larger flux of thermal waters moving from the deep aquifer upwards into the shallow volcanic aquifer, there is contamination of the relatively fresh shallow waters by the saline thermal waters. Therefore, the effect of continuous pumping from shallow wells must be carefully evaluated to avoid a decrease in quality of shallow groundwater as a consequence of the excessive mixing with thermal waters.

In the case of pumping from flowing artesian wells, the existing vertical gradient decreased due to the extraction of deep thermal water. This resulted in a decline in discharge at the thermal springs near the deep wells. The potential for contamination increases because of the decrease or reversal in the hydraulic gradient resulting in less upward movement of water, making the thermal waters vulnerable to contamination from the shallow volcanic aquifer. The over-extraction of thermal water is an important factor in determining the consequences of constructing flowing wells that penetrate the Tolfa Flysch groundwater reservoir. A limitation on the rate of pumping from deep wells must be considered, since reversal of the existing vertical gradient could cause thermal waters to be exposed to surface pollutants (Figure 10).

5 CONCLUSIONS

The contrasting land and groundwater usage patterns within the urban and rural settings of the Viterbo study area have distinct effects on the quality and quantity of the hydraulically interconnected aquifers found within the study area. Groundwater from the upper volcanic aquifer is used predominately for agricultural and domestic purposes whereas thermal waters found within the lower aquifer are used to supply hot spring pools. The future management of groundwater resources must take into account the complex hydrogeological regime of the study area, considering the emblematic problems of urbanisation of rapidly growing towns.

Results from this study show that urbanisation is the main impact to groundwater in the shallow volcanic aquifer. An increase in runoff is highlighted, but the large number of green areas ensures significant amounts of direct infiltration that somewhat offset the increase in runoff. The quality of shallow groundwater is affected by agricultural activity, as shown by the high content of nitrates. The effects of leakage from sewer and urban stream networks on groundwater quality must be further investigated.

In the surrounding rural (thermal) area of Viterbo, future groundwater planning must preserve the quality and quantity of the two main groundwater reservoirs. If present use of groundwater is continued and if the hydraulic interaction between the two aquifers is neglected, impacts to the hydrogeological equilibrium will be more evident in the rural area. Conjunctive-use of the shallow and deep groundwater resources must be planned to determine the level of water qualities required for irrigation and thermal spas purposes. A coordinated approach that takes into consideration the relationships between shallow and deep aquifers is planned so that the current divides to groundwater management will be resolved. These same issues concern many other towns and villages in central and southern Italy's volcanic regions (Piscopo et al., 2000), where escalating demands for domestic water irrigation and thermal spa centres is causing strain on the groundwater resources.

REFERENCES

Baldi, P., Decandia, F.A., Lazzarotto, A. and Calamai, A. 1974. Studio geologico del substrato della copertura vulcanica laziale nella zona dei laghi di Bolsena, Vico e Bracciano. *Mem. Soc. Geol. It.* 13: 575–606.

Boccaletti, M., Coli, M., Decandia, F.A., Giannini, E. and Lazzarotto, A. 1980. Evoluzione dell'Appennino Settentrionale secondo un nuovo modello strutturale. *Mem. Soc. Geol. It.* 25: 359–373.

Boni, C., Bono, P. and Capelli, G. 1986. Schema idrogeologico dell'Italia Centrale. *Mem. Soc. Geol. It.* 35: 991–1012.

Calamai, A., Cataldi, R., Locardi, E. and Praturlon, A. 1976. Distribuzione delle anomalie geotermiche nella fascia preappenninica tosco-laziale. *Simp. Intern. Sobre Energia Geotermica en America Latina, Città del Guatemala 16/23 ott. 1976*: 189–229.

Chiocchini, U., Madonna, S., Manna, F., Lucarini, C., Puoti, F. and Chimenti, P. 2001. Risultati delle indagini sull'area delle manifestazioni termominerali di Viterbo. *Geologia Tecnica & Ambientale* 1: 17–34.

Conforto, B. 1954a. Risultati della prima fase di ricerche di forze endogene nel Viterbese. *L'Ingegnere* 27 (1): 345–350.

Conforto, B. 1954b. Risultati della prima fase di ricerche di forze endogene nel Viterbese. *L'Ingegnere* 27 (1): 521–530.

IRSA. 1992. Metodi analitici per le acque. Istituto Poligrafico e Zecca dello Stato, Roma.

Lardini, D. and Nappi, G. 1987. I cicli eruttivi del complesso vulcanico cimino. *Soc. It. Min. e Petrol.* 42: 141–153.

Locardi, E. 1965. Tipi di ignimbriti di magmi mediterranei: le ignimbriti del vulcano di Vico. *Atti Soc. Tosc. Sci. Nat.* 72: 55–137.

Mattias, P.P. and Ventriglia, U. 1970. La Regione Vulcanica dei Monti Sabatini e Cimini. *Mem. Soc. Geol. It.* 9: 331–384.

Piscopo, V., Allocca, V. and Formica, F. 2000. Sustainable management of groundwater in Neapolitan volcanic areas, Italy. In Oliver Sililo et al. (ed.), *Groundwater: Past Achievements and Future Challenge*: 1011–1016. Rotterdam: Balkema.

Pistoni, S. 2004. Le acque termali di Viterbo: schema idrogeologico. Tesi di Laurea, Università degli Studi della Tuscia, Viterbo, Italy.

Ruggi, E. 2004. Le acque termali di Viterbo: problematiche di salvaguardia. Tesi di Laurea, Università degli Studi della Tuscia, Viterbo, Italy.

SIMN. 1950–1997. Annali Idrologici. Servizio Idrografico e Mareografico dello Stato, Ministero dei Lavori Pubblici, Roma.

Sollevanti, F. 1983. Geologic, volcanologic and tectonic setting of the Vico-Cimino area, Italy. *Journ. of Volc. and Geoth. Res.* 17: 203–217.

CHAPTER 18

Alarming rise in groundwater levels beneath the city of Jodhpur: an example of ground and surface water interaction in the Thar Desert of western India

B.S. Paliwal and Alka Baghela
Department of Geology, Jai Narain Vyas University, Jodhpur – 342005, India

ABSTRACT: Groundwater levels in western India's Thar Desert are declining at the startling rate of 3–4 m/y. In contrast, the water table in the city of Jodhpur is rising locally by 1–1.5 m/y, despite very low rainfall and three years of drought. In many places groundwater is barely a few centimetres below the surface and in some places water is seeping from the ground. Basements of buildings are flooded with water despite constant pumping. This problem first developed in 1997 when Himalayan water was brought to Jodhpur through the Indira Gandhi Canal (IGC). Studies show that human activities are responsible for the rising water levels. Due to the increased population and improved availability of water, supply to the city has risen dramatically. Poor drainage, leakage from aging water supply and sewer systems, coupled with the fact that the use of dug wells and step wells in the city has declined, all contribute to the extremely high groundwater conditions in Jodhpur.

1 INTRODUCTION

Declining groundwater levels are a common problem in many of the world's deserts, including the Thar desert in western India, and the problem is further exacerbated in densely populated areas (Paliwal, 1986, 1993, 1999; Paliwal and Shakuntala, 1993). Groundwater abstraction rates in the Thar desert have increased remarkably during the past decades (Paliwal and Sharma, 1999) due to advancements in the technology available to farmers and the electrification of villages. This has caused groundwater levels to decline at a rate of three to four metres per annum in many parts of the region, particularly to the west of the Aravalli Mountain range.

Surprisingly, in some isolated areas of the desert, such as the city of Jodhpur, the situation is reversed. In Jodhpur city there are places where groundwater levels are rising at an alarming rate of two to three centimetres per week and in several areas the water table has reached the surface (Table 1). As a result, dug wells, step wells, and ponds (Figure 1) have started overflowing, basements of houses and shops (Figure 2) within the central, walled city have been flooded and residential buildings are beginning to collapse.

Table 1. Increase in groundwater level in dug wells, tube wells and hand pumps.

S.No.	Area	No. of wells	Range (in m)	S.No.	Area	No. of wells	Range (in m)
1	Soor Sagar, Kali Beri	13	5	29	Manak Chowk, Gulab Sagar	14	3
2	Kaylana, Sodon Ki Dhani	6	5	30	Bagar Chowk	7	2.5
3	Masuria	3	6	31	Kalal colony to Kaga	7	4
4	Housing Board	3	4	32	Inside Nagauri gate	1	1.5
5	Housing Board	3	7	33	Fateh Sagar to Naya Talab	12	2.75
6	Masuria area	3	4	34	Udai Mandir area	1	1.5
7	Masuria area	1	3	35	Nai Sarak, Guljarpura	10	2.3
8	Pratap Nagar	1	3	36	Sojti gate to Ghoron ka Chowk	14	2.5
9	Soor Sagar, Raj Bag	6	3	37	Udai mandir to Merti gate	3	3
10	Outside Chandpole	3	4	38	Ratanada to Mohanpura	4	2.25
11	Inside Chandpole	16	4	39	Jodhpur railway station to Darpan	2	3
12	Kumhariya Kua, Bakra Mandi	8	4	40	Ratanada to Shiv Mandir	15	3.62
13	Inside Siwanchigate	12	3	41	PWD to Loco	6	3.25
14	Zakir Husain colony	7	2.5	42	Bhagat Ki Kothi	9	4
15	Chopasni Road	6	7	43	Madhuban Housing Board	4	2.75
16	Masuria	5	4	44	Airforce area	8	3
17	Shastri Nagr, Isaiyon ka Kabristan	1	1.5	45	Airforce to Pabupura	2	1.5
18	Nehru Park	2	7	46	Rasala road to Rai ka Bag	13	4
19	Sardarpura	1	9	47	Paota to Laxmi Nagar	11	12.75
20	Sardarpura	1	9	48	Paota polo	2	7.5
21	Siwanchigate	11	4	49	Inside Mahamandir	2	6
22	Jalorigate to Kabootron ka Chowk	12	4.5	50	Mahamandir to Khetanadi	8	5
23	Bhishtion ka Bas, Sunaron ka Bas	9	4.5	51	BJS, Mohanpura, ZSA, Digari	12	3.5
24	Tripolia, Gorinda	12	3	52	Bhadwasia, Maderna colony	3	3
25	Nagauri Silawaton ka Bas, Bohron ki pole	16	3	53	Basni Tamboliya, Magra	10	4
26	Juni Mandi, Lakhara	11	3	54	Balsamand and Mandore to Mandalnath	11	3.3
27	Manak Chowk to Batte Sagar	12	2.5	55	Ram Mohalla and Mahamandir	4	2.5
28	Ghantaghar area	13	3	56	Soor Sagar Bhoortia	9	2

Figure 1. View of Gulab Sagar Lake which is filled with water even during the summer months.

Figure 2. Basement of a house in the old city flooded by rising groundwater levels.

2 CLIMATE AND PHYSIOGRAPHY

Climatically, Jodhpur and the surrounding region can be classed as having an arid climate characterised by extreme temperature variations and scant erratic rainfall. The average daytime temperature is 40°C with highs of 50°C in the summer to an average of 23°C during the winter with evening lows of −1°C (average 5°C). The average rainfall for the area is 380 mm/year and the rate of evaporation is very high (potential evapotranspiration 1850 mm/year). About 80% of the monsoon precipitation occurs during July and August.

The physiography of the Jodhpur area is greatly influenced by the underlying geologic formations that consist of a number of flat-topped sandstone and rhyolite hills that run in a northeast-southwest direction. Horizontally bedded sandstones form erosional flat topped hills whereas hills composed of rhyolite are topographically more rugged. A continuous high ridge of rhyolite separates the Kailana and Takhat Sagar lakes from the rest of the area. At places, the rhyolite hills are capped by a sandstone unit a few metres thick, exhibiting a clear unconformity between these two units.

Jodhpur city is located on a buried pediment and has an average elevation of about 260 masl. The hills exhibit a partially developed dendritic drainage pattern that becomes lost in the surrounding plains and agricultural land. The main drainage outlet is the Jojari River which lies to the south-east of the city and flows in a south-westerly direction.

3 GEOLOGY

The area around Jodhpur comprises a variety of rock types ranging from Proterozoic bedrock to recent alluvium and desert sands (Paliwal, 1999; Roy and Jakhar, 2002). The geology of the region and study area is shown on Figures 3 and 4. Palaeoproterozoic deformed metasediments of the Aravalli Supergroup (Paliwal, 1999; Paliwal and Rathore, 2000) form the basement. This unit is followed by 745 million year old intrusive and extrusive rocks of the Malani Igneous Suite (Malani Supergroup of Paliwal and Rathore (2000)), overlain in turn by rocks of the Marwar Supergroup (Neoproterozoic to Cambrian in age) particularly its lower part comprising the Sonia Shale Formation and Girbhakar Sandstone Formation of the Jodhpur Group. The surrounding flat terrain is occupied by recent alluvium and desert sands of the Thar. Most of the city and surrounding area show outcrops of Malani volcanics i.e. rhyolite flows, pyroclastics, welded tuffs, volcanic breccia and pyroclastics. However, in the eastern part of the city, the Mehrangarh Fort and the Umaid Palace have their foundations on sandstone of the Jodhpur Group. All around the city area sandstone of the Jodhpur Group is quarried for its ornamental value as a building stone.

4 HYDROGEOLOGY

Hydrogeological classification of the study area formations (Figure 4) is based on the information generated by the surficial geology study conducted by Paliwal (1999) and Paliwal and Rathore (2000), and the study of well logs. In addition, water levels were regularly monitored in a large number of dug wells, step wells, hand pumped wells, tube wells, and ponds.

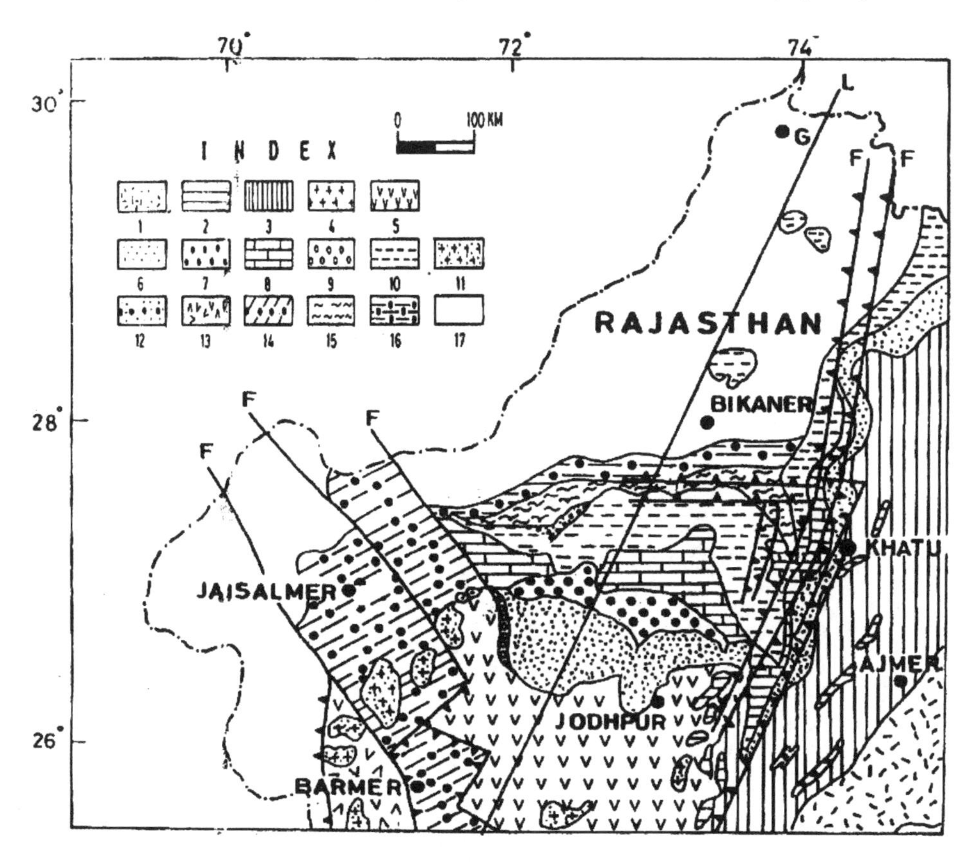

Figure 3. Regional geological map of western Rajasthan. 1 – Banded Gneissic Complex, 2 – Aravalli Supergroup, 3 – Delhi Supergroup, 4 – Erinpura Granite, 5 – Malani Igneous Suite, 6–10 – Marwar Supergroup (6–7 Sonia Formation and Girbhakar Formation of the Jodhpur Group, 8 – Bilara/Hanseran Group, 9 – Pokaran Boulder Bed, 10 – Nagaur Group), 11 – Jalore and Siwana Granites, 12 – Bap Boulder Bed, 13 – Deccan traps, 14 – Mesozoic sediments, 15–16 – Tertiary sediments, 17 – Quaternary sediments and sands of the Thar Desert, F–F – undifferentiated faults, G – Ganganagar, L – Lineaments (modified from Paliwal (1999)).

Three main hydrogeological units underlie the study area and include: (1) rhyolites, rhyolite porphyry, granite and pyroclastics of the Malani Supergroup, (2) basal conglomerates, sandstones and shales of the Jodhpur Group belonging to the Marwar Supergroup, and (3) recent alluvium. Groundwater in these three units occurs under unconfined conditions and is generally in a southward direction following the slope of the land surface. Topographic highs such as the one extending from Maherangarh Fort to Mandore and another exposed at the Umaid Palace, act as barriers to groundwater flow.

4.1 *Groundwater conditions in the Malani Supergroup*

Malani rhyolite and pyroclastics cover the entire city area and lie below the sandstone of the Jodhpur Group where the sandstone is exposed at the surface. The rhyolite is generally hard and compact and forms a poor aquifer with well yields ranging up to 20 m^3/day.

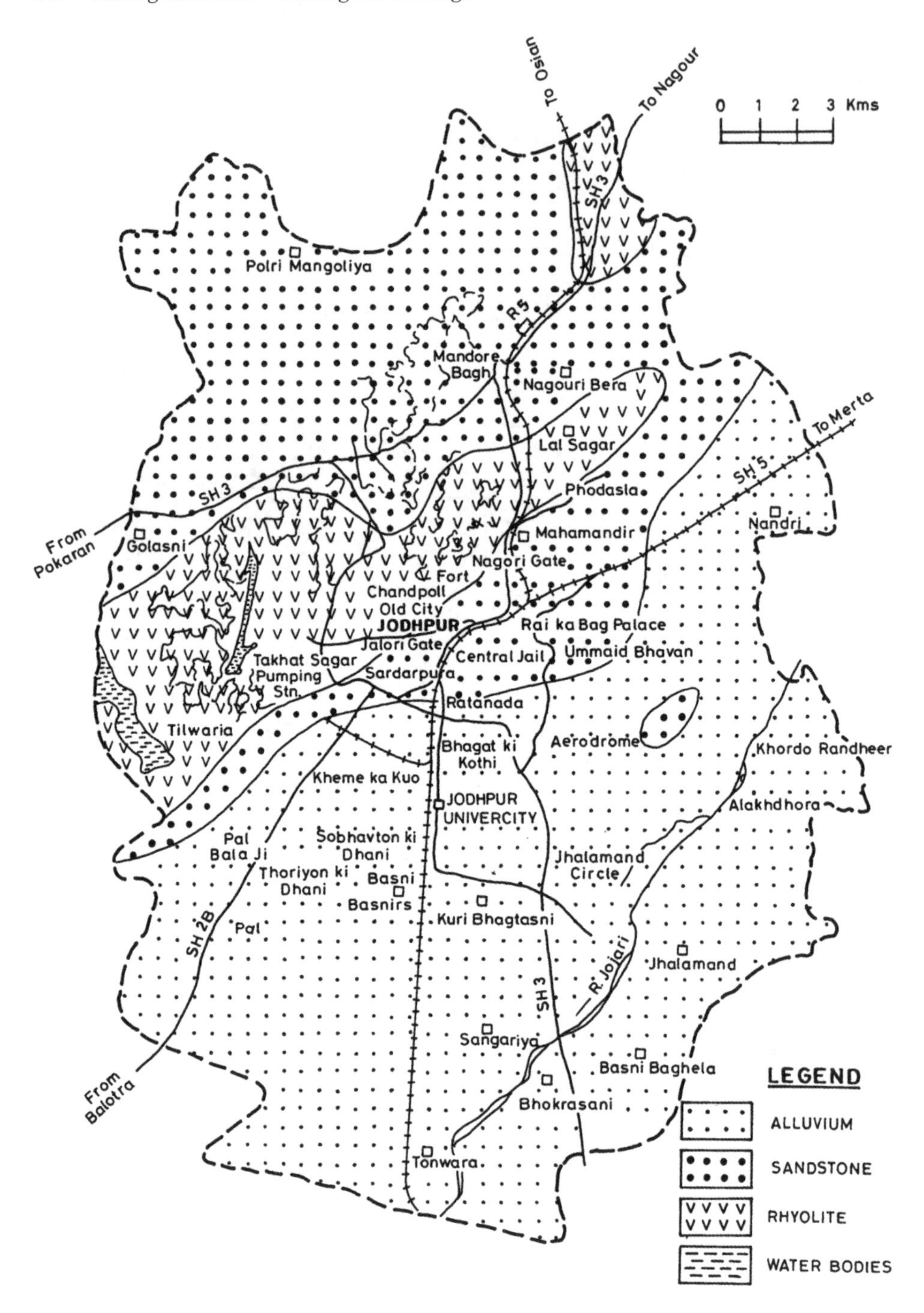

Figure 4. Geological map of the Jodhpur area.

However, in some places, the rhyolite exhibits secondary porosity and secondary permeability due to joints, fractures, and weathering that normally extends to a depth of 10–30 m. In these permeable zones, depth to the water table ranges from 20 to 40 m below ground level and the yield of dug wells and tube wells varies from 50 to 100 m^3/day. Water quality varies from potable to brackish in nature in the area around the Choupasni Housing Board, Shashtri Nagar, Sardarpura and Ratanada.

Along faults, shear zones and highly weathered zones the yield of groundwater in tube wells and dug wells is sometimes remarkably high. In the northern part of the area, groundwater also occurs in ash beds and clays where the discharge is very high but unfortunately the quality is brackish to saline.

4.2 *Groundwater conditions in the Marwar Supergroup*

The primary aquifer in the Marwar Supergroup is associated with the Girbhakar Formation, a medium to coarse-grained sandstone with some pebbly layers at the base. In some areas, notably in the north around Sursagar, Barli, Kali Beri, Chaukha and Golasani the sandstone is hard, compact and siliceous in nature with a low permeability. Water levels range from 7 to 35 m below the ground surface and the yield of dug wells and tube wells is low and varies from 10 to 25 m^3/day. Groundwater quality varies from potable to brackish.

Shale intercalated with medium grained sandstone occurs in the central part of the Jodhpur city covering the foothill side of the Maharangarh Fort, Umaid Palace, Poata, Ratanada and Masuriya areas. Because of the petrographic characters of the units in this area, groundwater occurs at a depth of 5–30 m and wells yield 25–75 m^3/day. The quality of the groundwater is brackish to saline with very high nitrate.

In the city area the overall structure is bowl-shaped resulting in shallow water levels along the foothill side of the sandstone and rhyolite hills. Water levels become deeper away from the plateau region. The water level in the walled city is very shallow.

4.3 *Groundwater conditions in the alluvium*

Recent alluvium forms an important hydrogeological unit southwest of Jodhpur covering Pal, Doli, Jhanwar, Gangana, Khema ka Kua and Chopasani. In this unit the water table normally ranges from 45 to 65 m below ground surface and yields range from 100 to 150 m^3/day. The quality of the groundwater varies from potable to brackish in nature. Along the course of Jojari River, the groundwater level is shallow, typically 20–30 m deep and the quality is saline. Farther east, around Jhalamand village, the alluvium becomes only 5 m deep and at places the underlying pink and grey granites reach the surface. Around the Pal-Doli-Jhanwar areas, the groundwater is found at a considerable depth but the quality of the groundwater is potable. Here the thickness of the aquifer ranges from 30 to 50 m.

5 HISTORICAL BACKGROUND OF WATER SUPPLY IN THE CITY

There are no perennial rivers close to Jodhpur which means that historically, people depended solely on harvested rainwater and groundwater. Prior to independence in 1947, a large number of dug wells, step wells and surface water reservoirs (e.g. Gulab Sagar, Ummed Sagar, Lal Sagar, Takhat Sagar, Kailana, Baiji Ka Talao and Balsamand) were constructed to harvest rainwater. An impressive network of canals was built to divert rainwater from the surrounding hilly terrain to the city. These surface water reservoirs were interconnected with each other so that excess water could easily be transferred from one reservoir to another.

Due to increased demand for water, combined with unpredictable rainfall and the failure of earlier water schemes, a lift canal was commissioned in 1993 to bring Himalayan water to important reservoirs such as Kailana and Takhat Sagar lakes. In 1997 water from these reservoirs reached the city and surrounding area and in the following year the outdated Ransi Gaon Canal and Jawai Dam were decommissioned. Presently, water is supplied to Jodhpur and nearby villages by (1) Takhat Sagar and Kailana lake reservoirs; (2) Pal-Doli, Mathania and Ransi Gaon groundwater schemes; and (3) bore wells within the city. The total water supply to the city in 1997 was 118 000 m^3/day which increased to 172 000 m^3/day in 2001 and rose to 182 000 m^3/day in 2004.

6 PLANNING OF THE TOWN

Constructed in 1459 by Rao Jodha, Jodhpur became the new capital of the state of Marwar. At this time the entire population of the city resided around the Meharangarh Fort, confined within the walled city. As the population grew the city expanded towards the flat terrain and a number of the small townships surrounding the city slowly merged into the city.

As Jodhpur grew, industries were built and more residential areas developed. In the 20th century major new developments included the largest housing project in Asia, the Kuri-Bhagtasani Housing Board Colony to the south of the city. The population continues to increase so that from an initial few thousand people in the fifteenth century, Jodhpur has now become a city of approximately 1.2 million people.

7 RISING GROUNDWATER LEVELS IN THE JODHPUR CITY AREA

In recent years there has been growing evidence of rising groundwater levels as witnessed by the following:

- Overflowing of surface water reservoirs, step wells and hand pumps (Figure 1).
- Water logging in low lying areas.
- Basements of houses and shops in some areas becoming flooded with groundwater despite all efforts to dewater them (Figure 2).
- Blockage of soak pits in the old city area.
- Increased groundwater contamination.
- Contamination of domestic water supply as a result of damage to pipelines caused by rising groundwater levels.
- Damage to residential buildings (Figure 5).

8 CAUSE OF THE RISING WATER LEVELS

Rising groundwater levels are evident throughout the city. For example, a step well popularly known as Gorinda Baori (Figure 6) was originally dug to a depth of 50 m in 1660 in

Figure 5. A house damaged in the old city due to rising groundwater levels.

Figure 6. Gorinda Baori – an old step well overflowing because of non-utilization of its water after commissioning of a new water supply scheme.

the heart of the city at the Tripolia Market. The water level at the time of construction can be inferred to be approximately 30–40 m below ground level as workers would not have been able to work below the water level. Today the water level in Gorinda Baori step well is close to the surface indicating a water level rise of over 30 m.

It seems that the present situation is not due to short term changes but is the long term result of continuously rising groundwater levels. The groundwater level in the city has likely been rising for many years, but the problem became noticeable only when groundwater started entering the foundations of buildings. In recent times, houses in the old city area are starting to collapse and ponds, lakes and step wells are beginning to overflow. In 1997 the issue came to a head when the local media highlighted the problem and the public demanded action. As a result the city administration called in experts from various organizations to examine the problem.

Hydrogeological information was collected by both the Central Ground Water Board, Government of India as well as the State Ground Water Department, Government of Rajasthan. In particular, groundwater and geological data from the past two decades (Paliwal, 1986, 1993, 1999; Paliwal and Rathore, 2000; Pareek, 1981, 1984) were examined in the light of the present issue. The unfortunate outcome of these investigations was that the Government of India and the Government of Rajasthan held different views regarding the possible causes of rising groundwater levels in Jodhpur.

8.1 *Subsurface flow via E-W lineaments*

The first view contends that the water is introduced via a series of re-opened sub-vertical parallel faults and fractures (lineaments) trending in the NE-SW and NW-SE directions. These lineaments were formed by Neotectonic activity in the region and are well recognized conduits for subsurface flow. A large number of surface water reservoirs in the area have been developed by constructing dams across these faults to intercept flow.

It is thought that some of these sub-vertical faults, particularly those trending in the E-W direction, pass beneath the Kailana and Takhat Sagar lakes and extend to the city area. It is argued that water from these lakes seeps through these E-W lineaments and recharges the groundwater of the city area and that the situation became aggravated when the Himalayan water was first introduced into these lakes after being lifted from the Indira Gandhi Canal in 1997. A total of 118 000 m^3 of surface water was stored in the Kailana and Takhat Sagar lakes and a water depth of between 12 and 18 m was maintained all year round. It is suggested that in previous years the water in these lakes had been maintained at a lower level (water depth ranging between 3 and 12 m) and that the fault system had been blocked and rendered ineffective by the accumulation of sediment. However, when water levels were raised, in 1997, water was able to seep through the upper parts of the fault system via fractures and joints that were not filled with sediment.

To investigate this hypothesis a detailed study of lineaments and joint patterns in the rocks of the Jodhpur city area was first undertaken involving data from the Indian Remote Sensing (IRS) Satellite. The IRS-1C PAN (panchromatic) data at 5.8 m resolution and LISS-III (Linear Imaging and Self Scanning) data at 23.5 m resolution were merged digitally and the FCC (false colour composite) prints of bands 234 BGR (blue, green and red) were generated on a 1:20 000 scale. To extract information about the lineaments, hydrogeological features and the geomorphology of the area, the hybrid product of the

LISS-III and PAN data were interpreted and a lineament map of Jodhpur city was prepared.

To investigate the flow of water along the lineaments, relative water level elevations were determined for (a) surface water reservoirs, (b) the areas affected by the problem of rising groundwater level in the city, and (c) the sites of the dug wells, step wells, tube wells, hand pumps and piezometers. In the city area much of the land surface varies from 200 to 226 masl which is over 20 m lower than the average elevation of the Kailana and Takhat Sagar lakes (245.7 masl). Interestingly, however, many areas severely affected by shallow groundwater levels are found along the north-south trending rhyolite ridge that separates the city from the lakes and has a ground level averaging 340 masl. These areas are significantly elevated above the level of the lakes and include Kriya Ka Jhalara (303 masl), Vyas Park (310 masl), and Padamsar-Raanisar (350 masl).

To further investigate the sources of the shallow groundwater in the city, chemical, biological and isotope analyses were carried out with the help of the Isotope Application Division of the Bhabha Atomic Research Centre, Mumbai and the Ground Water Department, Jodhpur. Water samples from hand pumps, basements, the two lakes, filter houses, dug wells, step wells and ponds were collected in the months of December 1998 and April 1999. These samples were analysed for oxygen-18, deuterium, tritium and various chemical parameters.

Electrical conductivity (EC) and major ion concentrations in shallow seepage water entering basements (BW) are mostly less than those in water samples collected from the nearby hand pumps and dug wells and quite different from those observed in Kailana lake (samples J1-1 and J1-25) and groundwaters from the surrounding area.

The plot of $\delta^{18}O$ versus $\delta^{2}H$ (Figure 7) indicates that the water samples from Kailana lake (lake and filter houses) are depleted in stable isotopes $\delta^{18}O$ and $\delta^{2}H$ as compared to other surface water bodies. The samples from basements fall on a mixing line between lake water and groundwater. In comparison, pond water samples are highly enriched in $\delta^{18}O$ and $\delta^{2}H$, showing the evaporation effect. Samples from basements which remained filled with water for months show higher $\delta^{18}O$ and $\delta^{2}H$ values than those from freshly filled basements.

Tritium in the lake water and the water collected from the lake filter houses varies from 9 to 12 TU (tritium units) (Figure 8), slightly higher than basement samples which range from 7 to 10 TU. Tritium in samples from hand pumps ranges from 2 to 7 TU and 5 to 11 TU for the samples collected in the months of December 1998 and April 1999, respectively. Tritium results support the argument that water in flooded basements is a mixture of lake water and groundwater. Water samples collected from a spring located in the Vyas Park area show depleted values of $\delta^{18}O$ and $\delta^{2}H$ and a relatively high tritium content indicating a strong component of lake water. However, direct seepage from Kailana lake to Vyas Park cannot be possible because of the significantly higher elevation of the Vyas Park area.

Samples of groundwater collected from the Chopasni and Golansni areas in December 1998 fall on the meteoric water line ($\delta^{2}H = 8\delta^{18}O + 10$) and are enriched in $\delta^{18}O$ and $\delta^{2}H$ as compared to the lake water (Figure 9). Tritium values of groundwater from the Chopasni and Golansni areas are randomly distributed and vary from 0.5 to 8 TU. The isotope data provide no evidence to suggest a contribution of lake water to the groundwater of these areas.

To further investigate the possibility of seepage through E-W lineaments connecting Kailana lake with the problem area, two 125 m deep investigation boreholes were drilled.

Table 2. Chemical analysis of groundwater samples collected from Jodhpur.

Sample No.	Source	Na^+ (ppm)	K^+ (ppm)	Ca^{++} (ppm)	Mg^+ (ppm)	Cl^- (ppm)	SO_4^- (ppm)	NO_3^- (ppm)	F^- (ppm)	pH	HCO_3^- (ppm)	EC (mS/cm)
J1-1	Surface water	8	2	42	5	14	14	1.6		7.01	90	233
J1-2	Surface water	75	9	76	34	113	48			6.85	350	770
J1-3	Surface water	66	15	64	12	85	19			7.57	260	710
J1-4	Surface water	14	4	48	1	28				7.2	120	300
J1-5	Surface water	14	5	42	7	28		15	0.28	6.38	200	410
J1-6	Groundwater spring	110	35	102	29	142	106	86	0.3	6.64	300	1130
J1-7	Surface water	270	176	134	44	312	252	126		7	526	2470
J1-8	Basement sample	158	31	176	41	184	232	229	0.2	7.1	350	2060
J1-9	Basement sample	224	289	106	11	262	420			8.5	110	2340
J1-10	Hand pump	339	344	108	57	369	516			7.1	280	3220
J1-11	Basement sample	144	143	132	17	156	194	128	0.28	7.1	370	1760
J1-12	Hand pump	618	35	388	163	822	572	1270	0.2	6.32	258	5310
J1-13	Dug well	256	274	118	62	277	423	398	0.2	6.9	304	2630
J1-14	Basement sample	177	180	90	29	156	232	153	0.36	7.3	354	1670
J1-15	Surface water	10	2	42	13	28	29	0.4		6.99	100	270
J1-16	Surface water	172	53	62	39	227	67	54	0.8	7.14	380	1420
J1-17	Surface water	138	55	74	38	277	120	30	0.8	7.5	380	1760
J1-18	Basement sample	188	199	110	39	248	213	362	0.12	7.6	230	2330
J1-19	Hand pump	348	328	164	71	440	461	627	0.64	6.8	378	3630
J1-20	Basement sample	360	356	192	63	468	615	460	0.2	6.88	350	3260
J1-21	Hand pump	428	360	220	126	567	567	930	0.56	6.83	324	4640
J1-22	Basement sample	708	321	44	41	638	509	448	0.28	9.1	540	4120
J1-23	Basement sample	378	285	316	36	469	818	286	0.28	7.4	360	3750
J1-24	Hand pump	558	78	192	100	638	461	555	0.56	6.7	324	4150
J1-25	Surface water	9	2	40	7	21	10	0.2	0.2	7.1	108	260
J1-26	Surface water	8	2	38	6	21		0.6	0.2	7.5	100	260
J1-27	Dug well	110	12	64	28	71	43	2.4	0.28	7.2	420	940
J1-28	Hand pump	160	3	86	39	241	67	82	0.28	6.42	260	1410
J1-29	Hand pump									6.2	258	750
J1-30	Hand pump	215	16	92	44	135	115	233	0.36	6.46	400	1590
J1-31	Hand pump	348	2	118	15	355	127	268	2.32	6.62	363	2450
J1-32	Dug well	82	2	118	40	135	79	144		7.1	280	1190
Testwell-1 (32 M)	–	86	33	62	19	85	65	143	0.04		235	950
Testwell-1 (125 M)	–	135	5	54	28	142	77	166	0.16		255	1180
Testwell-2 (20 M)	–	150	27	148	21	163	3	314	0.32		455	1670
Testwell-2 (125 M)	–	125	15	138	30	156	17	308	0.24		465	1560

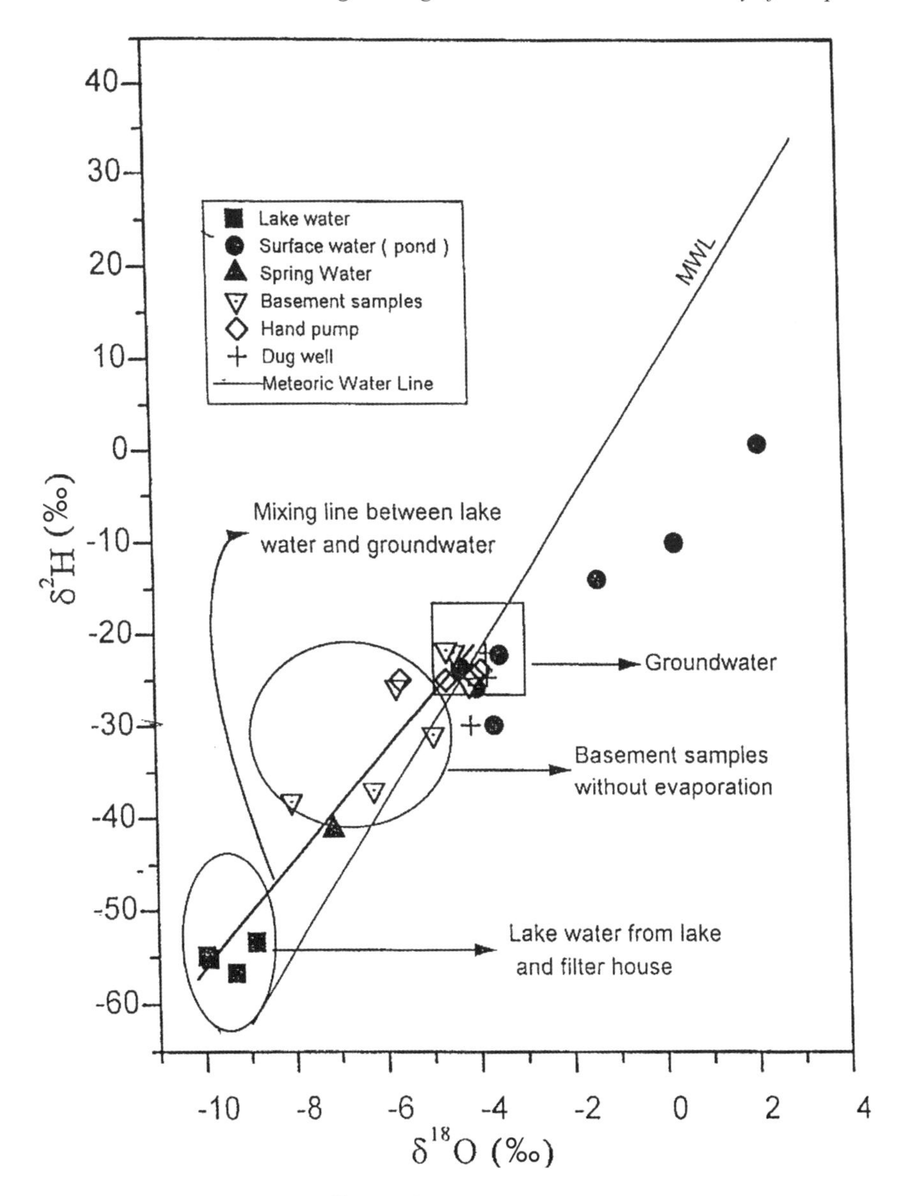

Figure 7. Figure showing plot of δ^{18}O vs. δ^2H water for samples from Jodhpur City.

Depths to water level, discharge of groundwater and the chemical quality of groundwater from these two boreholes refuted the contention that rising groundwater levels in the city are caused by seepage through east-west lineaments but confirmed the likelihood that subsurface flow was occurring along lineaments towards the south, where dug wells and tube wells have a good quality and high yield of groundwater.

To examine the possibility that elevated water levels in Kailana and Takhat Sagar lakes promote the movement of water along east-west fractures that had not been filled with

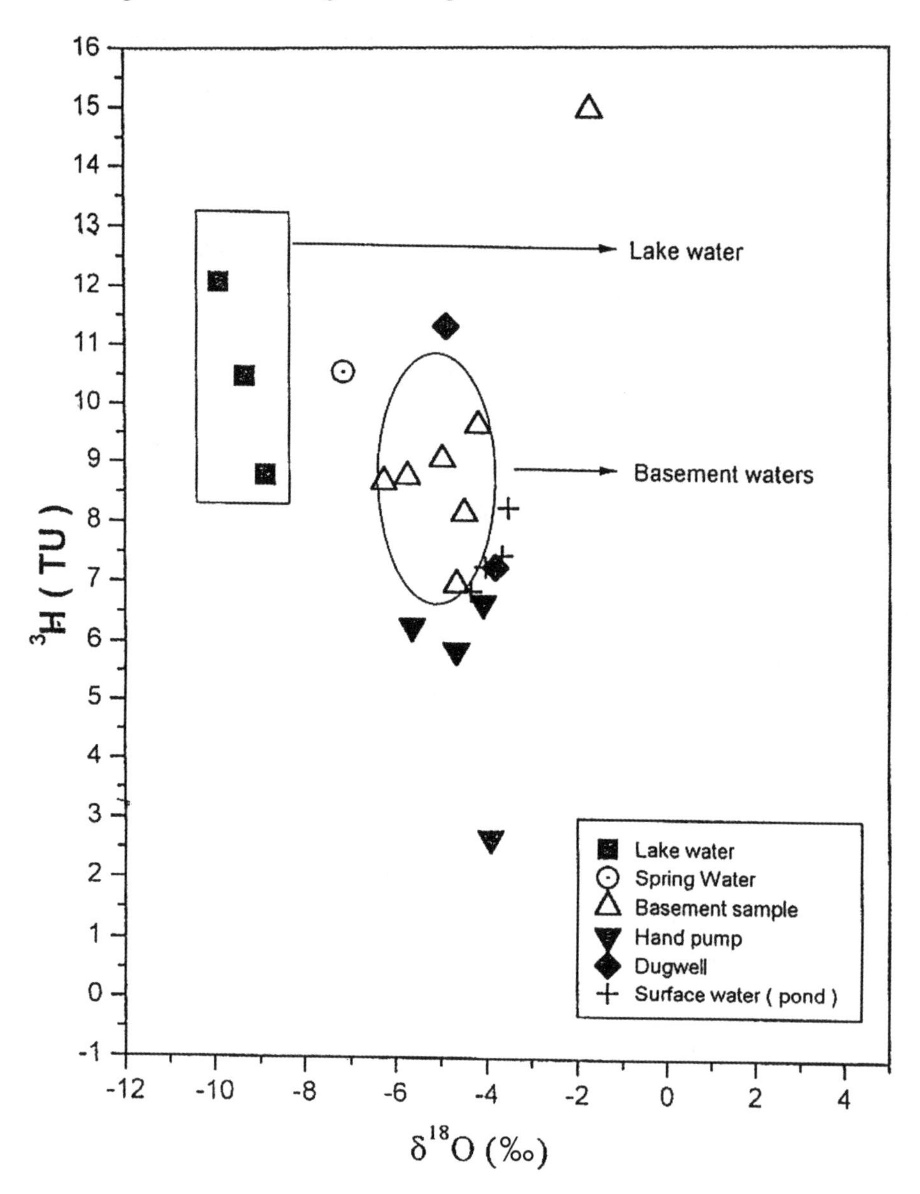

Figure 8. Tritium in lake water samples.

sediment, the water depth in the two lakes was lowered from 17.68 to 11.88 m between November 6th, 2001 and February 11th, 2002. Despite such a drastic reduction in lake levels, no significant change was observed in the subsurface. The problem continued to be severe and basements remained filled with water despite all efforts to dewater them. Surface water reservoirs continued to overflow and dug wells and step wells maintained an upward trend in their water levels. It can be concluded from this work that rising water levels in Jodhpur are not directly associated with the east-west trending lineaments.

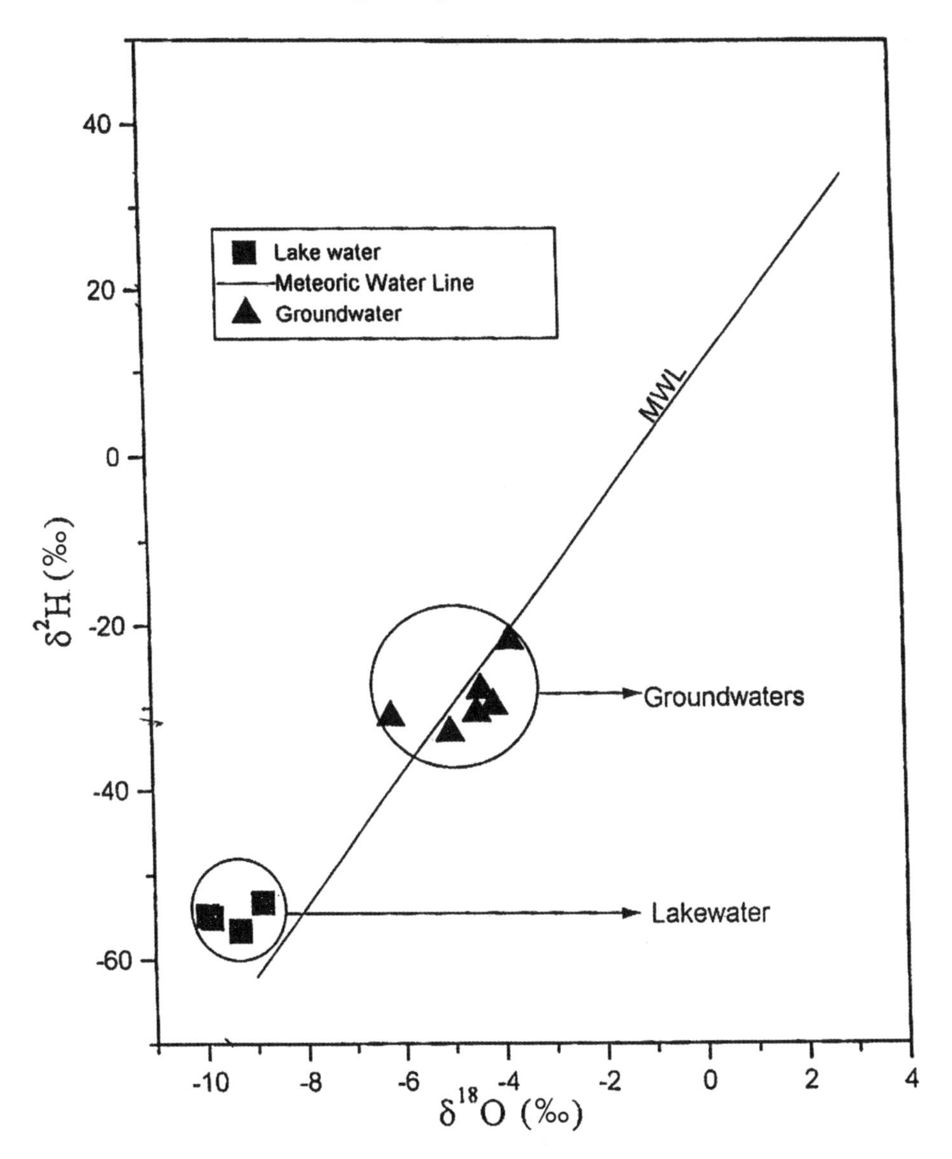

Figure 9. Diagram showing isotope concentrations for different locations.

8.2 *Excess net inflow of water*

The contending hypothesis is that rising water levels simply reflect an imbalance between the inflow and outflow of water in parts of the city. This was investigated by conducting a subsurface "water balance" for the central part of Jodhpur and examining the water balance in the context of long term water level fluctuations, rainfall, the presence of lineaments and the population density. Figure 10 shows water table contours and the electrical conductance of groundwater in the Jodhpur region. Groundwater flow directions are clearly indicated. Changes in the level of the groundwater beneath the area are shown in Figure 11.

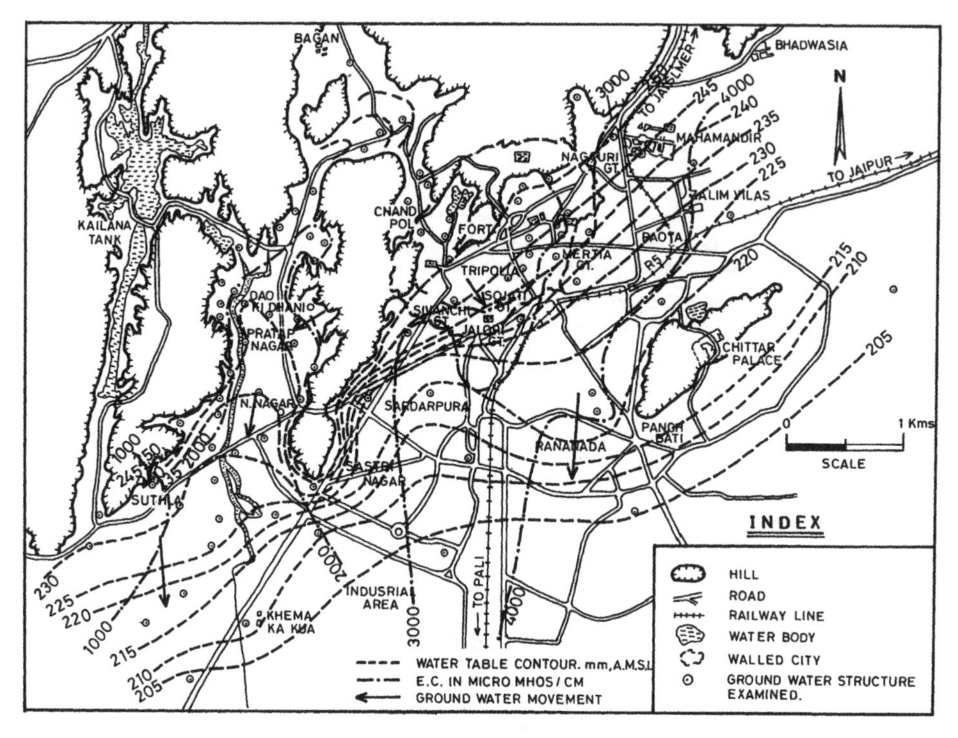

Figure 10. Map of Jodhpur showing water level contours and distribution of specific conductance.

8.2.1 *Water inflow*

The subsurface inflow of water to the centre of Jodhpur is contributed by five sources:

- Leakage from water supply pipelines (of which 80–90% would normally be converted into wastewater after consumption). Sources of water are shown in Table 3 and total 25 000 m^3/day.
- Rainwater infiltration through the ground. Rainfall between June and August raises the water table by about 0.7 m.
- Seepage from large surface water bodies.
- Contribution from wastewater through open drains.
- The underground flow of water.

8.2.2 *Water outflow*

Sewer and wastewater systems are found throughout the centre of Jodhpur. The outflow from this area is through sewers which ultimately discharge into open drains at Stadium-Darpan Cinema and Sojati Gate (behind the Anand Cinema). The results of discharge gauging studies are shown in Tables 4 and 5. Total daily wastewater discharge is calculated as 17 000 m^3/day.

8.2.3 *Water balance*

Of the 25 000 m^3/day of water used in the centre of the city, only 17 000 m^3/day or 65.6% is being returned as sewage outflow. This is between 15 and 25 percentage points less than the 80–90% that would normally be anticipated and represents a considerable volume of additional water. This additional water must inevitably recharge the groundwater system and contribute to rising groundwater levels.

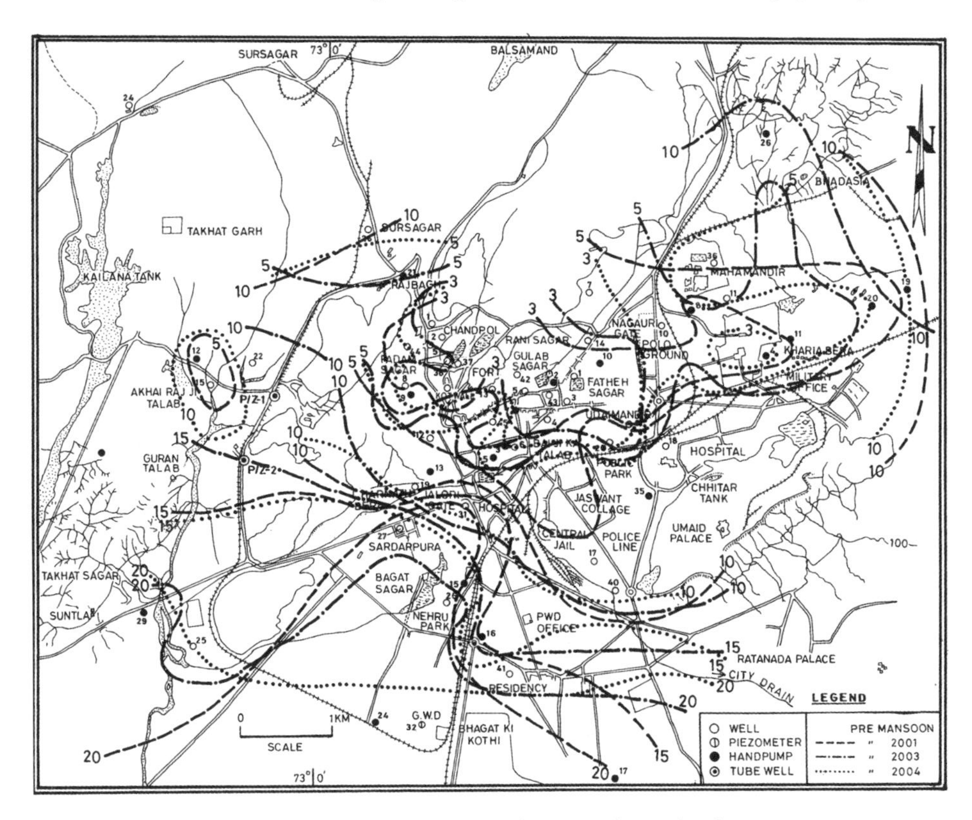

Figure 11. Map of Jodhpur city showing the changing groundwater level.

Table 3. Water supply for the centre of Jodhpur.

S.No.	Source	Quantity (m^3/day)
(i)	From Bal Kishan Ki Ghati (Vyas Park) Chouki through 450 mm, 300 mm, 350 mm lines from Baiji Ashram	14 000 to 15 000 (450 from 6 local sources)
(ii)	Fateh Sagar Chouky through 375 mm line	9100 to 10 000
(iii)	Girdhar Mandir ESR (elevated storage reservoir)	1800 to 2700
	Total	25 000 to 27 000 Average 26 000

9 CAUSES OF RISING WATER LEVELS IN JODHPUR CITY

Based on the present investigation, problems with shallow rising groundwater levels are due to the following:

1. The unique geomorphological setting of central Jodhpur – a bowl like structure, that conveys recharge to the central city area and creates shallow water table conditions.
2. Water supply to Jodhpur city which has increased considerably in response to the rising population. Leakage from the water supply system has created an additional source of aquifer recharge.

Table 4. Outflow of surface water recorded at Darpan Cinema observation point on the 11th and 12th of September 2003.

S.No.	Time (hrs)	Area of flow			Velocity observed			Discharge quantity (m^3/sec)
		Width (m)	Depth (m)	Area (m^2)	Distance (m)	Time (sec)	Velocity (m/sec)	
1	1000	2.1	0.21	0.441	27	90	0.3	0.132
2	1200	2.1	0.24	0.493	27	50	0.54	0.266
3	1400	2.1	0.19	0.399	27	60	0.45	0.179
4	1600	2.1	0.15	0.315	27	105	0.26	0.080
5	1800	2.1	0.20	0.420	27	63	0.428	0.179
6	2000	2.1	0.12	0.262	27	115	0.234	0.061
7	2200	2.1	0.15	0.262	27	115	0.234	0.061
8	2400	2.1			27			0.05
9	0200	2.1			27			0.05
10	0400	2.1			27			0.236
11	0600	2.1	0.20	0.525	27	60	0.45	0.236
12	0800	2.1	0.22	0.462	27	50	0.54	0.249

Table 5. Outflow of water in a drainage channel near ADB office recorded on the 11th and 12th of September 2003.

S.No.	Time (hrs)	Area of flow			Velocity observed			Discharge quantity (m^3/sec)
		Width (m)	Depth (m)	Area (m^2)	Distance (m)	Time (sec)	Velocity (m/sec)	
1	1000	2.1	0.11	0.213	13.2	55	0.24	0.51
2	1200	2.1	0.17	0.357	13.2	50	0.264	0.094
3	1400	2.1	0.12	0.252	13.2	90	0.146	0.36
4	1600	2.1	0.13	0.283	13.2	95	0.138	0.039
5	1800	2.1	0.10	0.220	13.2	115	0.114	0.025
6	2000	2.1	0.11	0.241	13.2	70	0.188	0.045
7	2200	2.1	0.11	0.241	13.2	75	0.176	0.042
8	2400	2.1	0.11	0.241	13.2	75	0.176	0.042
9	0200	2.1	0.11	0.252	13.2	40	0.33	0.083
10	0400	2.1	0.12	0.252	13.2	40	0.33	0.083
11	0600	2.1	0.17	0.315	13.2	30	0.44	0.138

3. The drainage and sewer system in the city area are quite old. Seepage from these open drains and sewer lines continues throughout the year and contributes to aquifer recharge.
4. Due to the increased availability of fresh water, the Government and private users have stopped using the existing groundwater resources which has generated surplus groundwater which has added to the problem.
5. Geologically, massive rhyolite forms the basement rock and does not have deep fractures or secondary permeability. As a result, water does not percolate down to deep and distant levels. The intermittent shale layers in the sandstone also hinder the movement of seepage water away from the area.

10 RECOMMENDED REMEDIAL MEASURES

Based on a detailed study of different aspects of the issue, the following measures are recommended to solve the problem of a rise in the water table in Jodhpur:

- The drainage system should be improved so as to minimize the opportunities for seepage losses that add to the groundwater system.
- Heavy pumping is required from sites such as Fateh Sagar, Gulab Sagar (Figure 1), Gorinda Baori (Figure 6), Ranisar and Padamsar etc. This water should be exported from the city to nearby rural areas where it can be used in agriculture.
- Withdrawal of groundwater must be restored on a large scale. If possible, a dual water supply system, based on available groundwater and surface water should be introduced to the city area.
- Lastly, it should be understood that the present problem is the result of human activity and society as a whole should be educated to change its attitude towards the utilization, protection and conservation of water for future generations (Paliwal, 1989).

ACKNOWLEDGEMENTS

The authors wish to acknowledge the assistance provided by the following organizations: Ground Water Department, Government of Rajasthan, Public Health and Engineering Department, Government of Rajasthan, Central Ground Water Board, Government of India, Bhabha Atomic Research Centre, Mumbai, Indian Space Research Organization, Municipal Corporation, Jodhpur and Jai Narain Vyas University, Jodhpur. We also wish to thank Dr. P.S. Rathore, Dr. Gyanesh Lashkari and Dr. Gopal Rathi of Department of Geology, Jai Narain Vyas University, Jodhpur for their help in preparing the text. Special thanks go to Mrs. S. Paliwal, Mr. B. Paliwal, Mrs. G. Paliwal and Mr. Shubham.

REFERENCES

Paliwal, B.S. 1986. Source of salinity in soil and surface and sub-surface waters and its bearing on the evolution of the Thar Desert, India. Transactions XIII Congress International Society of Soil Science. August 13–20, 1986, Hamburg, Germany (Abstract) Vol. 4, 1532p.

Paliwal, B.S. 1989. Status of water resources in developing countries in the twenty first century and the need of spreading awareness in the society through education – a case study from the Republic of India. Proceedings of International Seminar on Education and Training in Water Resources in Developing Countries. December 4–8, 1989. Aurangabad, India. Central Board of Irrigation and Power, New Delhi, India. pp. 347–353.

Paliwal, B.S. 1993. Geology of Rajasthan – a summarium. In: T.S. Chauhan (Ed.) Natural and Human Resources of Rajasthan. Scientific Publishers, Jodhpur, India. pp. 1–26.

Paliwal, B.S. 1999. Geological Evolution of Northwestern India. Scientific Publishers, Jodhpur, India. 414p.

Paliwal, B.S. and Shakuntala. 1993. Ground water resources of Rajasthan In: T.S. Chauhan (Ed.) Natural and Human Resources of Rajasthan. Scientific Publishers, Jodhpur, India. pp. 113–124.

Paliwal, S.C. and Sharma, D.C. 1999. Critical regions of ground water in western Rajasthan. In: B.S. Paliwal (Ed.) Geological Evolution of Northwestern India. Scientific Publishers, Jodhpur, India. pp. 293–304.

Paliwal, B.S. and Rathore, P.S. 2000. Neoproterozoic Volcanics and Sediments of Jodhpur – a reappraisal. In: K.C. Gyani and P. Kataria (Eds.) Proceedings National Seminar on Tectonomagmatism,

Geochemistry and Metamorphism of Precambrian Terrains. Department of Geology, M.L.S. University, Udaipur, India. pp. 75–96.
Pareek, H.S. 1981. Basin configuration and sedimentary Stratigraphy of western Rajasthan, Journal Geological Society of India. Vol. 22, pp. 517–527.
Pareek, H.S. 1984. Pre-Quaternary Geology and mineral resources of north western Rajasthan. Memoir Geological Survey of India, Calcutta, Vol. 115, 99p.
Roy, A.B. and Jakhar, S.R. 2002. Geology of Rajasthan (Northwest India) – Precambrian to Recent. Scientific Publishers, Jodhpur, India. 421p.

CHAPTER 19

Groundwater modelling to evaluate the risk of aquifer depletion due to a construction site in an urban area in Basel, Switzerland

C. Miracapillo
Civil Engineering, University of Applied Sciences NW-Switzerland

ABSTRACT: Groundwater drawdown due to simultaneous pumping at different locations and for different purposes can strongly influence groundwater flow. In particular, low water table levels due to construction sites can create a risk for over-exploitation, as the required pumping rate may not be met if sufficient hydraulic head is not available. A two dimensional numerical model is used to assess groundwater levels. Based on the predicted groundwater level, a comparative risk analysis for over-exploitation of the wells of the groundwater users around the construction site is carried out for management purposes. Two main factors affect the precision of the calculated drawdown: the simulation of vertical impermeable walls (not fully penetrating) in a two-dimensional horizontal model and the representation of the real geometry of the wells in the mesh of the discretised model domain. Special attention is given to the simulation of partially penetrating walls in the numerical model.

1 INTRODUCTION

The subterranean environment in urban areas is heavily exploited due to increasing traffic and the intensive and heterogeneous use of urban land. This leads developers to resort to underground alternatives for an increasing number of road networks and other infrastructure such as subways, tunnels, and water pipes. These often penetrate deep into aquifers and permanently modify the groundwater flow, sometimes acting as underground dams.

Construction below the water table often requires the drawing down of the water table to a prescribed level so as to enable construction to be completed in a dry environment. As a consequence of drawing the water table down at the construction site, the cone of depression extends widely around the construction site, causing the depletion of aquifers in neighbouring areas and eventually interfering with the pumping activities of other groundwater users in the region (Miracapillo, 2004, 2003).

The case presented here refers to a situation of conflicting needs: groundwater drawdown is required to provide dry conditions for workers and for building operations, but at pumping wells located near the construction site, a lowering of groundwater levels poses the risk of depleting well yields. The risk that the required pumping rates in these neighbouring wells cannot be obtained due to interference by pumping at groundwater drawdown at the construction site needs to be assessed. In order to do this, a comparative risk analysis was conducted to find out which wells are most exposed to such a risk.

2 PROJECT DETAILS

The project is located in an area of Basel, Switzerland, where industrial and commercial activities co-exist in a highly urbanised zone of the city. In this area there are also several groundwater users who exploit the aquifer using pumping wells. The construction site is located at the intersection of two major roads where a tunnel is being built as part of a national highway. The project area is situated on the west side of the Rhine River, which is an important hydrological component that acts as a receiving stream. Groundwater flows into the Rhine from the south-west to north-east, through deposits of gravel and sand of different sizes. An industrial area is located along the west bank of the river which includes a wall that is parallel to the river and which extends to the bottom of the aquifer. This wall acts as an impermeable barrier, forcing the groundwater flux to redirect to the north and south of the wall, thus preventing it from entering freely into the Rhine.

3 COMPARATIVE RISK ANALYSIS

The risk (R) of financial damage to groundwater users can be evaluated upon the probability (P) of damage occurring multiplied by the gravity (G) of the damage itself.

$$R = P \cdot G \quad (1)$$

In this case, harm to groundwater users occurs when they are forced to acquire water from the city water supply system rather than from their wells, thus incurring higher costs. These costs, G, are directly proportional to the amount of water required from the city water supply. The probability of harm to groundwater users, P, is also dependent on the drawdown of the water table, as the more drawdown that occurs, the greater the probability of harm.

$$G = f(Q, \ldots) \quad (2)$$

$$P = f(s, \ldots) \quad (3)$$

where Q is the required pumping rate and s is the total drawdown due to the hydrological conditions (s_h) and to the pumping at the construction site (s_c).

4 THE MODEL

A two-dimensional groundwater simulation model was constructed using a program based on the finite difference method (ASM, Aquifer Simulation Model). The model domain is defined through boundary conditions: first order boundary conditions at the constant head boundary on the western side of the model, second order boundary conditions for the outside stream lines on the northern and southern edges of the model as well as the impermeable boundary in the east, and third order boundary conditions at the outflow boundary simulating the interaction between the Rhine and the groundwater, north and south of the eastern impermeable boundary (Figures 1a, b). The location of the groundwater monitoring stations and of the groundwater users are shown on Figures 2a, b.

The model domain is divided into cells 5 m × 10 m. The aquifer bottom (Figure 3) was calculated using an interpolation code (krigging) with data from borehole profiles available at the University of Basel's Geological and Paleontological Institute (Miracapillo, 2001).

The numerical simulation was carried out under steady state conditions, using mean monthly values of pumping and recharging rates for the wells, which are derived from the data delivered by the Department for Energy and the Environment in Basel. The aquifer recharge was evaluated based on the monthly values of rain and on the land use (Figure 4). Mean monthly values are also used for the water level in the Rhine. The leakage coefficient is treated as a calibration parameter.

Pumping tests in the model domain were performed and values between 0.9 and 3.8 mm/s for hydraulic conductivity were found. Based on the results of the pumping tests the model was divided into zones of unique K-values. These are assumed as first estimation values and are adjusted during calibration to create a more realistic match between the observed groundwater level and the corresponding calculated levels at the observation points (Figure 5).

5 UNDERGROUND CONSTRUCTION IN URBAN AREAS

Underground construction projects often require vertical impermeable walls (fully or partially penetrating) made of metal or concrete, which effect the groundwater flow on both a local and regional scale by:

- reducing the permeability of the soil limiting the possibility of groundwater exploitation,
- acting as a groundwater dam and inducing the water level to rise upstream,
- causing abrupt deviations of the groundwater flows, and
- creating death zones, where water remains immobile in the ground.

In the present study attention is devoted to the industrial area, where the soil is intensively exploited and to the construction site, which is delimited by vertical walls.

5.1 *The industrial area*

The industrial area is to the west of the river bank wall. Most of the present buildings and infrastructure do not interfere with groundwater circulation as they do not extend deep enough down to reach groundwater. Nevertheless, underground foundations of destroyed buildings that are deep enough to reach groundwater still exist alongside foundations, canals, pipes, inspection tunnels and other subterranean infrastructures relating to existing industries (Miracapillo, 1991). In these areas the pumping tests show local values for hydraulic conductivity which, at some places, are 3 or 4 times lower than those for other parts of the model domain.

In this area, K-value zones are defined according to the degree of local underground exploitation, while the K-values themselves are treated as calibration parameters by matching calculated with observed piezometric levels.

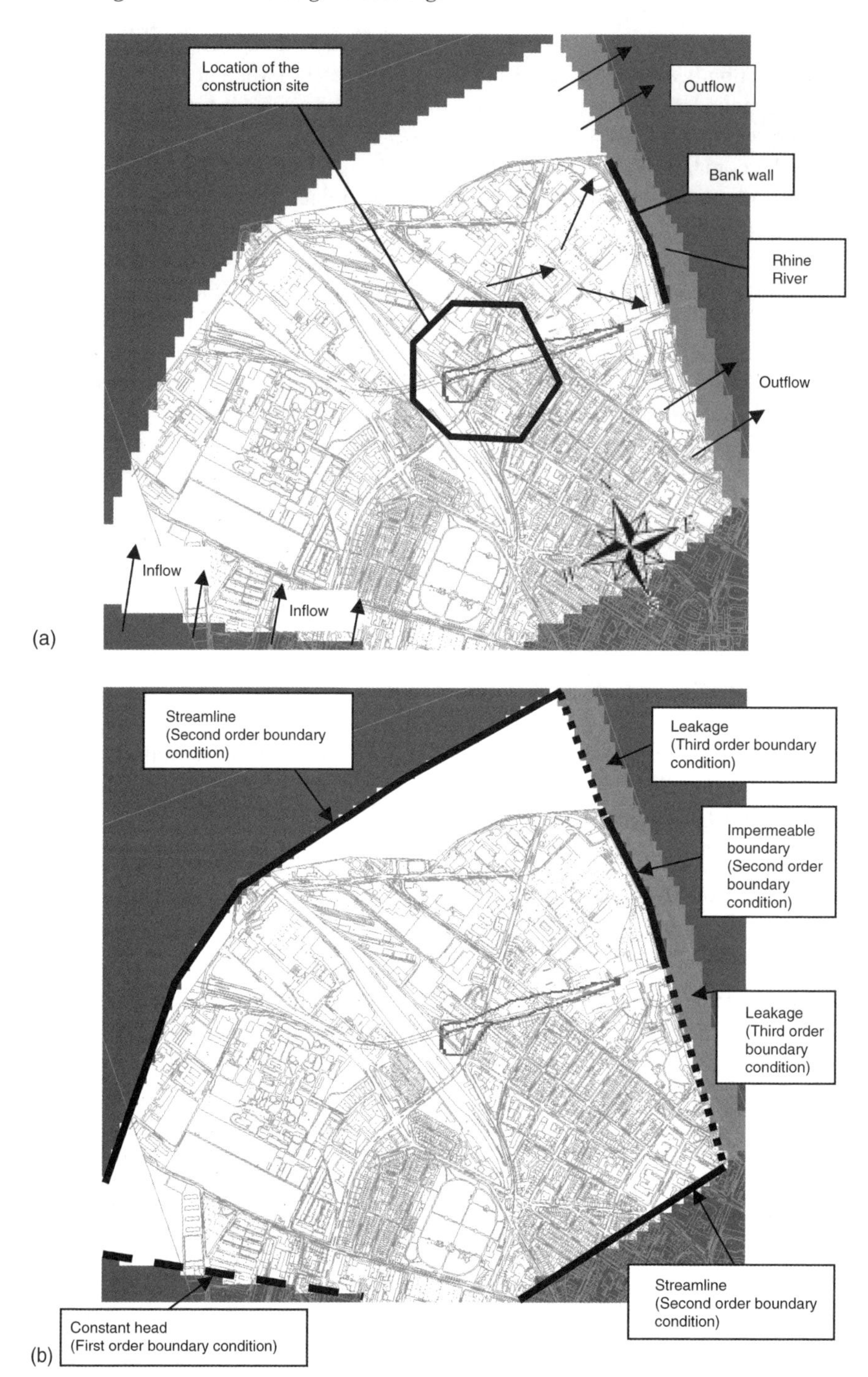

Figure 1. The project area (a) and the model domain (b).

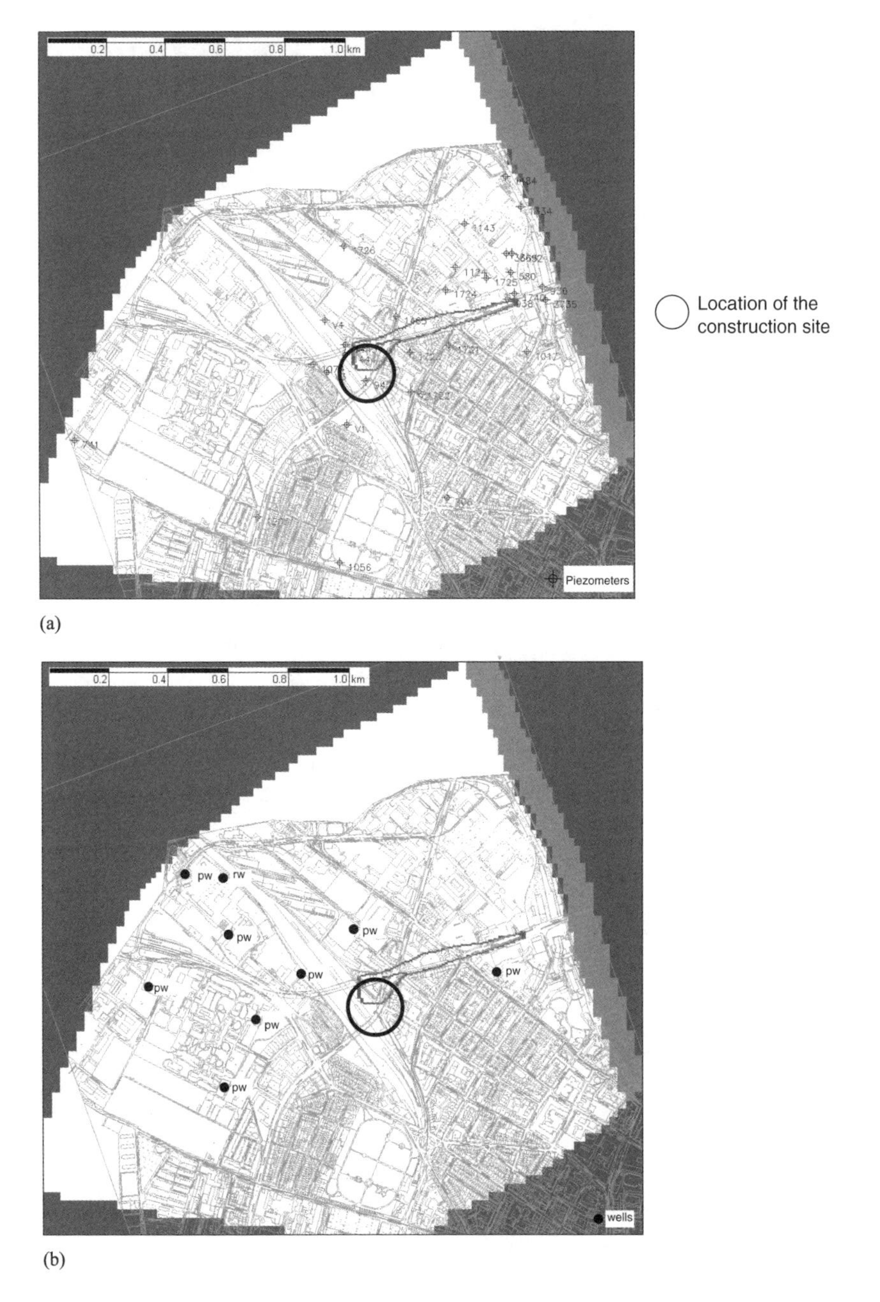

Figure 2. Location of the piezometers (a) and of the wells (b). pw = pumping well, rw = recharging well.

5.2 *The construction site*

The walls of the construction site are normally constructed using two kinds of bore piles (Figure 6):

- bore piles (with and without reinforcement bars and of different lengths),
- bore piles with an opening (a movable element or slide allowing the water to flow through).

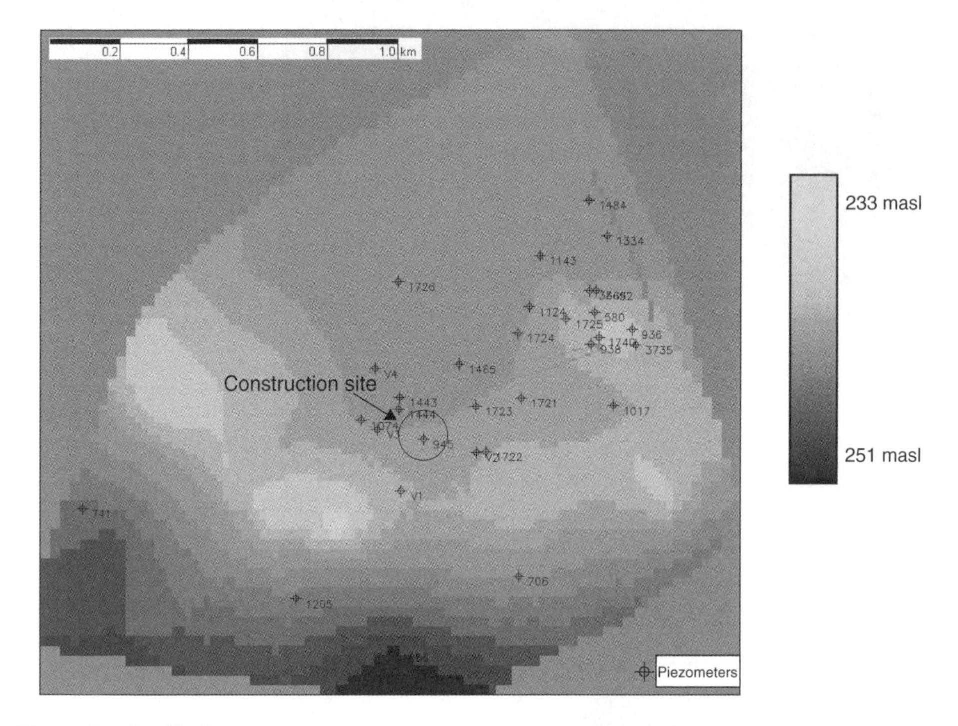

Figure 3. Aquifer base.

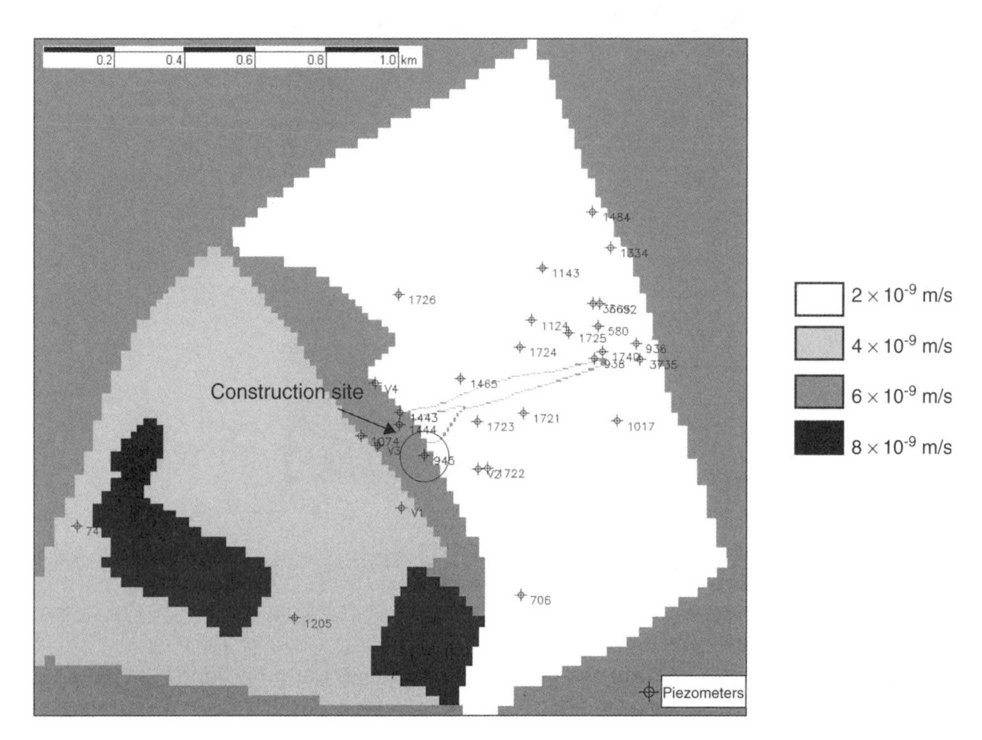

Figure 4. Groundwater recharge.

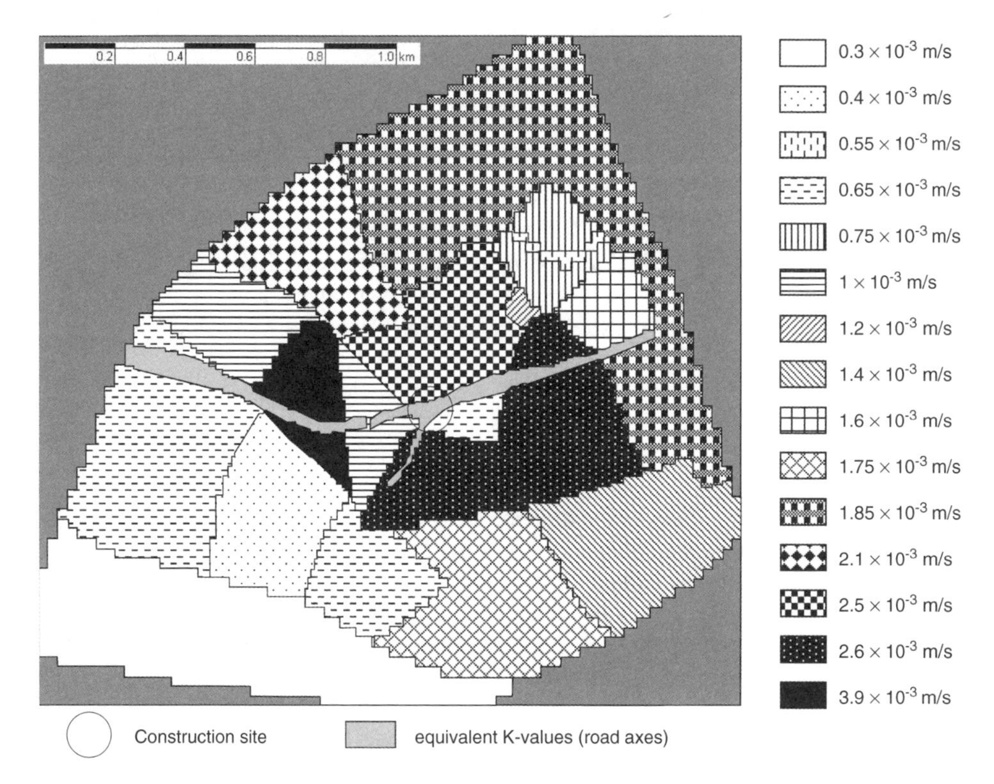

Figure 5. Hydraulic conductivity.

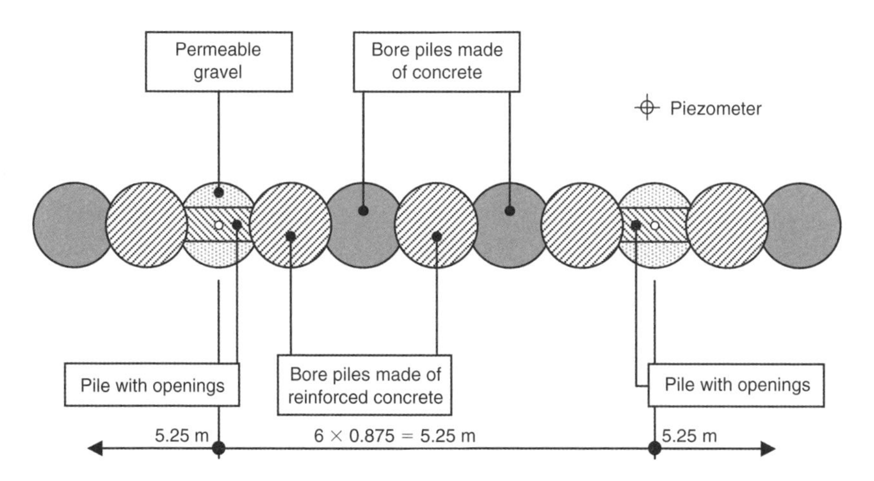

Figure 6. Vertical walls at the construction site.

The openings are kept closed during the construction phase in order to reduce the impact of pumping at the construction site on the regional groundwater flow (Buwal, 1998). The openings are kept open after the construction phase in order to reduce local impact of the underground walls to the groundwater fluxes. These underground structures, which reduce the infiltration surface vertically, are considered in the two-dimensional model using a fictitious value for hydraulic conductivity in the corresponding cells. The hydraulic conductivity that is proportional to the reduction of the infiltration area is then calculated. Thus, the walls of the construction site are simulated using an equivalent K-value, K_e, for the corresponding cells,

$$K_e = cK \tag{4}$$

where $c = H_0/H_{to}$, K = is the hydraulic conductivity of the zone, H_0 = free depth under the piles (including the openings if necessary) and H_{tot} = total depth of the aquifer.

Underground constructions which reach the aquifer bottom are simulated with impermeable cells (inactive cells).

6 RESULTS

The numerical simulation allows us to describe the groundwater flow field for the construction phase. Piezometer lines are shown in Figure 7a. The trend of the piezometer lines show that the groundwater flow is strongly influenced by the pumping at the construction site (cone of depression) and by the presence of vertical penetrating walls along the road axes (grey cells in Figure 5). It can be seen that the cone of depression extends asymmetrically. For example, the piezometer line at 245 masl shows a discontinuity by the intersection with the walls at the construction site.

Figure 7b shows the risk factor, R, calculated as the product of the required amount of water, Q, and the drawdown, s, due to the pumping at the construction site under low groundwater level conditions (worst case scenario).

$$R = Q \times s \tag{5}$$

Higher risks are related to wells with higher pumping rates or affected by bigger draw down or as a consequence of the combination of the two factors.

The risk factor in this context is not a measure of the absolute risk. Nevertheless it is of significance for a comparative study, since it allows a comparison of the different situations of the groundwater users with respect to the risk of aquifer depletion (due to the pumping at the construction site) and consequently of the additional costs incurred.

7 FUTURE DEVELOPMENTS

The simulation of walls with K_e values and the simulation of wells with pumping cells both represent sources of errors in the results. Two separate approaches allow the range of errors to be quantified:

- development of a three-dimensional model,
- and development of local two-dimensional models, one for each production well.

The former would allow a realistic simulation of the openings and of the free spaces under the walls, whereas the latter would allow a refinement of the grid down to the scale of the well diameter.

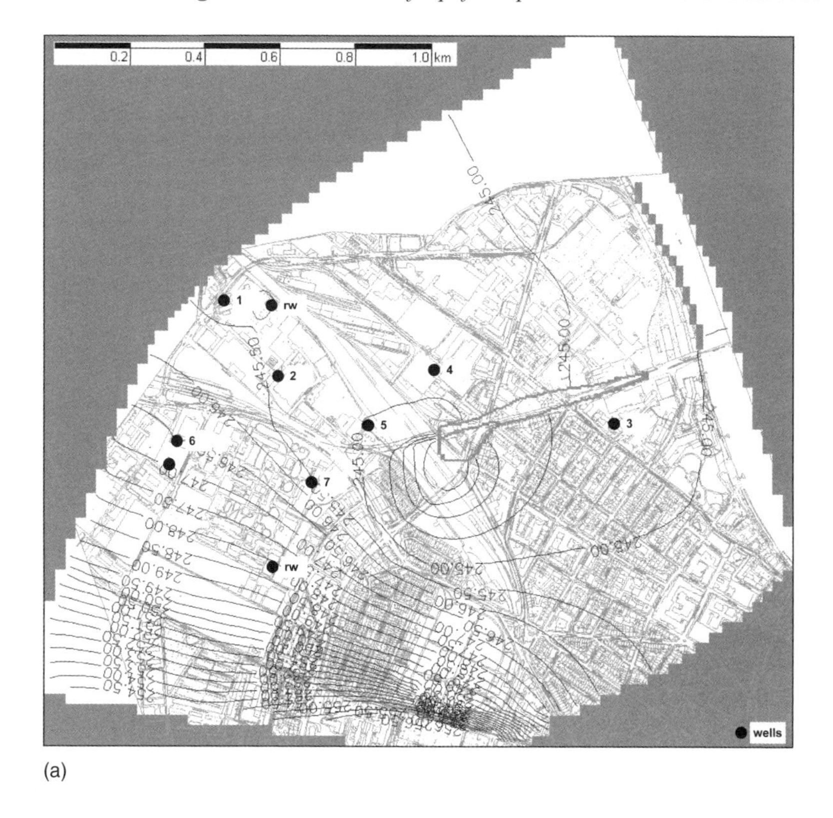

(a)

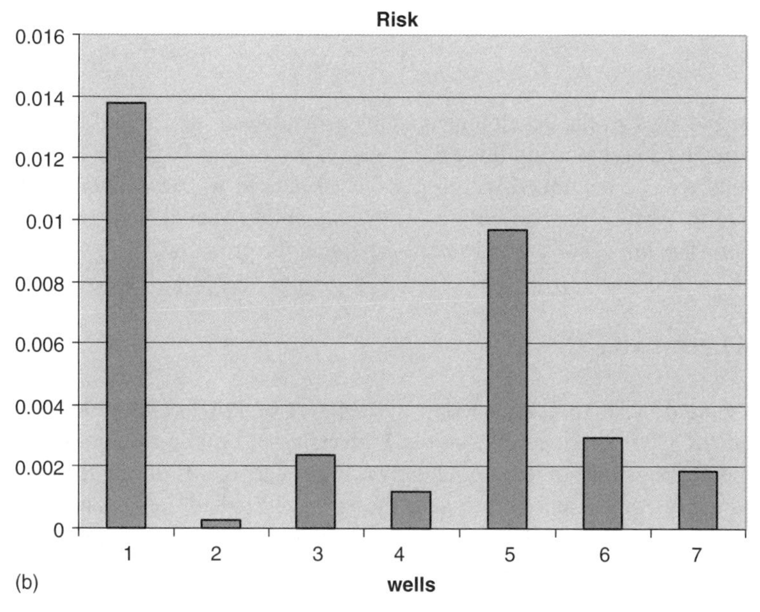

(b)

Figure 7. Groundwater flow field (a) and the risk factor (b).

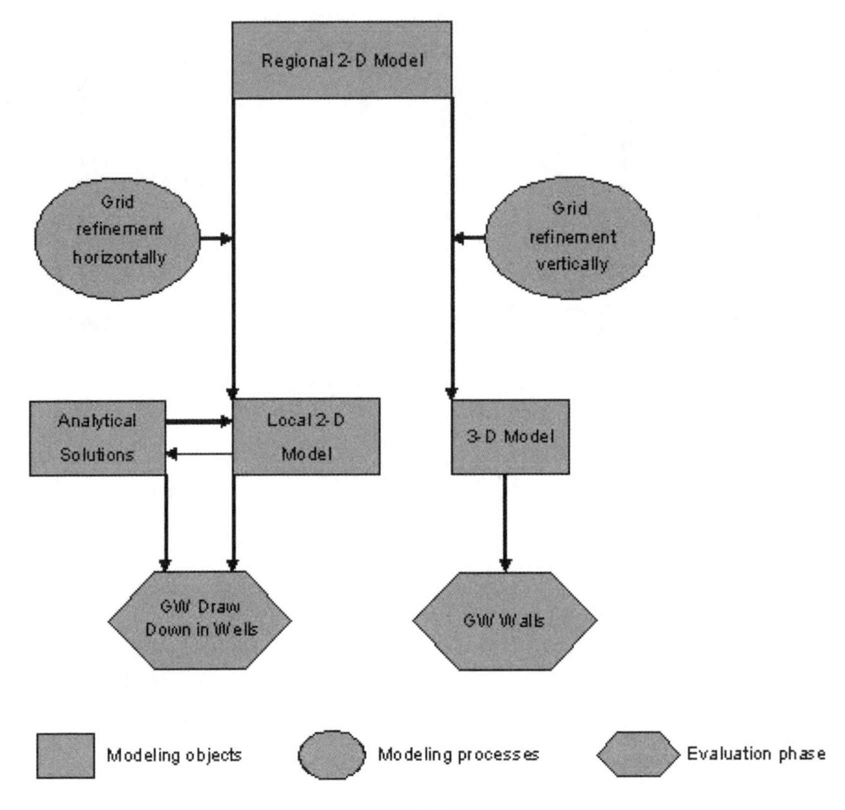

Figure 8. Development of the groundwater model.

A schematic view of the development of the groundwater model and of the modelling processes involved is shown in Figure 8.

This study will be continued within a project financed by the University of Applied Sciences North-West Switzerland in Basel: "Sustainability of civil engineering projects in urban areas – The impact of groundwater barriers on the groundwater flow."

ACKNOWLEDGEMENTS

The author would like to acknowledge the support of Prof. Gonsowski, Head of the Department of Civil Engineering, at the University of Applied Sciences North-West Switzerland in Basel. Part of the modelling work was carried out during the collaboration of the author as researcher and lecturer with the group of Applied Geology at the Geological Institute of the University of Basel. The author would also like to acknowledge the collaboration with the Geotechnical Institute in Basel for carrying out the groundwater monitoring program in the project area. The support of the Department of Energy and the Environment in Basel was invaluable in providing and interpreting the data. In addition the financial and technical support of the Department of Roads and Highways was essential for the results of this study.

REFERENCES

Buwal. 1998. Wegleitung zur Umsetzung des Grundwasserschutzes bei Untergebauten, *Mitt. Schweiz. Ges f. Boden-u. Felsmechanik, 1996.*

Miracapillo, C. 1991. Groundwater management in the region of Basel, *Lecture in NDK in angewandten Erdwissenschaften, 18. Blockkurs, 24–29 September, 1991, Kurszentrum Schloss Münchenwiler.*

Miracapillo, C. and Huggenberger, P. 2001. NT-Anschluss Luzernerring, offene/geschlossene Wasserhaltung Untersuchungen zur Auswahl der Variante, *Bericht GPI, BS-154 (7.12.01).*

Miracapillo, C. 2003. Criteria for Risk Evaluation in Groundwater Management Projects: a comparative study, *Proc. Symp. Hydrology Days, Fort Collins, Colo., March 31–April 2, 2003.*

Miracapillo, C. 2004. Groundwater modelling for monitoring purposes in construction projects, *Proc. Symp. Hydrology Days, Fort Collins, Colo., 10–12 March, 2004.*

CHAPTER 20

Urban groundwater resources: a case study of Porto city in northwest Portugal

M.J. Afonso[1], H.I. Chaminé[1], J.M. Carvalho[1], J.M. Marques[2], A. Gomes[3], M.A. Araújo[3], P.E. Fonseca[4], J. Teixeira[5], M.A. Marques da Silva[5] and F.T. Rocha[5]

[1]*Dep. de Engenharia Geotécnica, Instituto Superior de Engenharia do Porto, Rua do Dr. António Bernardino de Almeida, 431, 4200-072 Porto, Portugal (E-mail: mja@isep.ipp.pt); and Centro de Minerais Industriais e Argilas (MIA), Universidade de Aveiro, Portugal*
[2]*Dep. de Engenharia de Minas e Georrecursos (DEMING), Instituto Superior Técnico, Lisboa, Portugal*
[3]*Dep. de Geografia (GEDES), Faculdade de Letras da Universidade do Porto, Portugal*
[4]*Dep. de Geologia (LATTEX), Faculdade de Ciências da Universidade de Lisboa, Portugal*
[5]*Dep. de Geociências, Universidade de Aveiro (MIA, ELMAS), Portugal*

ABSTRACT: Regional geological, morphotectonical and hydrogeological mapping in the sustainable management of groundwater resources is demonstrated for the Porto metropolitan area in northwest Portugal. Porto City is the second most important city on the Portuguese mainland and supports about 1 million inhabitants. Thirty-five groundwater samples collected from springs, dug-wells and boreholes were analysed for major ion parameters to investigate: i) their hydrochemical character, ii) their suitability for use and iii) the contaminant sources influencing water resource quality. The results obtained show that despite natural low level salinity derived from atmospheric transport of salts from the Atlantic Ocean, groundwaters underlying the Porto metropolitan area are generally suitable for both potable and irrigation uses. Locally, groundwater quality is compromised by human activities, most notably related to the intense urbanisation and local agricultural activities, the latter most predominant in the NW and NE sectors of the city.

1 INTRODUCTION

Urbanisation can have a profound effect on water resources and the hydrological cycle. The urban subsurface is a network of pipes, conduits, and other structures that serve to alter the natural hydraulic conductivity of the geologic materials. These features can also provide pathways for the movement of urban-sourced contaminants into underlying aquifers. A consequence of such complexities is that urban groundwater resources are frequently prone to uncontrolled exploitation and to degradation resulting from indiscriminate effluent and waste disposal practices (e.g. Legget, 1973; Foster, 1996; Custodio, 1997; Lerner, 1997; Morris et al., 1997; Foster et al., 1999). Research in earth sciences can help resolve such problems, but a common difficulty is that ground and surface water resources are rarely managed conjunctively despite their strong natural interdependence. Faced with increasing worldwide pressure on water resources due to escalating demand, contamination and climatic change, it is becoming evident that integrated multidisciplinary approaches must be adopted to address the scientific issues related to water resources (e.g. Lerner, 1997; Morris et al., 1997; Aureli, 2002; Barrett, 2004).

This paper synthesises, from a multidisciplinary perspective, the nature and importance of surface water/groundwater interactions in the Porto metropolitan area of northwest Portugal. Porto city is the second most important city on the Portuguese mainland and supports about 1 million inhabitants in a 1000 km^2 area. The work is necessary to ensure that future demand for water resources can be met in the face of rapid urban, industrial, and agricultural growth. Local urban aquifers are recharged by meteoric waters and, under natural conditions this water discharges to local surface water systems or to the sea (Zaadnoordijk et al., 2004). The aquifers are recognised as vulnerable to depletion and contamination, and there is an urgent need to manage water resources in an equitable, sustainable, and ethical manner (e.g. Custodio, 1997; Barrett, 2004).

2 A BRIEF HISTORY

Porto City is located on sloping granitic hills on the banks of the Douro River, in Northern Portugal. It is one of the oldest cities in Europe and dates back to the days of the Suevi prior to the 6th Century, when it was known as *Portucale* (Harbour of Cale). It grew in importance following its conquest of 868 AD, eventually becoming the motivational centre behind the Christian re-conquest of the Iberian Peninsula (Oliveira Marques, 1972). Much of the original city was built in the 12th century and the architectural and historical attributes of its old neighbourhoods led Porto to be recognised by UNESCO as a World Heritage Site in 1996.

Until the end of the 13th century, most groundwater systems located across the Portuguese mainland could be considered natural, and had not been seriously degraded by human activity (Carvalho, 2001). Groundwater resources were utilised only locally, and where minor problems occurred they were generally resolved through natural regeneration. It was not until the turn of millennium (most notably, the last few decades of the 20th century), that groundwater resources became progressively more endangered, both in quantity and quality. The problem has been of particular concern in dense Southern Europe conurbations, such as those in the northern/central part of the Portuguese mainland where steep slopes leave little flat land suitable for urban development.

3 GEOMORPHOLOGICAL AND GEOTECTONICAL FRAMEWORK

The study area corresponds to the Porto metropolitan area, adjacent to the Atlantic Ocean. It is located in a complex geotectonic domain of the Iberian Massif, on the so-called Ossa-Morena Zone and Central-Iberian Zone boundary (Ribeiro et al., 1990), alongside the western border of the Porto–Tomar–Ferreira do Alentejo dextral major shear zone (Chaminé et al., 2003a,b; Ribeiro et al., 2003).

Topographically, the Porto region (*sensu latu*) is represented by a planar littoral platform dipping gently to the west and culminating around 120 masl (metres above sea level). To the east, the surface is bounded by a series of ridges ranging from 250 to 300 masl. Deeply incised river valleys interrupt the flatness of the surface, particularly the Douro river valley which is tectonically controlled. Evidence of neotectonic activity (Araújo et al., 2003) includes: (a) numerous major faults affecting the uppermost fluvial deposits of the littoral platform and (b) outcrops of the same marine unit occurring at various elevations, forming an irregular pattern with a general trend dipping from north to south.

Several fractures cross the region; most are associated with the NNW-SSE set which dominates the region. A more discrete yet extensive fracture set oriented NE-SW has also been identified (Chaminé et al., 2003a; Araújo et al., 2003). Discontinuities are generally vertical to sub-vertical. In the latter fault system, dextral strike-slip faulting is associated with transpressive kinematics triggered by the post-orogenic collapse of the structure along the ancient Porto–Coimbra–Tomar thrust planes. These processes generated a multitude of ENE-WSW to NE-SW regional fault systems which provide tectonic control on the drainage network (Araújo et al., 2003) (Figure 1).

The regional geotectonic framework of the Porto metropolitan area (e.g. Sharpe, 1849; Delgado, 1905; Barata, 1910; Rosas da Silva, 1936; Carríngton da Costa, 1958; Oliveira et al., 1992; Almeida, 2001; Fernández et al., 2003; Chaminé et al., 2003a,b; and references therein) comprises a crystalline fissured basement complex which is strongly deformed and overthrusted by Late Proterozoic/Palaeozoic metasedimentary rocks and granites. The substratum complex is mainly composed of phyllites, black schists, garnetiferous quartzites, mica schists, migmatites and gneisses, whereas the sedimentary cover rocks are dominated by post-Miocene alluvial and Quaternary marine deposits. The igneous rocks include pre-orogenic and syn-orogenic Variscan suites, which comprise a large component of granitic rocks.

The crystalline bedrock of Porto city consists of granites in the eastern part and a gneiss-mica schist complex in the west (Figure 2). A major fault zone, Porto–Coimbra–Tomar shear zone (Chaminé et al., 2003a,b), trending NNW-SSE, defines the boundary between these two geological units. Variscan granitic rocks, representing the Porto granite facies and the Ermesinde porphyritic facies, underlie the Porto site (*s. str.*).

The Porto basement consists of a greyish (yellowish weathered surface) two-mica, coarse granite. The granite is generally weathered ranging from fresh-rock to residual soil up to depths of over 100 m (e.g. Begonha and Sequeira Braga, 1995; Begonha, 2001; Russo et al., 2001; Gaj et al., 2003; COBA, 2003). Most of the chemical palaeoweathering took place during Cenozoic times under tropical/subtropical conditions (Araújo et al., 2003).

4 REGIONAL HYDROGEOLOGY

The main hydrogeological subdivisions of the northern Portugal region are defined by the major active fault zones (e.g. Régua-Verin fault, Bragança-Manteigas fault, Porto–Coimbra–Tomar shear zone, Douro-Beira shear zone) (Brum Ferreira, 1991; Cabral, 1995; Chaminé et al., 2003a; Ribeiro, 2002). In the study area, the mean annual rainfall to the east of the Douro-Beira shear zone is over 1300 mm, but it is lower towards the coast, reaching 1150 mm in the urban area of Porto city (Afonso, 2003).

Hydrogeological data are scarce for the study area (see details in Afonso, 1997, 2003; Afonso et al., 2004, 2005), so the proposed regional hydrogeological classifications correspond broadly with the main geological features (Table 1):

(i) sedimentary cover (post-Miocene), including alluvium and fluvial deposits;
(ii) metasedimentary rocks (upper Proterozoic-Palaeozoic), which include schists, greywackes, quartz-phyllites and quartzites; and
(iii) granitic rocks (Variscan and/or pre-Variscan), including two mica granites, biotite granites; gneisses, migmatites and gneissic granites.

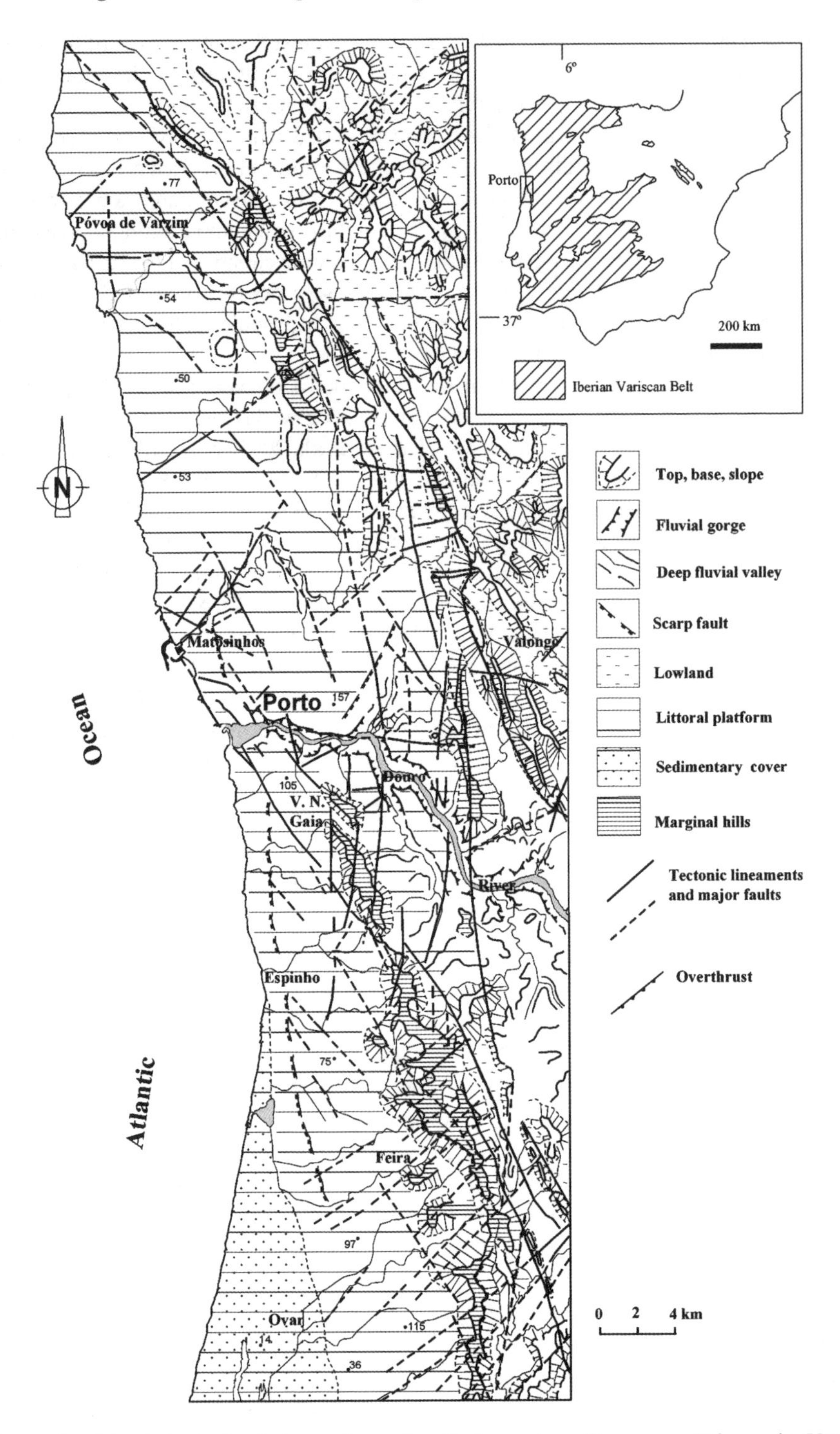

Figure 1. Morphotectonic features from the Porto metropolitan area (Póvoa de Varzim–Porto–Feira), NW Portugal (adapted from Araújo et al., 2003 and Chaminé et al., 2003a).

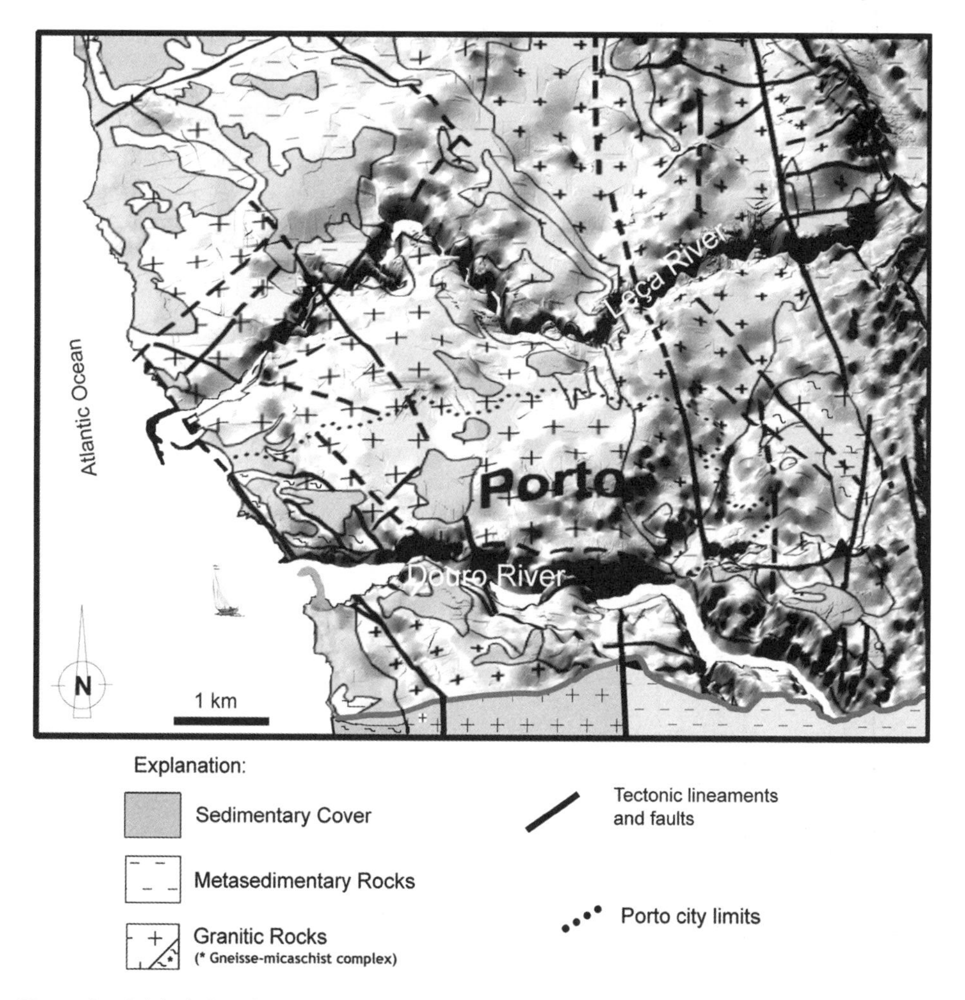

Figure 2. Digital elevation model (DEM) with an overlay of the geology for the Porto urban area (geological basement adapted from Chaminé, 2000; Araújo et al., 2003; Chaminé et al., 2003a). DEM of the studied area was generated by kriging digitized contour lines at a 1:25,000 scale. Ground resolution is 10 m. Shadowed image of the DEM is artificially illuminated from the west.

The geological, morphostructural and climatological conditions of the Porto metropolitan area strongly influence the distribution of groundwater resources (Carvalho, 1996; Monteiro, 1997; Carvalho et al., 2003, 2005). An assessment of regional hydrogeological units was carried out using the same lithological and structural framework used to define the regional geological units, and a regional hydrogeological map (Figure 3) was developed (Struckmeier and Margat, 1995; Assaad et al., 2004). Almost all aquifers in this region are associated with fissured hard rock and normally comprise weathered material that may extend to considerable depth. The zone of weathering has an important influence on the extent to which recharge reaches underlying aquifers.

Groundwater flowpaths are mainly governed by secondary permeability features such as faults, fractures and fissures, locally enhanced by weathering to produce discontinuous

Table 1. Regional hydrogeological units and related features in the Porto metropolitan area (see Figure 3).

Regional hydrogeological groups	Hydrogeological units	Hydrogeological features																
		Connectivity to the surface drainage network			Prevalent permeability		Weathering				More suitable exploitation structures		Geological risk of failure (MCI*, m/l/s)			Long-term well capacity (L/s)**		
		With	Without	Possible	Porous	Fissured	Low thickness	High thickness	Clayey	Sandy	Dug-wells, galleries and springs	Boreholes	Very high MCI > 120	High 80 < MCI < 120	Low MCI < 80	Very low Q < 1	Low 1 < Q < 2	High Q > 2
Sedimentary cover	Sands and alluvium	x			x		n.a.	n.a.	n.a.	n.a.	x				x		x	
	Sandstones and conglomerates	x			x		n.a.	n.a.	n.a.	n.a.	x				x		x	
Metasedimentary rocks	Quartz-phyllites, mica schists and black shales			x		x	x	x		x		x		x	x		x	
	Quartzites and slates		x	x		x	x			x		x		x	x		x	
	Schists, graywackes and metaconglomerates			x		x		x	x			x	x	x	x	x	x	
Granitic rocks	Granite, medium to coarse grain, with megacrystals			x		x		x		x	x		x	x		x		
	Granite, medium to fine grain, essentially biotitic			x		x		x		x	x		x	x		x		
	Gneisses and migmatites			x		x	x	x		x	x		x			x		

* MCI [Metres Capacity Index] in a given area, total drilled metres in one or several wells to obtain 1L/s; ** median long-term well capacity.

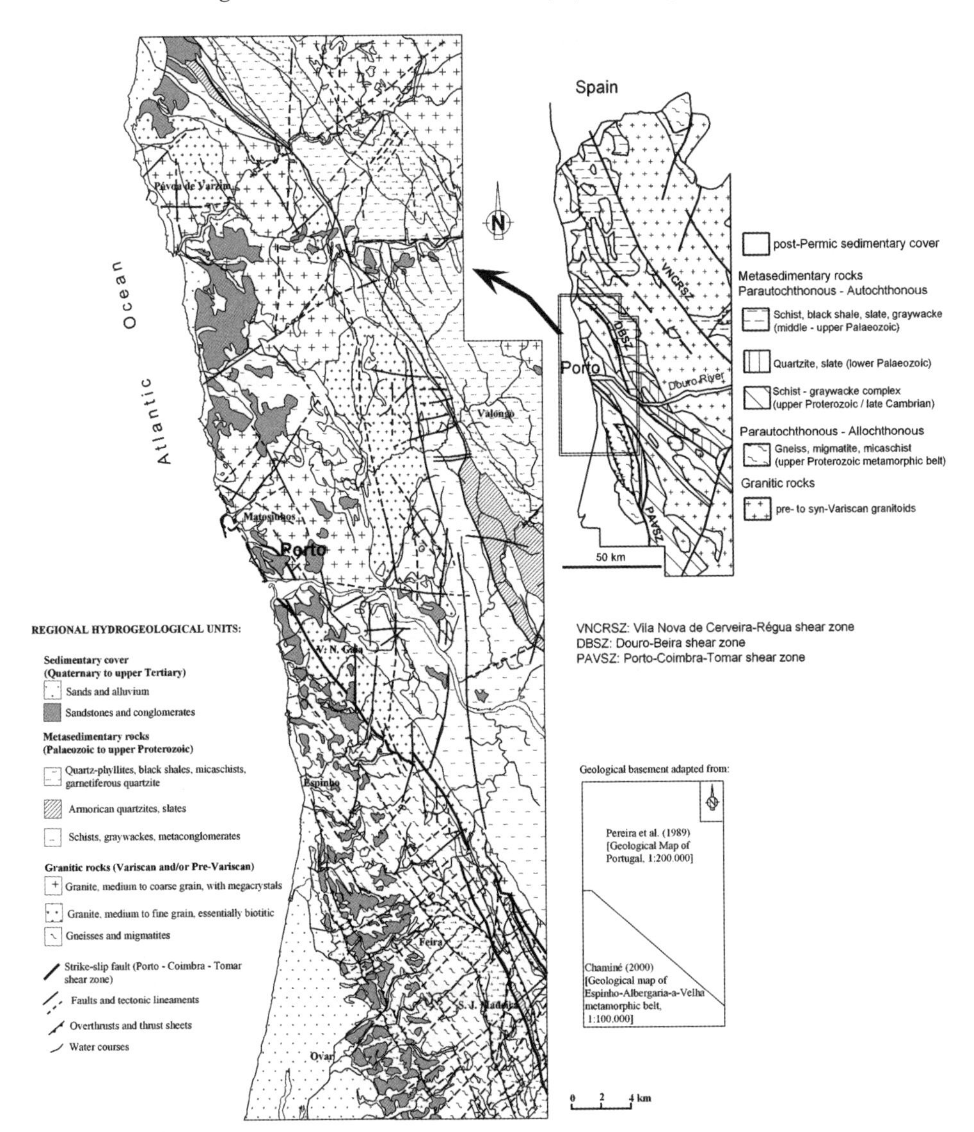

Figure 3. Regional hydrogeological outline from the Porto metropolitan area, northern Portugal (see Table 1 for hydrogeological features).

productive zones. While fracture connectivity is weak at the regional scale, it is nevertheless clear, that in the Variscan Iberian Massif, lithology and structure play a major role on the productivity of regional geological units and related water wells (Carvalho, 1996; Afonso et al., 2004; Carvalho et al., 2005). To assess well productivities, Carvalho (1993) and Carvalho et al. (2003, 2005) used data available from local well drillers to produce a Metres Capacity Index (MCI), defined for a given area as: MCI = Σ total drilled metres/ Σ total yield (L/s) for wells with long-term yields of $>$0.5 L/s. The MCI (see Table 1) is affected by the morphology, the thickness and nature of the weathering layer, the well penetration, and the presence of filonian rocks (e.g. quartz veins) and fault gouges (Pereira, 1992; Carvalho, 1996).

5 HYDROCHEMICAL CLASSIFICATION OF GROUNDWATERS

During hydrogeological investigations of urban areas, chemical analyses of waters collected from springs, dug-wells and boreholes can provide important information concerning water quality, contaminant sources and the degree of mixing between surface waters and groundwaters. In the Porto city area a complicating factor is the adjacent coastline and the possibility that elevated groundwater salinity may be associated with the sea.

Data from 35 sampling points: 28 boreholes (mean depth of 103 m), 4 dug-wells and 3 spring-collection chambers. Sample locations are shown in Figure 4. Three fieldwork campaigns were performed (November, 1995, February, 1996 and July, 1996), during which time, temperature (°C), pH, electrical conductivity (μS/cm) and Eh (mV) were measured on site while major ions were analysed in samples sent to the laboratory. A summary of results for the February, 1996 study is presented in Table 2.

Most of the sampled groundwaters show near-neutral pH values and medium to low electrical conductivities (Figure 5). The highest pH values are mainly found in borehole waters, probably reflecting a higher interaction with the local geology. The highest conductivity values (~1000 μS/cm) were obtained during the February sampling program (Afonso, 1997). The increased water mineralisation noted during this time was indicative of surface water–groundwater mixing, whereby the surface water introduces ions (e.g. SO_4, Cl and NO_3) of anthropogenic origin. The use of water temperature as an indicator of depth of groundwater circulation is not appropriate here since, for many of the sampling sites, the groundwater flows through a network of pipes/conduits before arriving at the sampling point.

Groundwater circulating in the deeper granitic aquifers exhibits a wide range of chemical signatures which are difficult to explain simply on the basis of water-rock interaction processes. Analyses of samples collected during the July, 1996 campaign are plotted on a Piper diagram in Figure 6. The broad scatter of points, notably in the anion field, suggests multiple processes are operating including ion exchange and mixing with shallow waters contaminated by anthropogenic sources.

Table 2. Groundwater analyses from the Porto city area (February, 1996 fieldwork campaign). Concentrations in mg/L, temperature (T) in °C and electrical conductivity in μS/cm.

	No. of samples	Mean	Minimum	Maximum	Standard deviation
T	25	15.5	9.6	17.3	2.0
pH	25	6.13	4.55	7.45	0.62
Conductivity	25	489	201	1011	204
Na	25	43.9	9.1	103.0	20.6
K	25	6.7	2.0	16.2	4.3
Ca	25	29.3	5.6	60.8	13.5
Mg	25	10.6	1.5	22.8	5.5
HCO_3	25	58.6	2.4	125.7	33.4
Cl	25	62.9	10.6	188.5	38.5
SO_4	25	51.1	5.8	114.6	29.6
NO_3	25	38.4	0.8	138.6	38.2

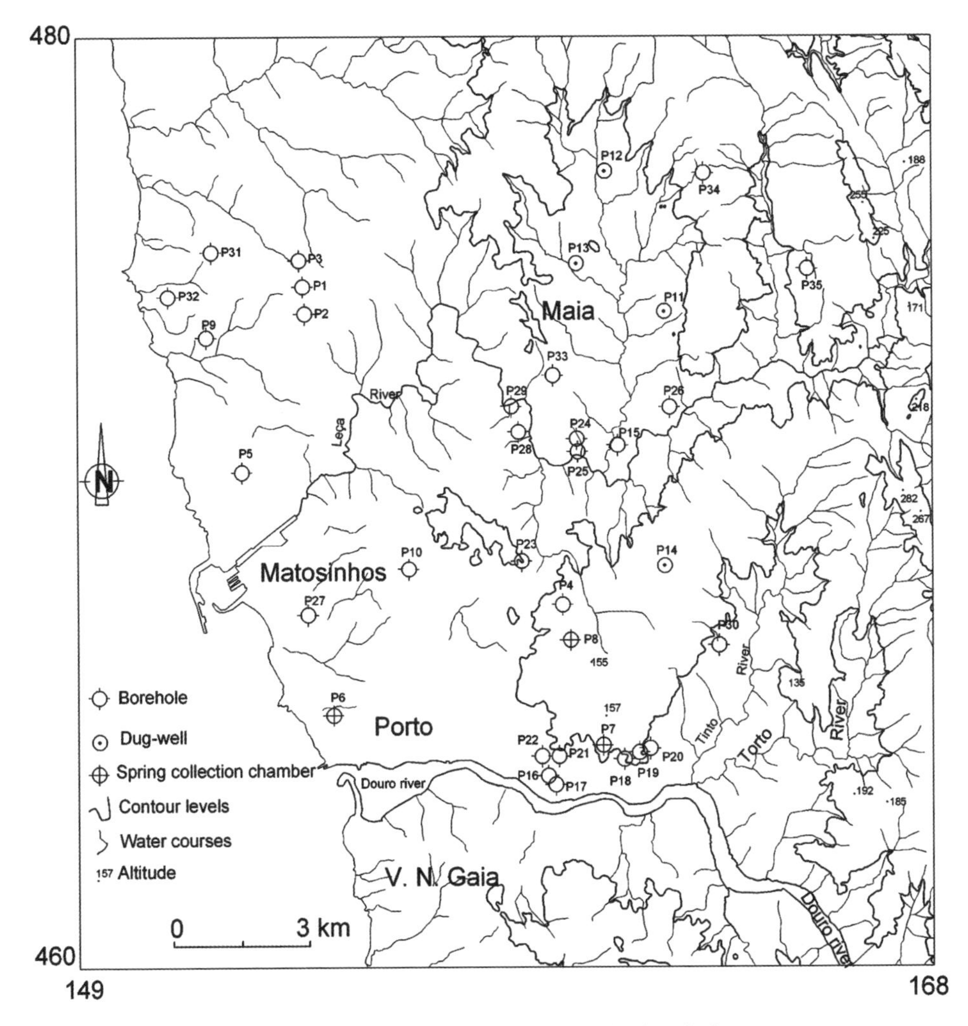

Figure 4. Location of the sampling points for hydrogeochemical analysis.

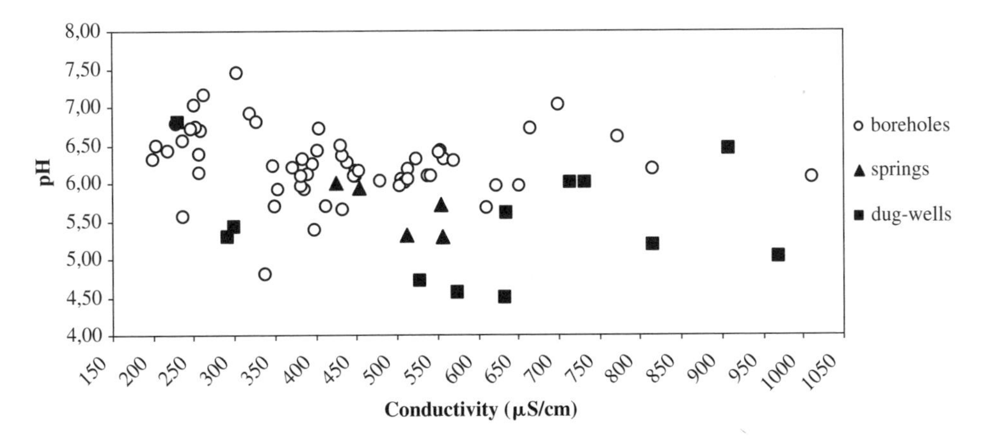

Figure 5. Conductivity *vs.* pH for water samples collected during the three fieldwork campaigns.

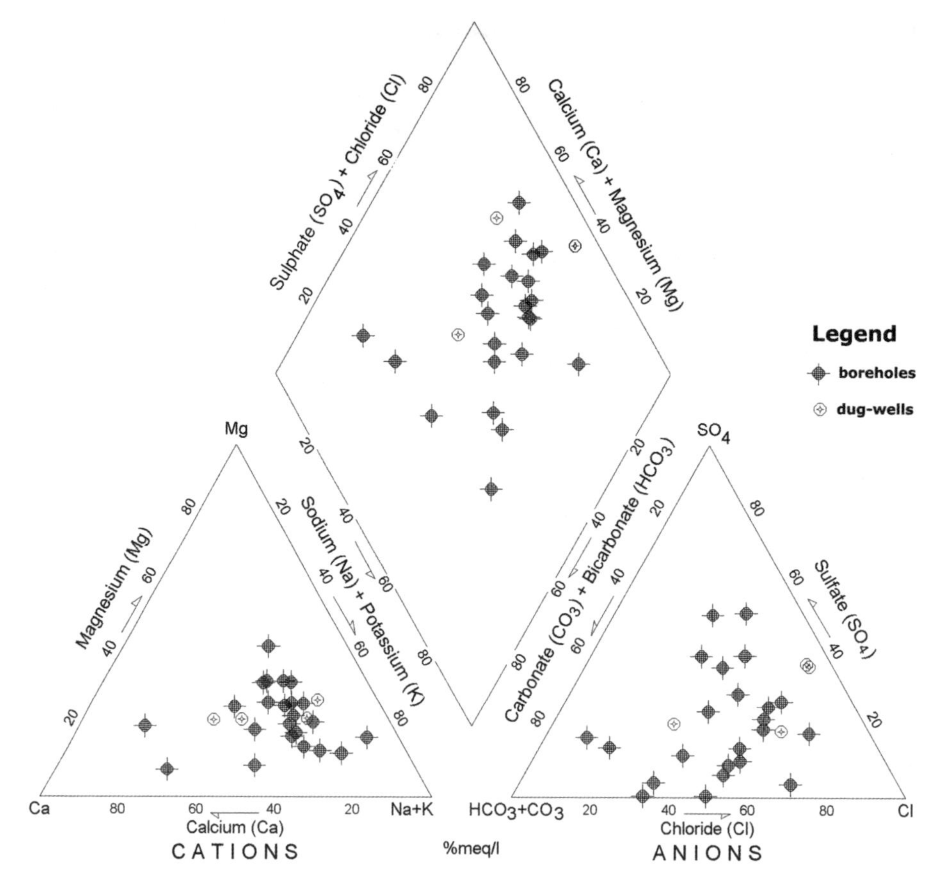

Figure 6. Piper diagram for water samples collected during the July, 1996 fieldwork campaign.

5.1 *Natural sources of mineralisation*

Natural sources of mineralisation include; (i) rock-water interaction and (ii) the airborne transport of salt from the adjacent Atlantic Ocean. Local hydrogeological conditions discourage the direct intrusion of seawater.

The atmospheric transport of sea spray is a well-known phenomenon in coastal areas. The droplets and evaporative particulates are transported by wind and deposited inland. Salt deposition is a function of topography and the exposure of the land surface to prevailing winds (Lorrai et al., 2004). Studies performed by Gustafsson and Franzén (2000) show that salt deposition rates decrease rapidly over the first 2–5 km from the coast, followed by a slower rate of reduction further inland.

Statistical and graphical analyses of the available data favour airborne salt deposition as the primary source of elevated sodium and chloride. For example, a weak correlation between HCO_3^- and Na^+ ($r = 0.17$) does not support the hydrolysis of plagioclase as a primary source of mineralisation, while a strong correlation between electrical conductivity and Na^+ ($r = 0.81$; data from July, 1996) and conductivity and Cl^- ($r = 0.75$; data from July, 1996), indicates a sodium chloride source.

Table 3. Chemical composition of rain waters (mean values) from the Porto region. Taken from Begonha et al. (1995) and Begonha (2001). Concentrations in mg/L.

HCO_3	Cl	SO_4	NO_3	PO_4	Na	K	Ca	Mg	NH_4	SiO_2
2.8	9.5	4.1	0.65	0.02	5.6	0.3	1.8	0.7	0.17	0.24

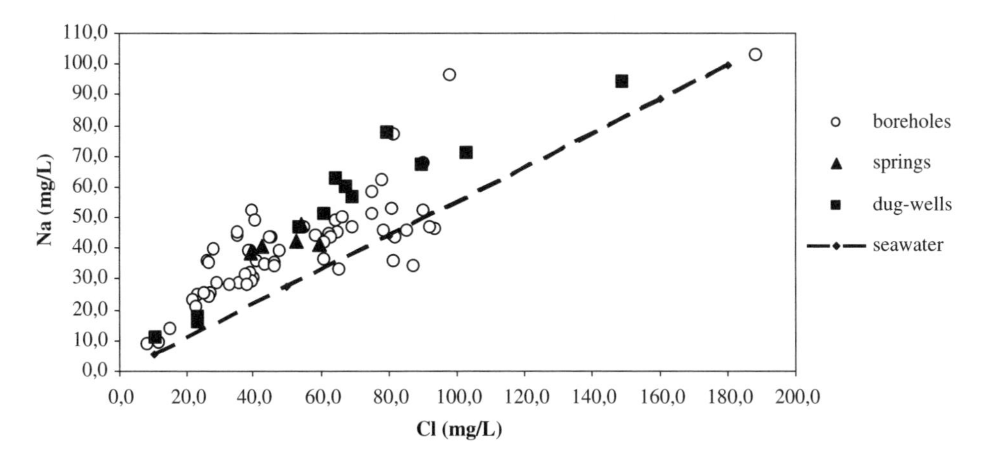

Figure 7. Cl *vs.* Na (mg/L) scatter diagram for groundwater samples collected during the three fieldwork campaigns within the Porto region.

Table 3 shows the chemical composition of rain waters from the Porto region. The ratio of Na to Cl in local precipitation is very close to the ratio found in seawater and is only slightly less than the ratio observed in much of the groundwater (Figure 7). After deposition of seawater salts in the soil zone, evaporation processes would be responsible for elevating the ion concentrations to those typically observed in the groundwater (Lorrai et al., 2004). The fact that the sodium concentration in the groundwater is slightly higher than would be expected from a pure NaCl source suggests that hydrolysis of plagioclase may provide an ancillary source of sodium.

5.2 *Groundwater contamination due to anthropogenic impacts*

Although groundwaters from the Porto city study area are of generally good quality and only slightly mineralised from natural processes, there is nonetheless evidence of anthropogenic pollution albeit, at relatively low levels. This is not unusual in rapidly growing cities which are host to numerous pollutant sources (Foster, 1997). Examples include:

- high nitrogen concentrations (normally nitrate, but sometimes ammonium) related to indiscriminate use of fertilizers, animal wastes and leaking sewer systems,
- high concentrations of chloride, commonly associated with waste water and landfills,
- high sulphate and borate concentrations, from detergents and construction waste,
- elevated dissolved organic carbon, and soluble manganese and/or iron and/or bicarbonate (as a result of oxidation of waste organic matter).

Ion scatter plots involving Ca^{2-}, SO_4^{2-} and Cl^- are shown (Figures 8 and 9, respectively) for the Porto city groundwater. Ca^{2+} and SO_4^{2-} are significantly elevated with respect to

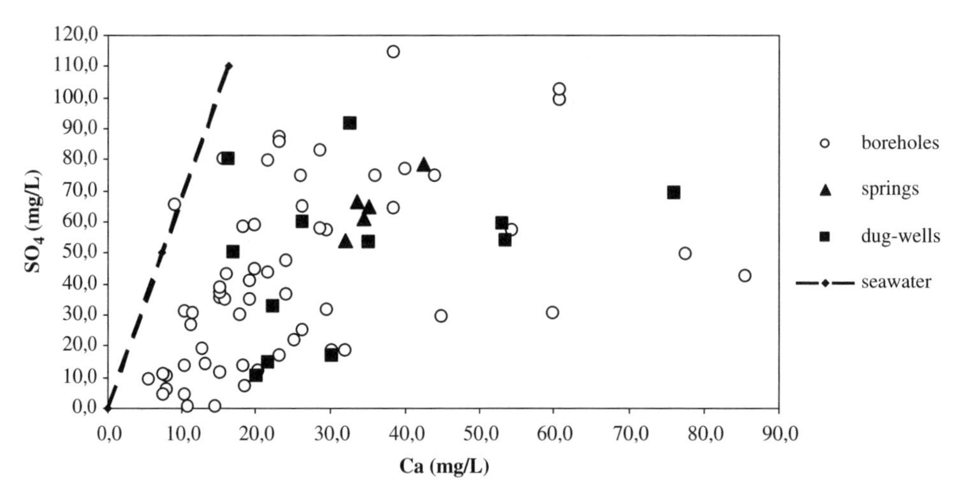

Figure 8. Ca *vs.* SO_4 (mg/L) scatter diagram for groundwater samples collected during the three fieldwork campaigns within the Porto region.

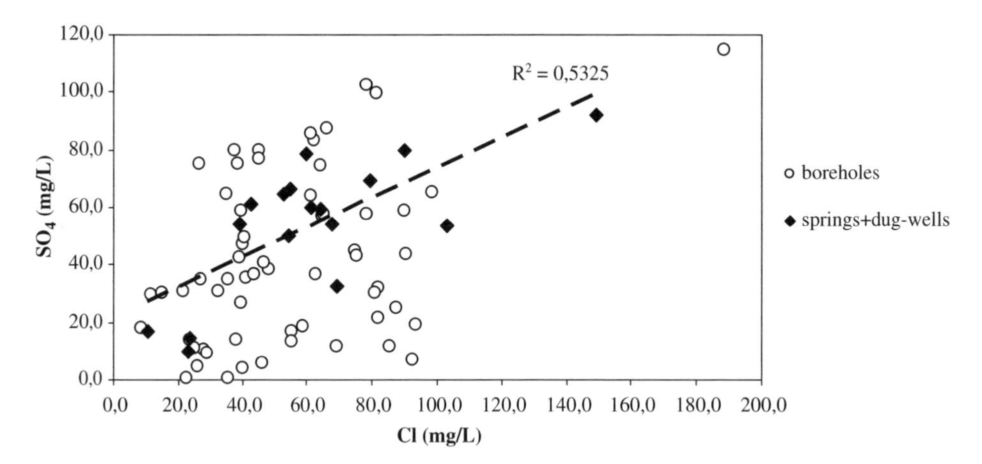

Figure 9. Cl *vs.* SO_4 scatter diagram for groundwater samples collected during the three fieldwork campaigns within the Porto region.

the concentrations anticipated from a simple seawater source, with SO_4^{2-}/Cl^- ratios in groundwater (commonly 1 to 1) being seven times higher than the ratio found in seawater (0.14 to 1) and Ca^{2+}/Cl^- ratios (typically 0.5 to 1) over twenty times higher than the ratio found in seawater (0.022 to 1). The probable source of sulphate, and to a lesser extent calcium, is air pollution (Aires-Barros, 1991; Begonha et al., 1995; Begonha, 2001) which is comprised of atmospheric gaseous SO_2 and particulate matter. SO_4^{2-}/Cl^- and Ca^{2+}/Cl^- ratios in local precipitation (Table 3) are approximately 0.5 to 1 and 0.2 to 1 respectively, which are significantly higher than ratios found in seawater, and approach the ratios found in Porto groundwater. Additional subsurface sources of calcium and sulphate found in solution may include construction waste containing gypsiferous material, although it should not

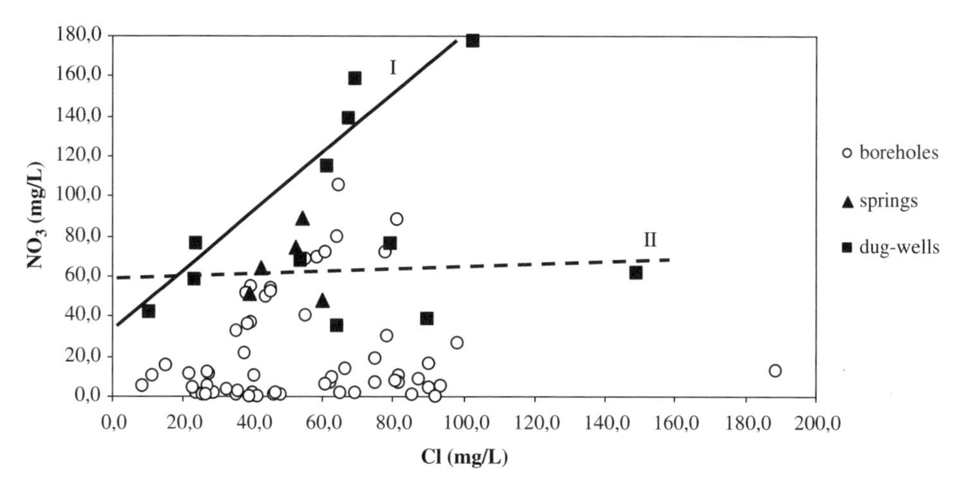

Figure 10. Cl *vs.* NO_3 scatter diagram for groundwater samples collected during the three fieldwork campaigns within the Porto region.

be discounted that the elevated calcium concentrations could also be partially attributed to ion exchange, the calcium being released from clays by the infiltration of sodium-rich water.

Finally, many of the Porto city groundwaters show elevated nitrate that in some cases exceed water quality guidelines. A Cl *vs.* NO_3 scatter plot is shown in Figure 10 and reveals two trends. Trend line I corresponds to the case where Cl and NO_3 are likely derived from the same source, and trend line II for which the Cl and NO_3 sources appear to be different. Typically, wastewater is a source of both Cl and NO_3 while sea water and agricultural fertilizer would be regarded as sole sources of the Cl and NO_3, respectively. Significantly, agricultural activity is relatively common in the NW and NE sectors of the city.

6 CONCLUSIONS

The value of regional geological, morphotectonical and hydrogeological mapping for the sustainable management of groundwater resources is demonstrated for the Porto metropolitan area in northwest Portugal. Hydrochemical studies involving major ion parameters were used to investigate (i) the hydrochemical character of groundwater resources, (ii) the suitability of groundwaters for use and (iii) the contaminant sources influencing water resource quality. A summary of conditions is shown in Figure 11. The results reveal that groundwaters underlying the Porto metropolitan area are slightly mineralised due to the atmospheric transport of salts derived from the nearby Atlantic Ocean and natural rock-water interaction. However, they are generally suitable for both potable and irrigation uses. Locally, groundwater quality is compromised by human activities which cause additionally elevated concentrations of calcium, sulphate nitrate and chloride. In some cases, concentrations of nitrate exceed the water quality guideline. The contamination is associated with the intense urbanisation and local agricultural activities, the latter most predominant in the NW and NE sectors of the city.

In the future, special emphasis will be put on integrating geochemical knowledge with isotopic analysis of precipitation and groundwater to increase the understanding of urban

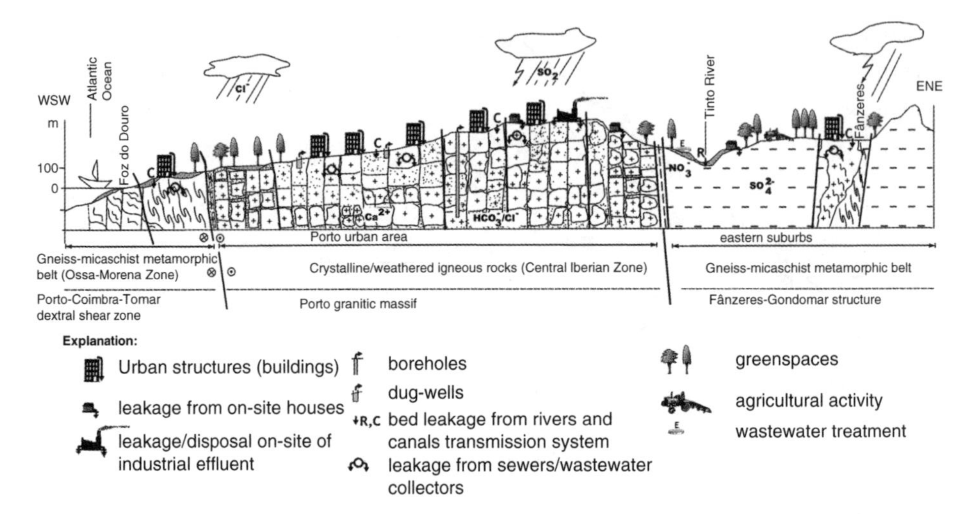

Figure 11. Conceptual hydrogeological cross-section illustrating the relationship of the main Porto city urban groundwater features (not to scale; geological and geomorphological background adapted from Chaminé et al., 2003a; Araújo et al., 2003).

water quality and quantity issues. The isotopic techniques to be employed include δ^2H and $\delta^{18}O$ for water and $\delta^{15}N$ and $\delta^{18}O$ for nitrate. In addition the vulnerability of groundwater sources will be assessed using environmental ^3H (Gonfiantini et al., 1998), and in some cases ^{14}C, thus allowing aquifer protection areas to be established.

ACKNOWLEDGEMENTS

This study was performed under the scope of the GROUNDURBAN (POCTI/CTE-GIN/59081/2004) project granted by the Portuguese Foundation for Science and Technology (FCT). This work was supported partially by grants from the GEODYN (POCTI-ISFL/5/32, Lisbon), FCT-SFRH/BPD/3641/2000 (Aveiro) and TBA (POCTI/CTA/38659/2001, Porto). We acknowledge Prof. Ken Howard and an anonymous reviewer for constructive reviews that helped to improve the clarity of the manuscript.

REFERENCES

Afonso, M.J.C. 1997. *Hidrogeologia de rochas graníticas da região do Porto.* Faculdade de Ciências da Universidade de Lisboa. 150 pp. (unpublished M.Sc. Thesis).

Afonso, M.J.C. 2003. Hidrogeologia de rochas graníticas da região do Porto (NW de Portugal). *Cadernos Laboratório Xeolóxico de Laxe*, A Coruña, 28: 173–192.

Afonso, M.J., Chaminé, H.I., Gomes, A., Teixeira, J., Araújo, M.A., Fonseca, P.E., Carvalho, J.M., Marques, J.M., Marques da Silva, M.A. and Rocha, F.T. 2004. Cartografia geológica e geomorfológica estrutural da área metropolitana do Porto: implicações na gestão dos recursos hídricos subterrâneos. *Xeográfica, Revista de Xeografía, Territorio e Medio Ambiente*, Santiago de Compostela, Spain, 4: 101–115.

Afonso, M.J., Espinha Marques, J., Marques, J.M., Carreira, P., Carvalho, J.M., Marques da Silva, M., Samper, J., Borges, F.S., Rocha, F.T., Fonseca, P.E., Gomes, A., Araújo, M.A., Teles Vieira, G.,

Mora, C., Teixeira, J., Almeida, P.G. and Chaminé, H.I. 2005. Hydrogeology of hard-rocks from two key-sectors in the Portuguese Iberian Massif: examples from Porto urban area and Serra da Estrela mountain region. In Lopo Ferreira, J.P. and Vieira, J. (eds.) *Proceedings The Fourth Inter-Celtic Colloquium on Hydrology and Management of Water Resources – Water in Celtic Countries: Quantity, Quality and Climate Variability*, Univ. Minho, Guimarães, LNEC-IAHS (Cd-Rom edition).

Almeida, A. 2001. Caracterização geoquímica e geocronológica do granito de duas micas sintectónico do Porto (NW de Portugal). In Lago, M., Arranz, E. and Galé, C. (eds.) *Proceedings III^{er} Congreso Ibérico de Geoquímica/VIII Congreso de Geoquímica de España*. Instituto Tecnológico de Aragón, Zaragoza, pp. 311–315.

Aires-Barros, L. 1991. *Alteração e alterabilidade de rochas*. INIC, Lisboa. 384 pp.

Araújo, M.A., Gomes, A.A., Chaminé, H.I., Fonseca, P.E., Gama Pereira, L.C. and Pinto de Jesus, A. 2003. Geomorfologia e geologia regional do sector de Porto–Espinho (W de Portugal): implicações morfoestruturais na cobertura sedimentar cenozóica. *Cadernos Laboratório Xeolóxico de Laxe*, A Coruña, 28: 79–105.

Assaad, F.A., LaMoreaux, P.E., Hughes, T.H., Wangfang, Z. and Jordan, H. 2004. *Field methods for geologists and hydrogeologists*. Springer-Verlag. 420 pp.

Aureli, A. 2002. What's ahead in UNESCO's International Hydrological Programme? (IHP VI 2002–2007). *Hydrogeology Journal*, 10: 349–350.

Barata, J.M.P. 1910. *Contribuição para o estudo das rochas do Porto*. Faculdade de Philosophia Natural. Coimbra. 59 pp.

Barrett, M.H. 2004. Characteristics of urban groundwater. In Lerner, D.N. (ed.) *Urban groundwater pollution*, A.A. Balkema, Lisse, 24: 29–51.

Begonha, A. and Sequeira Braga, M.A. 1995. A meteorização do granito do Porto. In Borges, F.S. and Marques, M. (eds.) *IV Congresso Nacional de Geologia. Memórias Mus. Lab. Min. Geol. Fac. Ciênc. Univ. Porto*, 4: 171–175.

Begonha, A., Sequeira Braga, M.A. and Gomes da Silva, F. 1995. A acção da água da chuva na meteorização de monumentos graníticos. In Borges, F.S. and Marques, M. (eds.) *IV Congresso Nacional de Geologia. Memórias Mus. Lab. Min. Geol. Fac. Ciênc. Univ. Porto*, 4: 177–181.

Begonha, A. 2001. Meteorização do granito e deterioração da pedra em monumentos e edifícios da cidade do Porto. *Colecção monografias, FEUP Edições*, Porto, 2: 1–445.

Brum Ferreira, A. 1991. Neotectonics in Northern Portugal: a geomorphological approach. *Zeitschrift für Geomorphologie*, Berlin-Stuttgart, N.F., 82: 73–85.

Cabral, J. 1995. Neotectónica em Portugal Continental. *Memórias Inst. Geol. Min.*, Lisboa, 31: 1–256.

Carríngton da Costa, J. 1958. A geologia da região portuense e os seus problemas. *Boletim Academia Ciências Lisboa*, 30: 36–58.

Carvalho, J.M. 1993. Groundwater exploration in hard rocks for small scale irrigation in Trás-os-Montes, Portugal. In Sheila and David Banks (eds.) *Hydrogeology of hard rocks. Memoirs 24th Congress Int. Ass. Hydr.*, Oslo, 24(2): 1021–1030.

Carvalho, J.M. 1996. Mineral water exploration and exploitation at the Portuguese Hercynian massif. *Environmental Geology*, 27: 252–258.

Carvalho, J.M. 2001. As águas subterrâneas no abastecimento de núcleos urbanos no norte de Portugal. *Tecnologia da Água*, 4, 1: 4–18.

Carvalho, J.M., Chaminé, H.I. and Plasencia, N. 2003. Caracterização dos recursos hídricos subterrâneos do maciço cristalino do Norte de Portugal: implicações para o desenvolvimento regional. In *A Geologia de Engenharia e os Recursos Geológicos: recursos geológicos e formação. Volume de Homenagem ao Prof. Doutor Cotelo Neiva* (Portugal Ferreira, M., coord.), *Imprensa da Universidade, Série Investigação*, Coimbra, 2: 245–264.

Carvalho, J.M., Chaminé, H.I., Afonso, M.J., Marques, J.E., Medeiros, A., Garcia, S., Gomes, A., Teixeira, J. and Fonseca, P.E. 2005. Productivity and water cost in fissured-aquifers from the Iberian crystalline basement (Portugal): hydrogeological constraints. In López-Geta, J.A., Pulido Bosch, A. and Baquero Úbeda, J.C. (eds.) *Water, mining and environment Book Homage to Professor Rafael Fernández Rubio*. Instituto Geológico y Minero de España, Madrid, pp. 193–207.

Chaminé, H.I. 2000. *Estratigrafia e estrutura da faixa metamórfica de Espinho-Albergaria-a-Velha (Zona de Ossa-Morena): implicações geodinâmicas*. Faculdade de Ciências da Universidade do Porto. 497 pp. (unpublished PhD Thesis).

Chaminé, H.I., Gama Pereira, L.C., Fonseca, P.E., Noronha, F. and Lemos de Sousa, M.J. 2003a. Tectonoestratigrafia da faixa de cisalhamento de Porto-Albergaria-a-Velha-Coimbra-Tomar, entre as Zonas Centro-Ibérica e de Ossa-Morena (Maciço Ibérico, W de Portugal). *Cadernos Laboratório Xeolóxico de Laxe*, A Coruña, 28: 37–78.

Chaminé, H.I., Gama Pereira, L.C., Fonseca, P.E., Moço, L.P., Fernandes, J.P., Rocha, F.T., Flores, D., Pinto de Jesus, A., Gomes, C., Soares de Andrade, A.A. and Araújo, A. 2003b. Tectonostratigraphy of middle and upper Palaeozoic black shales from the Porto–Tomar–Ferreira do Alentejo shear zone (W Portugal): new perspectives on the Iberian Massif. *Geobios*, 36(6): 649–663.

COBA – Consultores de Engenharia e Ambiente, SA. 2003. *Notícia explicativa da Carta Geotécnica do Porto*. 2th edition, COBA/FCUP/Câmara Municipal do Porto, 230 pp.

Custodio, E. 1997. Groundwater quantity and quality changes related to land and water management around urban areas: Blessings and misfortunes. In Chilton et al. (eds.) *Proceedings of the 27th IAH Congress on Groundwater in the urban environment: Problems, processes and management*. Balkema, Rotterdam, 1: 11–22.

Delgado, J.F.N. 1905. Contribuições para o estudo dos terrenos Paleozóicos. *Comun. Comm. Serv. Geol. Portg*., Lisboa, 6: 56–122.

Fernández, F.J., Chaminé, H.I., Fonseca, P.E., Munhá, J.M., Ribeiro, A., Aller, J., Fuertes-Fuentes, M. and Borges, F.S. 2003. HT-fabrics in a garnet-bearing quartzite from Western Portugal: geodynamic implications for the Iberian Variscan Belt. *Terra Nova*, 15(2): 96–103.

Foster, S. 1996. Groundwater quality concerns in rapidly-developing cities. In Guswa, J.H. (ed.) *Hydrology and hydrogeology of urban and urbanizing areas. American Institute of Hydrology*, St. Paul. pp. MIU12–MIU26.

Foster, S. 1997. The urban environment. Evaluation of hydrological changes and their consequences. *Proceedings of an International Symposium on Isotope Techniques in the Study of Past and Current Environmental Changes in the Hydrosphere and the Atmosphere*. International Atomic Energy Agency, Vienna, pp. 321–338.

Foster, S., Morris, B., Lawrence, A. and Chilton, J. 1999. Groundwater impacts and issues in developing cities: an introductory review. In Chilton, J. (ed.) *Proceedings of the 27th IAH Congress on Groundwater in the urban environment: Selected cities profiles*. Balkema, Rotterdam, 21: 3–16.

Gaj, F., Guglielmetti, V., Grasso, P. and Giacomin, G. 2003. Experience on Porto: EPB follow-up. *Tunnels and Tunnelling International*, pp. 15–18.

Gonfiantini, R., Frohlich, K., Araguás-Araguás, L. and Rozanski, K. 1998. Isotopes in groundwater hydrology. In Kendall and McDonnel (eds.) *Isotope tracers in catchment hydrology*, Elsevier, pp. 203–246.

Gustafsson, M.E.R. and Franzén, L.G. 2000. Inland transport of marine aerosols in southern Sweden. *Atmospheric Environment*, 34: 313–325.

Legget, R.F. 1973. *Cities and geology*. McGraw-Hill, New York. 579 pp.

Lerner, D.N. 1997. Too much or too little: recharge in urban areas. In Chilton et al. (eds.) *Proceedings of the 27th IAH Congress on Groundwater in the urban environment: Problems, processes and management*. Balkema, Rotterdam, 1: 41–47.

Lorrai, L., Fanfani, L., Lattanzi, P. and Wanty, R.B. 2004. Processes controlling groundwater chemistry of a coastal area in SE Sardinia (Italy). In Wanty and Seall II (eds.) *Proceedings of the International Symposium on Water-Rock Interaction*, 1: 439–443.

Monteiro, A. 1997. *O clima urbano do Porto: contribuição para a definição das estratégias de planeamento e ordenamento do território*. Colecção Texto Universitários de Ciências Sociais e Humanas. Fundação Calouste Gulbenkian/Junta Nacional de Investigação Científica e Tecnológica. 486 pp.

Morris, B.L., Lawrence, A.R. and Foster, S.D. 1997. Sustainable groundwater management for fast-growing cities: mission achievable or mission impossible? In Chilton, J. et al. (eds.) *Proceedings of the 27th IAH Congress on Groundwater in the urban environment: Problems, processes and management*. Balkema, Rotterdam, 1: 55–66.

Oliveira, J.T., Pereira, E., Ramalho, M., Antunes, M.T. and Monteiro, J.H. [coords.]. 1992. *Carta Geológica de Portugal escala 1/500000*, 5ª edição. Serviços Geológicos de Portugal, Lisbon.

Oliveira Marques, A.H. 1972. *History of Portugal. Vol. 1: from Lusitania to Empire*. Columbia University Press, New York. 507 pp.

Pereira, E., Ribeiro, A., Carvalho, G.S., Noronha, F., Ferreira, N. and Monteiro, J.H. [coords.]. 1989. *Carta Geológica de Portugal, escala 1/200000*. Folha 1. Serviços Geológicos de Portugal, Lisbon.

Pereira, M.R. 1992. Importância dos filonetes de quartzo na pesquisa de água subterrânea em rochas cristalinas. *Geolis*, Lisboa, 6(1/2): 46–52.

Ribeiro, A., Quesada, C. and Dallmeyer, R.D. 1990. Geodynamic evolution of the Iberian Massif. In Dallmeyer, R.D. and Martínez-García, E. (eds.) *Pre-Mesozoic Geology of Iberia*, Berlin, Heidelberg. Springer-Verlag. pp. 397–410.

Ribeiro, A. 2002. *Soft plate and impact tectonics*. Springer-Verlag, Berlin, Heidelberg. 324 pp.

Ribeiro, A., Marcos, A., Pereira, E., Llana-Fúnez, S., Farías, P., Fernandéz, F.J., Fonseca, P.E., Chaminé, H.I. and Rosas, F. 2003. 3-D strain distribution in the Ibero-Armorican Arc: a review. *Ciências da Terra (UNL)*, Lisboa, Nº Esp. V (CD-Rom, VI Congresso Nacional de Geologia): D62–D63.

Rosas da Silva, D.J. 1936. *Granitos do Porto*. Provas de Agregação, Porto. 63 pp.

Russo, G., Pescara, M., Kalamaras, G. and Grasso, P. 2001. A probabilistic approach for characterizing the complex geologic environment for design of the Metro do Porto. In Teuscher, P. and Colombo, A. (eds.) *Proceedings of the AITES-ITA 2001 World Tunnel Congress: Progress in tunnelling after 2000*, Milano, Italy. Pàtron Editore, Bologona, 3: 463–470.

Sharpe, D. 1849. On the Geology of the neighbourhood of Oporto, including the Silurian coal and slates of Vallongo. *Quart. Journ. Geol. Soc. London*, 5: 142–153.

Struckmeier, W.F. and Margat, J. 1995. Hydrogeological maps: a guide and a standard legend. *International Association of Hydrogeologists*, Hannover, 17, pp. 1–177.

Zaadnoordijk, W.J., van den Brink, C., van den Akker, C. and Chambers, J. 2004. Values and functions of groundwater under cities. In Lerner, D.N. (ed.) *Urban groundwater pollution*, A.A. Balkema, Lisse, 24: 1–28.

CHAPTER 21

Water resources management in Taiz, Yemen: a comprehensive overview

Essam S.A. El Sharabi
Geology Department, Faculty of Science, Taiz University, Yemen

ABSTRACT: This paper provides a comprehensive overview of the water situation in Taiz, Yemen and investigates the causes of a water crisis in the area. Taiz is a prominent and rapidly developing area showing much socioeconomic potential which could potentially be hampered if the National Water Resource Authority (NWRA) is not given more authority over Taiz's water resource management strategies. An evaluation of current water resource management strategies in the city of Taiz reveals an urgent need to enhance the responsibilities of the national water resource authority (NWRA). In order to meet the demands of an overwhelming water crisis, it is essential to develop an integrated water resource management scheme.

1 INTRODUCTION

The primary purpose of water resource management is to ameliorate hydrologic extremes and increase the reliability of water related-services. It is inevitable that demand for fresh water will continue to rise; however, the finite nature of water resources leads to water shortages in arid areas which can place severe constraints on economic development. Of particular concern in Yemen are the following (Ward, 1998):

- rapid population growth,
- rainfall variability,
- pollution of aquatic systems,
- water supply problems, and
- the lack of a comprehensive national water policy and management strategy.

Other studies have highlighted issues relating to bacteriological pollution in the city of Taiz (Metwali, 2002). The critical water situation in Taiz has received the attention of both governmental and scientific organizations because, regardless of differing socio-economic conditions, development of the area depends on the availability of water.

2 STUDY AREA SETTING

2.1 *Location*

The city of Taiz is the third largest city in Yemen, located approximately 200 km south of Sana'a and about 90 km east of the Red Sea. The area under investigation (Figure 1) is situated

Figure 1. Topographic map of the area of study.

in the foothills and slopes surrounding Jabal Saber Mountain (Jabal Saber) at elevations ranging from 1,100 to 1,600 metres above sea level (masl).

Jabal Saber is the dominant morphological feature of the area and rises up to more than 3,000 masl. Hydromorphologically, the study area can be classified into two major units, these are:

1) Highland areas
2) Basin and plains

These two hydromorphic areas have very different runoff and infiltration characteristics. The highlands receive the majority of precipitation as well as recharge from the upland mountains and foothill slopes, while the basin and plains areas include features such as wadis (ephemeral streambeds, only active during the rainy season) and alluvial plains.

The city of Taiz is growing rapidly, with population censuses carried out in 1986 and 1994, showing the population of Taiz increasing from 172,439 to 317,157, respectively. Today the population has reached approximately 592,158 inhabitants with an estimated annual growth rate of 5.72%. Projections estimate that by 2020, the population in Taiz will be 1,154,608.

Table 1. Maximum and minimum air temperatures during summer and winter.

	Summer	Winter
Maximum air temperature range	32.7°C (June) 32.1°C (August)	26.9°C (January) 27.8°C (February)
Minimum air temperature range	18.9°C (June) 18.2°C (August)	12.2°C (January) 12.9°C (February)

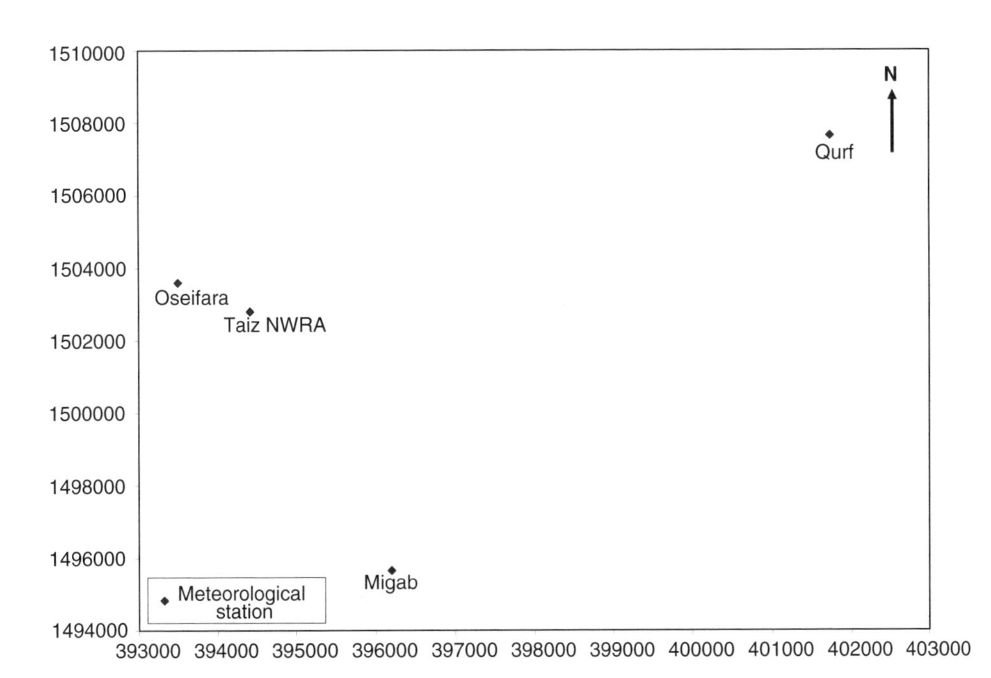

Figure 2. Location map of the meteorological stations in the area of study.

The total settlement area of the city of Taiz amounted to about 3200 ha in 2000 and estimates of expected growth extend that area to 5,200 ha in 2020 (NWASA, 2000). Projecting the rates of population increase from the year 2000 through 2004, the approximate settlement area of the city is as much as 4,000 ha. Therefore, the population density in the study area is about 145 inhabitants/ha.

3 CLIMATIC CONDITIONS

3.1 *Air temperature*

Air temperatures vary between summer and winter seasons, as displayed in Table 1. The difference between the highest maximum air temperature and the lowest minimum air temperature in the area of study during the year is 14.5°C.

3.2 *Precipitation*

Precipitation data are collected from four meteorological stations in the study area (Figure 2).

Table 2. Average monthly rainfall mm/month and average annual rainfall mm/year.

Station	Month												Average annual rainfall
	01	02	03	04	05	06	07	08	09	10	11	12	
Taiz NWRA	11.7	12.4	58.32	43.6	49.38	58.78	48.1	112.04	123.56	102.92	9.14	358	633.5
Migab	12.6	8.0	47.8	76.4	91.9	93.7	94.4	177.0	184.7	145.9	7.8	14.1	918.3
Qurf	5.4	7.3	43.7	42.4	62.1	91.5	62.4	93.4	67.7	53.7	6.9	5.5	542.0
Oseifara	4.86	8.36	31.61	61.96	75.02	66.43	54.6	101.1	97.2	73.7	11.1	4.7	590.6

Note: The measurement period for Taiz, Migab, and Qurf stations was from 1999 to 2003, while for the Oseifara station it was from 1990 to 2003.

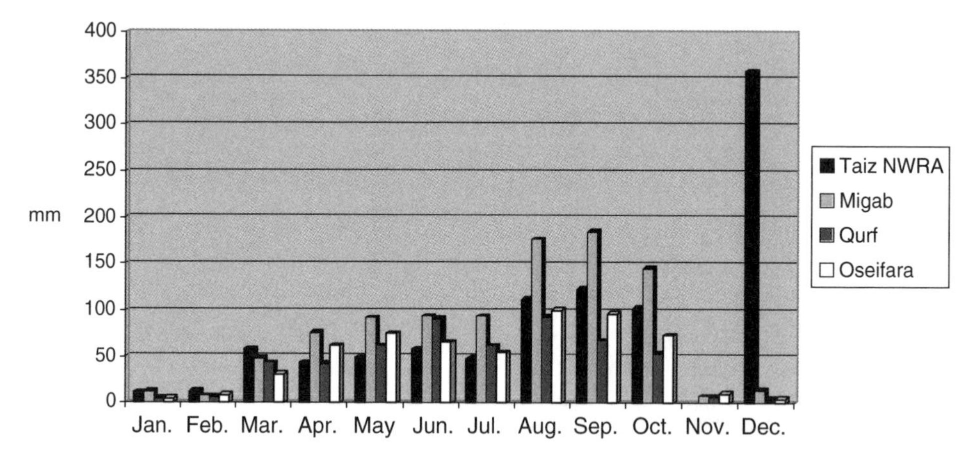

Figure 3. Average monthly rainfall during different periods.

The data collected from these stations provide effective coverage for characterizing rainfall in the various parts of Taiz. Average annual rainfall is approximately 671.1 mm/year, while the total annual rainfall ranges between 542 mm/year (Qurf station, 1999–2003) and 918.3 mm/year (Migab station, 1999–2003) (Table 2 and Figure 3). Meteorological data show that the annual rainy season in the study area is confined to the period from March to October with a period of significantly increased precipitation from July to October. The dry season lasts 4 months (November through February) during which there is no precipitation, except for a few storms.

The highland station of Migab receives a significantly greater amount of precipitation (900 mm on average) and the entire region shows considerable variation from year to year and from place to place depending on the morphology and the elevation. The maximum annual rainfall during the period 1999–2003 (all stations) was 1062 mm while the minimum annual rainfall during the same period was 448.5 mm (Table 3 and Figure 4).

Table 3. Annual rainfall in mm from 1999 to 2003.

Station	Year				
	1999	2000	2001	2002	2003
Taiz	608.1	596.2	621.8	836.8	707.6
Migab	864.1	1062	1034	814.1	714
Qurf	–	633	556	517.2	448.5
Oseifara	617	581.8	651	591	677

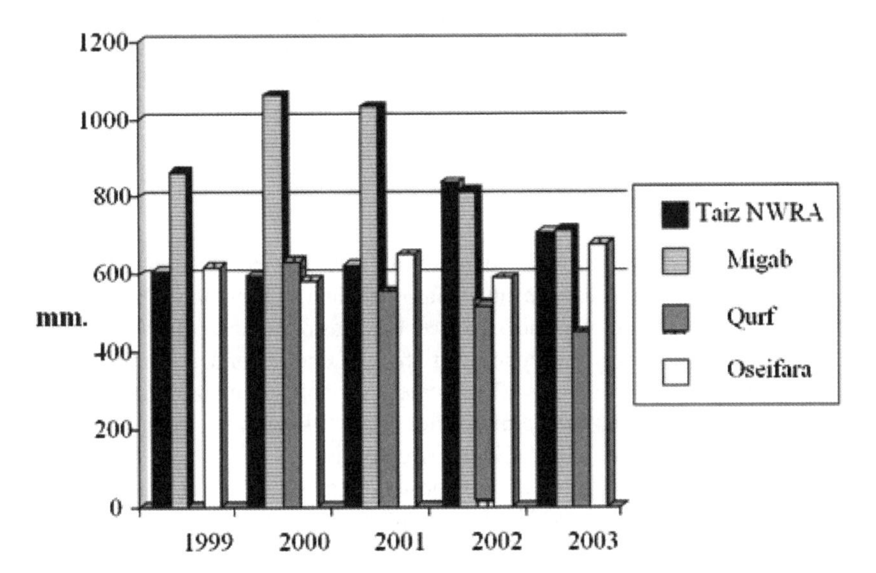

Figure 4. Annual rainfall from 1999 to 2003.

3.3 *Evaporation*

Evaporation is driven by the local climatic regime and is particularly sensitive to solar radiation intensity and wind velocities. In Taiz, potential evaporation rates are extremely high, and exceed 6 mm/day (Table 4). Not surprisingly, the evaporative rates are highest during the summer season due to hot temperatures.

Surface water bodies experience substantial evaporation, including the Amirah dam reservoir and Buraihi wastewater lagoons as well as ephemeral standing water left from rainfall events. Evaporation also takes place from vegetated lands. Groundwater, on the other hand, experiences negligible evaporation due to the depth of the water table.

4 WATER RESOURCES

The existing water supplies for Taiz come from well fields located north and northeast of the city, in addition to a number of urban wells. Groundwater is the main source of water to the city and is extracted from both a shallow and a deep aquifer in the area. The only natural surface water source is found within wadi features, during short periods in the summer season.

Table 4. Average monthly potential evaporation (mm/day) from 1982 to 1987.

Station	Month											
	01	02	03	04	05	06	07	08	09	10	11	12
Oseifara	6.79	6.53	7.63	7.27	7.8	7.13	7.94	7.44	6.86	6.85	6.98	7.59

4.1 *Surface water sources*

Temporary flow of surface runoff from rainfall is the only source of ephemeral surface water which may be collected using rainwater harvesting techniques. The surface water which flows into the wadis is diverted for irrigation by means of excavated channels called Sawaaqi. The estimated volume of the surface runoff flow in wadis (Al Haima, Dhabab, Hawban, Central Zone and Upper Wdi Rasyan) is about 9 Mm^3/year (NWRA, 2000).

4.2 *Groundwater sources*

Rainfall is the major and only renewable source of fresh groundwater in the area. Three aquifers can be distinguished in and around the city of Taiz including a Quaternary alluvium aquifer, a Tertiary fractured volcanic aquifer, and a Cretaceous sandstone aquifer (Tawillah Sandstone).

The alluvial aquifer forms the uppermost shallow aquifer and is composed of non-consolidated Quaternary sediments with extremely variable grain-size ranging from boulders to silt. The Quaternary alluvium aquifer exists along wadis and depressions. It receives recharge from floods and overlying irrigated areas.

The Tertiary volcanic rock aquifer is composed of fractured basalts separated by more compact volcanic rocks and clays. These volcanic rocks are mainly located in the south-west of Yemen while flood basalts are more predominant in the outermost south and south-western parts of Yemen (Kruck et al., 1996). Generally, this type of aquifer is not a very productive source as it tends to provide low yields and poor quality water.

The Cretaceous sandstone aquifer (Tawillah Sandstone) in the study area is not fully exploited except in the NE zone (Dhi Sufal) and by a number of wells on farms. Wells that tap into this aquifer generally have high yields and good quality water. Efforts to identify new water sources are focusing on this particular aquifer, but these efforts have been hampered slightly by both drilling complications and community resistance (NWRA, 2000).

It is estimated that about 47 Mm^3 is withdrawn annually from the three aquifers and about one third of this volume (15.6 Mm^3) eventually returns as indirect recharge. The total groundwater used per sector is as follows: irrigation (66%), industrial (7%), urban (17%), and rural (10%).

5 WATER CONSUMPTION

The estimated amount of water consumption by domestic, agricultural and industrial sectors varies depending on the data source, making it difficult to obtain an accurate estimation of consumption rates. The following section presents the best estimates of water use and consumption rates for Taiz.

Table 5. Summary of water supply, water use and losses through 1981–2003.

Year	Population	Water supply (m^3)	Water use (sale) (m^3)	Losses (%)
1981	121,393	2,256,000 (51 l/c/d)	1,808,898 (40.8 l/c/d)	20
1982	130,206	2,256,000 (47.5 l/c/d)	2,009,886 (42.3 l/c/d)	11
1983	139,659	2,630,000 (51.6 l/c/d)	2,233,207 (43.8 l/c/d)	15
1984	149,798	3,924,000 (71.8 l/c/d)	2,893,951 (52.9 l/c/d)	26
1985	160,674	6,039,650 (103 l/c/d)	3,036,770 (51.8 l/c/d)	50
1986	172,439	6,036,770 (96 l/c/d)	3,263,928 (51.9 l/c/d)	46
1987	186,096	5,717,780 (84.2 l/c/d)	2,900,501 (42.7 l/c/d)	49
1988	200,835	4,836,390 (66 l/c/d)	2,905,260 (39.6 l/c/d)	40
1989	216,741	4,704,550 (59.5 l/c/d)	3,275,262 (41.4 l/c/d)	30
1990	233,907	5,537,790 (64.9 l/c/d)	3,220,510 (37.7 l/c/d)	42
1991	252,432	5,572,986 (60.5 l/c/d)	3,466,558 (37.6 l/c/d)	38
1992	272,425	5,525,640 (55.6 l/c/d)	3,286,605 (33.1 l/c/d)	41
1993	294,001	5,443,610 (50.7 l/c/d)	3,462,589 (32.3 l/c/d)	36
1994	317,157	6,027,640 (52.1 l/c/d)	3,478,148 (30 l/c/d)	42
1995	342,276	5,547,060 (44.4 l/c/d)	3,063,062 (24.5 l/c/d)	45
1996	365,311	5,267,699 (39.5 l/c/d)	2,656,473 (20 l/c/d)	50
1997	389,897	4,988,338 (35 l/c/d)	3,054,566 (21.5 l/c/d)	39
1998	416,137	5,041,864 (33.2 l/c/d)	3,905,051 (25.7 l/c/d)	23
1999	444,143	5,391,264 (33.3 l/c/d)	4,218,470 (26 l/c/d)	22
2000	474,033	6,186,156 (35.8 l/c/d)	4,092,877 (23.7 l/c/d)	34
2001	501,148	7,320,276 (40 l/c/d)	3,542,532 (19.4 l/c/d)	52
2002	529,813	7,427,046 (38.4 l/c/d)	3,699,056 (19.1 l/c/d)	50
2003	560,119	6,713,332 (32.8 l/c/d)	3,875,827 (19 l/c/d)	42

Growth rate 1981–1986 = 7.26%.
Growth rate 1986–1995 = 7.92%.
Growth rate 1995–2000 = 6.73%.
Growth rate 2000–2003 = 5.72%.

5.1 *Domestic water supply*

The main water sources for the domestic water supply system in Taiz are from wells located to the north and northeast of the city. Table 5 presents domestic water supply rates, including all losses due to leaks or other sources as well as the domestic use (water sale) in Taiz from 1981 to 2003.

From Table 5, the minimum and maximum domestic water supply rates, including estimated losses, are 2,256 Mm^3/year (47.5–51 l/c/d in 1982 and 1981) and 7,427 Mm^3/year (38.4 l/c/d in 2001), respectively, whilst the minimum and maximum domestic use (sale) are estimated to be 1,809 Mm^3/year (40.8 l/c/d in 1981) and 3,699 Mm^3/year (19 l/c/d), respectively. Estimates of losses range between 11% (1982) and 52% (2001). These losses can be categorized as administrative (illegal connections, incorrect readings and measuring of gauges, incorrectly measured production, or an unsuitable billing system) and technical (leaks of water from the surface and underground pipes).

Predicted population growth rates (non government census years) at the bottom of Table 5 were determined for each of the following time periods 1981–1986, 1986–1995, 1995–2000 and 2000–2003.

5.2 *Irrigation and industrial water supply*

The total water used for irrigation in Taiz is estimated to be approximately 66–68% of the total groundwater use (NWRA, 2000). Industrial use within the study area was comparatively small, using just under 7% of the total groundwater, however, this quantity is expected to increase due to industrial growth in Taiz.

6 WATER DEMAND

6.1 *Domestic water demand*

Increased water demand for domestic purposes in Taiz is related to the following factors:

- High rate of population growth (approximately 6% per year).
- Improving socio-economic status or increased affluence of the community.
- Increased population due to the return of approximately 11,000 expatriates to Taiz following the Gulf War.

The population is expected to reach 626,029 by 2005, 793,677 by 2010, 971,680 by 2015, and 1,154,608 by 2020. Two scenarios for estimating the domestic water demand from 1998 to 2020, are provided in Table 6. Scenario 1 is a low rate of demand (75 l/c/d) and in scenario 2 the demand rate fits the minimum human requirements of 120 l/c/d. In order to predict the water demand until 2020, the population growth rates provided at the bottom of Table 6 were derived using the same methods as those provided for Table 5.

6.2 *Irrigation and industrial water demand*

The expected demand for irrigation water will decrease or remain stable, assuming continued improvements to irrigation methods. However, major industrial development is expected to accelerate in the coming years so the ratio of water consumed to water

Table 6. Domestic water demand in Taiz city.

Year	Population	Water demand (m^3)	
		Scenario 1 (75 l/c/d)	Scenario 2 (120 l/c/d)
1985	160,674	4,398,845	7,037,521
1990	233,907	6,403,204	10,245,127
1995	342,276	9,369,805	14,991,689
2000	474,033	12,976,653	20,762,645
2005	626,029	17,137,544	27,420,070
2010	793,677	21,726,908	34,763,053
2015	971,680	26,599,740	42,559,584
2020	1,154,608	31,607,394	50,571,830

Growth rate 1985–1986 = 7.26%. Growth rate 2000–2005 = 5.72%.
Growth rate 1995–2000 = 6.73%. Growth rate 2010–2015 = 4.13%.
Growth rate 2005–2010 = 4.86%. Growth rate 2015–2020 = 3.51%.
Growth rate 1986–1995 = 7.92%.

returned as indirect recharge is also expected to increase. Current projections indicate that, the total consumption and demand for industries will be approximately 13 Mm^3/year and 27 Mm^3/year, respectively (NWRA, 2000).

Depending on the actual groundwater abstraction rates for different purposes, the estimated water demand deficit within the study area can be expected to worsen. Therefore, in order to alleviate water deficiencies both water supply and water demand strategies are needed.

7 WATER MANAGEMENT STRATEGIES

The previous sections highlight the need for water management efforts that work towards alleviating the current water supply crisis in Taiz. Increases in water supply can be attained by identifying new sources of water or finding new ways to reuse existing, non-consumed water supplies. In the past few years many projects and studies have been completed with the objective to improve the water budget in Taiz and increasingly people are becoming more aware of issues concerning water management. Important efforts resulting in increases to water supply have been achieved through:

- rain water harvesting and reuse,
- groundwater exploration and development projects to provide new wells to the east and west of the city,
- feasibility study for a brackish water desalination pilot plant,
- desalination of brackish water abstracted from Al Hawban and Al Hawjalah wells, and
- lining most of the wadi drainage courses.

At the same time, decreases in use patterns, creative reuse, and protection measures are also improving the water supply balance, some of these efforts include:

- reuse of treated domestic waste water for irrigation,
- increased usage of groundwater,
- surface runoff collection,
- water pollution avoidance,
- improved water policy management in the study area, particularly by avoiding groundwater depletion with improved pumping controls,
- decreasing the loss rate from domestic water supply networks which are currently estimated to be 50% of abstracted water,
- improving efficiency of irrigation systems, and
- public awareness and education about environmental concerns.

To be successful, water resource management must be flexible enough with its approach and tools and ensure that stakeholders are involved (Savenije, 1994).

8 SUMMARY AND CONCLUSIONS

This paper examines water resource management in Taiz and demonstrates the urgent need to enhance the implementation activities and jurisdiction of the National Water Authority.

The existing water supplies come from well fields located to the north and northeast of Taiz, in addition to a smaller number of wells drilled in the city. Groundwater supply is from both shallow and deep aquifer systems, which provide the primary source of water. Surface water supply is limited to intermittent wadi formation during the short summer period.

Considering the current water demand (sales between 1,809 and 3,699 Mm^3/year) for domestic supplies and the estimated water supply capacities (estimated between 2,256 and 7,427 Mm^3/year), which vary according to uncertainty about the quantity of losses prior to delivery, there may be a shortfall in Taiz's ability to provide water under certain conditions in the short term. As the city continues to grow, there is little doubt that the dependence upon groundwater supplies will increase and the potential for shortfalls may become even more exaggerated if efforts are not taken to alleviate potential deficits.

In recent years, many projects and studies have been completed to improve the water supply situation in the area, as well as conducting educational awareness campaigns for local residents. The most important projects include evaluations of rainwater harvesting and reuse, identification of new groundwater resources to the east and west of the city, a feasibility study for a pilot desalination plant, lining of most wadi drainage courses, retrofitting existing infrastructure to decrease the rate of water supply losses, and improving water management policies.

The measures currently under development are encouraging and show a level of commitment to finding solutions to the water crisis through a variety of methods. Continued investments to economic and educational resources will be needed for the city of Taiz to successfully meet its water needs and to assure a safe, secure future for its citizens.

The strategies of highest priority for assuring Taiz's water supply in the future are as follows:

- reduce infrastructure water losses through both administrative and technical measures,
- select region appropriate water harvesting and water conservation techniques (e.g. rain water harvesting, brackish groundwater treatment, and runoff storage during wet-periods),
- reuse treated waste water, both for irrigation and for recharging groundwater aquifers,
- implement artificial recharge to groundwater, a technique that is widely used in other arid and semi-arid regions where it, together with water conservation, forms a package of remedial action for dry and semi-arid regions (UNESCO, 1995),
- increase investment in water treatment at the Al Amirah reservoir,
- control and manage the new exploration of water wells and prevent random drilling, and
- continue public awareness campaigns that focus on water conservation.

There is no single solution for the water deficit facing the city of Taiz, but through the strategic use of several techniques, rapid improvements can be expected. Like many urban centers worldwide, the ability of scientists, engineers, citizens, and government managers to find creative means for addressing water resource problems may determine the future economic viability of Taiz.

ACKNOWLEDGMENTS

The author is deeply thankful to Engineer Adel K.S. Magaref, Director General of Taiz Water and Sanitation Local Corporation for his assistance with data acquisition.

REFERENCES

Kruck, W., Schäffer, U. and Thiele, J. 1996. Explanatory Notes on the Geological Map of the Republic of Yemen – Western Part. Geologisches Jahrbuch, Hannover.

Metwali, R.M. 2002. Groundwater Quality in Taiz City and Surrounding area, Yemen Republic. Arab Gulf Journal of Science Research, March 2002, Vol.20, Issue 1, pp. 50–54.

NWASA. 2000. Taiz Water Supply and Sanitation Project. Initial Development Plan, Second Edition, National Water and Sanitation Authority, Republic of Yemen, 45p.

NWRA. 2000. Water Resources Management Action Plan for the Taiz Region. National Water Resources Authority, Republic of Yemen, 71p.

Savenije, H.H.G. 1994. Water Resources Management Concept and Tools, IHF, Delft, the Netherlands, pp. 4 and 18.

UNESCO. 1995. Rainfall Management in the Arab Region. State of Art Report, ROSTAS, Cairo, 147p.

Ward, C. 1998. Practical Responses to Extreme Groundwater Overdraft in Yemen. International conference Yemen: The Challenge of Social, Economic and Democratic Development, University of Exeter, Centre for Arab Gulf Studies. April 1–4, 1998.

Index